Safety in Medication Use

Safety in Medication Use

Edited by

Mary P. Tully
The University of Manchester

Bryony Dean Franklin
UCL School of Pharmacy

CRC Press is an imprint of the
Taylor & Francis Group, an **informa** business

CRC Press
Taylor & Francis Group
6000 Broken Sound Parkway NW, Suite 300
Boca Raton, FL 33487-2742

Printed on acid-free paper
Version Date: 20150722

International Standard Book Number-13: 978-1-4822-2700-0 (Paperback)

Visit the Taylor & Francis Web site at
http://www.taylorandfrancis.com

and the CRC Press Web site at
http://www.crcpress.com

This book is dedicated to our respective daughters, Maeve and Caitlin (MPT) and Katie (BDF), who have had to do without their mothers more than once while we were writing this book.

Contents

Foreword

While they are highly beneficial in the aggregate, medications also appear to cause more harm than any other therapy. Thus, it is critical to use them as safely as possible in the delivery of health care.

This book, edited by Mary P. Tully and Bryony Dean Franklin, represents a welcome addition to the literature. It delivers an overview about both the theory and practice of medication safety as well as summarizes the international literature and delivers practical suggestions. It is aimed at health care professionals and others with an interest in patient safety and/or quality improvement in the field of medication use, as well as researchers in the field.

When I first started working in this area nearly 30 years ago, orders for medications were written on paper, and the Internet had not been invented. There was little automation in pharmacies, and the notion of using robots in clinical care seemed like science fiction. Things have evolved dramatically since then. We now know a great deal more about the harm that medications can cause and the many types of errors that can occur in the medication use process. But at the same time, we have learned an enormous amount about how to improve medication safety. We know that computerization of prescribing—when done well, and linked with robust decision support—can make that step of the process dramatically safer. Robots can dispense medications in hospitals and pharmacies with a very high degree of accuracy. Using barcodes for medications, patients and providers can result in major improvements at the dispensing and administration stages. Smart pumps can warn nurses if they try to administer dosages too fast or that are too high. Electronic tools can help with medication reconciliation, a complex and challenging process. We also have far greater understanding of systems design, human factors, psychology, sociotechnical approaches and educational theories, and importantly, how they can be applied to medication practices. Designed well, even "low-tech" systems can make a big difference.

But the reality is that most organizations have implemented only some of these approaches that can improve medication safety. In addition, how an organization implements them will drive, to a very great extent, which benefits are achieved (or not). Furthermore, very few organizations have made all of these approaches work together. It is as if there is an orchestra, and hospitals have a few of the instruments, but they have not yet learned to play together. To do this, humans as well as technologies will need to work together. This book is intended to provide a musical score—it will describe the many, many things you will need to do well if you want to make the sort of music that will improve medication safety where you work. It will not be an easy or simple journey, but on any such journey, it is exceptionally helpful to have a robust guide, and in this book you will have that.

David W. Bates, MD, MSc
Chief Quality Officer and Chief, Division of General Internal Medicine
Brigham and Women's Hospital
Professor of Medicine, Harvard Medical School
Boston, Massachusetts

Preface

The use of medication is probably the most common intervention in today's medical practice. Medication use occurs in many different settings and involves many different health care professionals, as well as patients and their carers. Modern-day medication has made many previously untreatable conditions treatable, contributing to our increasing longevity and quality of life. However, the increasing numbers of available medications, together with the increased complexity of medication use, comes at a price—errors can occur at any stage of the medication use process, and errors involving today's powerful drugs can potentially lead to serious harm. Particularly over the last 20 years, a considerable amount of research, development, and quality improvement work has focused on tackling these problems, with the goal of preventing avoidable harm relating to medication use. This book is intended to pull together key aspects of such work and is aimed at health care professionals, researchers, medication safety officers, and all those concerned with medication safety.

The concept of "medication safety" potentially encompasses a wide range of areas, from drug development and formulation through to patients' medication-taking behaviors at home, incorporating pharmacovigilance, antimicrobial stewardship, and illicit drug use. In this book, we focus on one aspect: preventable harm relating to medication use in clinical practice, generally referred to as medication errors. Other textbooks address other important aspects of medication safety, such as pharmacovigilance and patient adherence. We also recognize that safety is only one aspect of health care quality; other aspects, such as effectiveness, patient-centeredness, timeliness, efficiency, and equity are equally vital, but are not considered here.

The book is split into three sections. Within each of these sections, each chapter has been written by one or more authors from around the world, who were chosen because of their standing in their field. Section I sets the scene by describing the extent of the problems that can occur in each of the main stages of medication use. Next, Section II considers a range of perspectives: measurement, education, psychology, sociotechnical theories, systems design, quality improvement, and safety culture. Chapter authors present the relevant theories in these fields and apply them specifically to the topic of medication safety. Finally, Section III considers approaches to interventions and solutions. We recognize that solutions can be very context specific—what works in one country or one setting may not be appropriate in another. Each chapter in Section III therefore includes one or two commentaries from relevant experts, including patients (who are undoubtedly experts in medication use), who provide a different perspective to the chapter authors. These commentaries highlight the key issues in their author's particular context and/or profession, and point out where there are important differences or similarities that someone else from their context should take into account when reading the chapter.

Together, these three sections describe an international body of work that shows not only how widespread medication errors are, but also how we are developing and implementing interventions that can reduce them and thus improve patient safety.

Bryony Dean Franklin and Mary P. Tully

Acknowledgment

We would like to thank Joe Schaffer, who spent the summer working on the referencing.

Editors

Mary P. Tully, BSc, MSc, PhD, FFRPS, MRPharmS is a reader in pharmacy practice at Manchester Pharmacy School, University of Manchester, United Kingdom. Her research interests are two-fold. She has a research team working on the processes and outcomes of hospital and non-medical prescribing, particularly on ways of reducing prescribing errors, and has published widely in this area. In addition, she is co-lead for the Patient and Public Involvement Theme for the Health e-Research Centre within the Farr Institute, with an interest in public attitudes and opinions on the secondary use of electronic health data for research. She has an honorary contract at Salford Royal Foundation Trust Hospital, supervises numerous research students, and teaches professional ethics to undergraduate pharmacy students. She was a visiting professor at the University of Uppsala, Sweden in 2003–2008, has served as chair of three and vice-chair of two international conferences, and has been a member of the scientific committee of many other conferences.

Bryony Dean Franklin, BPharm, MSc, PhD, FFRPS, FRPharmS is director of the Centre for Medication Safety and Service Quality (CMSSQ), a joint research unit between Imperial College Healthcare NHS Trust and University College London School of Pharmacy. She is a professor of medication safety at UCL School of Pharmacy, chair of the Centre for Patient Safety and Service Quality at Imperial College Healthcare NHS Trust, a visiting professor at Imperial College, and a theme lead for the NIHR Imperial Patient Safety Translational Research Centre.

Bryony has been involved with medication safety research for nearly 20 years. She has published widely on methods for studying medication errors and the frequency and causes of prescribing, dispensing, and medication administration errors. She has particular research interests in the evaluation of various technologies designed to reduce error and how we can better involve patients in developing and evaluating safety-related interventions. Her role includes research and teaching, as well as clinical practice as a hospital pharmacist.

Contributors

Jos Aarts
Department of Biomedical Informatics
School of Medicine and Biomedical Sciences
University at Buffalo
Buffalo, New York

Derar H. Abdel-Qader
Training and Development in Middle East and North Africa (MENA)
International Group for Education Consulting
Amman-Jordan, Jordan

Gerry Armitage
School of Health Studies
University of Bradford and Bradford Institute for Health Research
Bradford Royal Infirmary
Bradford, United Kingdom

Darren Ashcroft
Manchester Pharmacy School
University of Manchester
Manchester, United Kingdom

Melissa T. Baysari
Centre for Health Systems and Safety Research
Australian Institute of Health Innovation
Macquarie University
Sydney, Australia

Todd Boyle
Gerald Schwartz School of Business
St. Francis Xavier University
Antigonish, Nova Scotia, Canada

Gordon Caldwell
Worthing Hospital
Brighton, United Kingdom

Pascale Carayon
Department of Industrial and Systems Engineering
University of Wisconsin-Madison
Madison, Wisconsin

Maria Cordina
Department of Clinical Pharmacology and Therapeutics
University of Malta
Msida, Malta

Anique de Bruin
Department of Educational Development and Research
Maastricht University
Maastricht, the Netherlands

Ailsa Donnelly
United Kingdom

Tim Dornan
Department of Educational Development and Research
Maastricht University
Maastricht, the Netherlands

and

Centre for Medical Education
Queen's University Belfast
Belfast, Northern Ireland

Bryony Dean Franklin
UCL School of Pharmacy
London, United Kingdom

Carolyn Gamble
Canada and United Kingdom

TDF	Theoretical Domains Framework
TDM	Therapeutic drug monitoring
U.K.	United Kingdom
U.S.	United States of America/United States
VA	Veterans Affairs
WHO	World Health Organization

Section I

Problems in the Medication Use Process

1 Introduction

Bryony Dean Franklin and Mary P. Tully

Medication errors are common across all health care settings. Some lead to significant patient harm; others do not. Even so, the latter can reduce confidence among patients in their health care and/or reduce efficiency by requiring work to be redone by the health care professionals involved. Importantly, all indicate the presence of weaknesses in the system and the possibility for improvement to better support safe practice.

This section comprises individual chapters summarizing the evidence on error rates, types, and causes relating to each step of the medication use process across a range of health care settings. We consider the main stages of prescribing and monitoring (Chapter 2), dispensing (Chapter 3), administration (Chapter 4), and professional communication (Chapter 5). While these stages are considered in separate chapters for ease of presentation, this distinction is somewhat artificial. When things go wrong, multiple stages in the medication use process are often involved, and errors originating in one stage can also continue into other stages before becoming apparent (Carayon et al. 2014a; Huckels-Baumgart and Manser 2014). From the patient's perspective, such distinctions are also likely to be irrelevant (Franklin 2014). To take a simple example from our own experience: a doctor prescribed amlodipine for a patient newly admitted to a hospital, but contrary to hospital policy, wrote it on the drug chart using the brand name of ISTIN, written in capital letters. The pharmacist checked the medication order, but did not clarify on the prescription chart that the generic name was amlodipine, again contrary to local policy. The nurse saw the letters ISTIN and interpreted this as ISMN, a possible abbreviation for isosorbide mononitrate, even though abbreviations were not permitted in the hospital concerned. A dose of isosorbide mononitrate was administered. Fortunately, both have anti-hypertensive properties and the patient did not suffer harm. However, even this example illustrates how problems at every stage of the system can contribute to things going wrong.

Before considering errors in each of these stages in more detail, we would like to highlight some potentially confusing terminology within medication safety, most of which seems to involve three-letter acronyms: ADRs, ADEs, and DRPs. These are explained in more detail in the relevant chapters, but here we explain some general principles. Adverse drug events (ADEs) refer to any *harm* relating to the use of a medication; this harm may be due to an unpredictable adverse drug reaction (ADR) at a routine dose or due to a medication error resulting in a ten-fold overdose. Medication errors are *preventable* events that lead to actual or potential harm. As above, many do not result in actual harm; those that do cause harm are ADEs. A related concept is drug-related problems (DRPs), which include a wider range of issues, such as non-adherence and product unavailability, as well as medication errors. Studies in the field of medication safety also differ as to whether they focus on process (did a medication error occur, regardless of whether or not harm resulted?) or outcome (did harm occur, and if so, was it due to a medication error?).

We would also like specifically to emphasize the importance of context in medication safety. Internationally, there is seemingly infinite variation in the ways in which medications are prescribed, stored, prepared, dispensed, administered, and documented. Such differences arise due to variations in legislative frameworks, health care systems, and available resources, as well as many other drivers. Wide variation also exists in terminology. Different terms are used in different ways, and the same term can mean different things in different settings. For example, the terms outpatient care, ambulatory care, primary care, and general practice have all been used in various health care settings to mean something similar: health care accessed directly by people themselves as their first point of contact. In some settings, the terms may be interchangeable; in others, they have subtle but important differences. In this book, we generally refer to primary care (meaning health care accessed directly by the patient as the first point of contact) and secondary care (meaning any hospital-based specialist care that requires referral from primary care), without differentiating further. There is also very different terminology as to what are variably known as care homes, nursing homes, long-term care facilities, skilled nursing facilities, aged-care facilities, convalescent homes, or rest homes. We generally refer to "nursing homes," but recognize that not all employ registered nurses and that terminology varies widely around the world. When reading any literature in this field, we would therefore caution against making any assumptions about the setting and advise reading the original papers cited, where necessary, in order to assess applicability elsewhere. Only then can you reassure yourself as to the applicability of those findings to your context.

2 Prescribing and Monitoring Medication

Penny J. Lewis

CONTENTS

KEY POINTS

- Prescribing errors are common across international health care settings.
- Some can lead to significant harm to patients, and all errors demonstrate some weakness in current systems.
- The causes of errors are multifactorial and as such are likely to require multiple interventions in order to reduce their burden.

INTRODUCTION

The dictionary definition of the verb "prescribe" is to "advise and authorize the use of (a medicine or treatment) for someone, especially in writing" (Oxford University

Press 2014). Prescribing medication is the most common intervention used by medical practitioners. The enormous technical advances in drug development over the past few decades have led to an explosion of new drug therapies to treat an ever-increasing number of defined illnesses. People are living longer, leading to a greater reliance on pharmaceutical interventions. This has led to a year-on-year rise in the number of medications prescribed. In England, in primary care alone, over one billion prescriptions were dispensed in 2013, a figure that has doubled since 2003 (Health and Social Care Information Centre 2014). In the United States, the equivalent number is 3.5 billion (IMS Institute for Healthcare Informatics 2013). The average number of prescription items per person per year in the United Kingdom is thought to be 18.7 (Health and Social Care Information Centre 2013) and within the U.K. National Health Service (NHS), prescribing is the biggest expenditure after staffing.

The act of prescribing medication has been placed under various legal controls by international governments. These laws, such as the U.K. Medicines Act and the U.S. Federal Food, Drug, and Cosmetic Act, restrict what can be prescribed and by whom. Until relatively recently, prescribing has been the exclusive right of medical doctors and dentists but, more recently, in some countries, other health care professionals, such as nurses, pharmacists, and physiotherapists, have been granted the right to prescribe within various models and limits of practice. Different countries, as well as having different governance arrangements, also vary in their systems and processes for prescribing. For example, in the United Kingdom, the NHS sets out further formal controls on prescribing, in addition to the Medicines Act, limiting what can be prescribed under its service, such as requiring standardized NHS prescription forms in general practice. Countries also vary in their adoption of electronic prescribing (e-prescribing) and computerized decision support systems.

The act of prescribing is not a singular, discrete event, but rather a complex process that can be subdivided into two main components—deciding on the medication and then preparing the prescription for dispensing and/or administration. When making the prescribing decision, the prescriber should have an understanding of the patient's current clinical condition, which is derived from physical examination, descriptions of symptoms, and assessments of biochemical and/or physiological parameters. This information is then combined with knowledge of the patient's past medical history and any co-morbidities to determine the most appropriate treatment option. This option, in turn, depends upon the current evidence, the availability of medications or other treatments, the side effect profile of medication and, ideally, the patient's preferences. Figure 2.1 provides an overview of the various influences upon a prescribing decision.

After the medication has been chosen, the prescriber then needs to prepare the prescription. There are differences in how this is done both between countries and between sectors in the same country. In the United Kingdom, for example, prescriptions in primary care are overwhelmingly ordered via an electronic health record system and either printed to give to the patient or transferred electronically to a pharmacy, whereas in some other countries, equivalent prescriptions are often handwritten. In secondary care in the United Kingdom, on the other hand, inpatient prescriptions are commonly handwritten on paper "drug charts," which combine the prescription with the medication administration record. In the United States, the

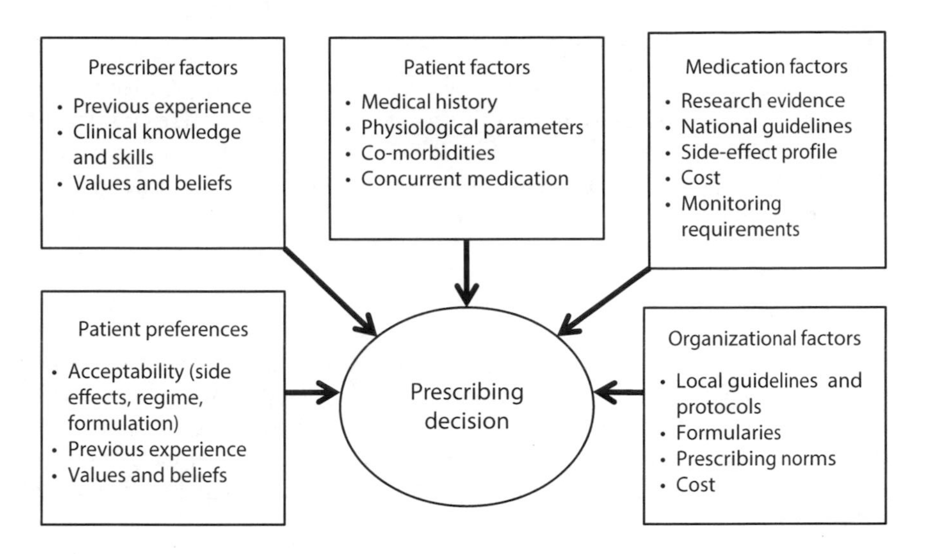

FIGURE 2.1 Influences on a prescribing decision.

prescription may be handwritten in the patient's medical record and then either transcribed by nursing staff into a paper or electronic medication administration record or may be entered directly by the prescriber into an electronic system. E-prescribing for all inpatients is more common in the United States, although still only available in less than half of hospitals (DesRoches et al. 2013), compared to only 13% of hospitals in England (Ahmed et al. 2013). This contextual information is relevant, as research into prescribing errors will be dependent upon the health care system in which it is undertaken—a topic covered in greater detail in Chapter 7.

Once the prescription has been prepared and the patient receives the medication, the prescriber must then monitor its effects to establish effectiveness and identify possible adverse drug reactions (ADRs). The frequency of this monitoring will vary depending on the particular medication prescribed.

The high volume of prescribing in health care, combined with the complexity of the task, makes prescribing a common source of error (Winterstein et al. 2004). Key studies of hospital prescribing report that preventable adverse drug events (ADEs) (i.e., harm due to medication use) occur most commonly at the ordering stage of the medication process (Bates et al. 1995b; Leape et al. 1995), a finding mirrored in general practice (Thomsen et al. 2007).

The potential implications of prescribing errors that reach the patient are substantial, with unnecessary and preventable harm to the patient being the most immediate outcome. Further repercussions might include hospital admission or increased hospital stay, loss of earnings, or even loss of life. ADEs (many of which are preventable) are estimated to cost the NHS £466 million ($847 million or €706 million) (2014) purely from the result of admission to a hospital (Pirmohamed et al. 2004), and further financial losses will be incurred following patients' legal action. A less quantifiable, but important impact of prescribing errors is a reduction in the public's trust in the health care system.

THE STUDY OF PRESCRIBING ERRORS

This section explores some fundamental concepts in relation to prescribing error study design. A lack of study uniformity makes this a difficult area to describe completely and succinctly; therefore, this section will cover the most common and widely accepted approaches adopted by rigorous research studies. Further detail on the measurement of errors is given in Chapter 7.

Prescribing Error Definitions

What constitutes a prescribing error is potentially subjective. Studies of prescribing errors that provide a definition (many do not) often develop their own. These vary in their inclusiveness and clarity, leading to difficulties in operational use (Lewis et al. 2009). Some studies do not provide a definition but provide a list of included errors. Particular differences can be found in whether or not definitions include proprietary prescribing, prescribing outside of a product license, and deviation from local formularies or guidelines. This lack of standardization is one of the many reasons that comparison across studies of prescribing errors is problematic.

A definition used frequently in the United Kingdom, although by no means universally (Lewis et al. 2009), is one developed by Dean and colleagues (2000). They used a two-stage Delphi technique to develop a practitioner-led definition of a prescribing error plus a list of examples of situations that should be included and excluded as errors. Internationally, the National Coordinating Council for Medication Error Reporting and Prevention's (NCC MERP) definition of *medication* errors is commonly used synonymously with the definition for *prescribing* errors (National Coordinating Council for Medication Error Reporting and Prevention 2011) and, in the United States, the American Society of Hospital (now Health-system) Pharmacists' (ASHP) definition of a medication error is sometimes applied (Anon. 1982).

When measuring prescribing error rates, careful consideration should be given to what is deemed an error and what might otherwise be considered suboptimal prescribing. It is useful to perform a quality check, reviewing individual errors to ensure that they fall within a study's definition. Review panels can provide some consistency and robustness to this process, especially when several data collectors are used. For example, a study of prescribing errors in a psychiatric hospital found that pharmacists collecting prescribing error data included omission of details such as allergy or patient identification number as prescribing errors (Stubbs et al. 2006). Although these were erroneous acts, they did not fall within the study's definition of a prescribing error and were therefore excluded by the review panel. Other studies require that the prescriber agrees that an error has been made before inclusion (Lesar et al. 1997b). Although discussion with the prescriber is appropriate and may reveal important information, this can be problematic if disagreement arises.

An overlapping concept with prescribing errors is that of inappropriate prescribing, which has sometimes been operationalized in similar ways. Appropriate prescribing, however, has been defined as the outcome of the process of decision-making that maximized net individual health gains within society's available resources (Buetow et al. 1996). Thus, it is possible for some inappropriate prescribing (such as prescribing an excessively expensive treatment) to not contain any errors. The terms are not synonymous therefore, and should not be used interchangeably.

Monitoring errors are often incorporated within studies exploring medication errors in general and therefore explicit definitions of monitoring errors are rare. The ASHP guidelines on preventing medication errors in hospitals include one such definition: "Failure to review a prescribed regimen for appropriateness and detection of problems, or failure to use appropriate clinical or laboratory data for adequate assessment of patient response to prescribed therapy" (American Society of Hospital Pharmacists 1993). Alldred and colleagues (2008) validated a definition of a monitoring error for application to older people in U.K. care homes. This was given as "when a prescribed medicine is not monitored in the way that would be considered acceptable in routine general practice. It includes the absence of tests being carried out at the frequency listed in the criteria for each medicine, with tolerance of +50%." This definition is clearly context specific, and transfer to other care settings and other countries may be difficult.

Data Collection Methods

Incident report review, document review, and self-report are the main methods used to detect prescribing errors in research studies. Using incident reporting systems to investigate the prevalence of prescribing errors leads to a major underestimation of the true error rate, as reporting rates are notoriously low. For these reasons, incident reports are not recommended for determining the prevalence of errors. Document review, in which trained researchers detect errors by examining drug charts, computerized prescribing logs, and medical notes, is most effective at detecting prescribing errors (Franklin et al. 2009). However, due to the time-consuming nature of this task, this approach to data collection is costly. A common approach to data collection is the self-report of prescribing errors identified by pharmacy staff. Detection of prescribing errors is part of the normal remit of a pharmacist and additional documentation of the errors they detect is a valuable and potentially cost-effective source of data.

Self-report of errors by pharmacists can be considered a process-based approach as errors are generally detected prior to reaching the patient. Studies that report only on errors that result in patient harm are considered to be outcome based; these would usually retrospectively review documents or incident reports in order to detect ADEs. Further details of these methods are given in Chapter 7.

Structured (Leape et al. 1995) and semi-structured (Dean et al. 2002a; Lewis et al. 2014) interviews and participant observation (Kopp et al. 2006) are common methods used to identify the causes of prescribing errors. Data on other characteristics, such as time of day and prescriber type, are often collected when gathering data on the prevalence of errors in order to explore associations between these factors and error rates (Folli et al. 1987; Hendey et al. 2005). This type of information is useful but can only be used to identify associations and not causality.

Prescribing Error Types and Severities

Prescribing errors are often grouped into decision-making errors (sometimes called clinical errors) and prescription writing errors (sometimes called clerical or technical errors). The former could include errors in the choice, dose, or frequency of medication to be prescribed, and the latter could include omission of the route of medication. Dosing errors are one of the most common types of prescribing errors (Lewis et al. 2009) and they are also potentially more serious than other types of prescribing errors (Bobb et al. 2004). Such errors can originate in the act of writing a prescription or in decision-making and, as with many errors, it is impossible to know which has occurred without discussion with the prescriber. Other commonly reported types of errors are incomplete prescription orders, omission of therapy, illegibility, errors in dosage interval, incorrect formulation, drug–drug interactions, and transcription errors (Lewis et al. 2009).

Many studies set out to classify the severity of errors that they detect. Overall, most prescribing errors are deemed minor or moderate both in hospitals (Dornan et al. 2009) and in general practice (Avery et al. 2013). However, different studies use different classification systems for severity, making comparisons across studies difficult (Lewis et al. 2009).

OVERVIEW OF THE NATURE AND PREVALENCE OF PRESCRIBING AND MONITORING ERRORS

This section provides a broad overview of key research findings from studies of the prevalence and nature of prescribing and monitoring errors. It should be noted that there is limited literature explicitly exploring monitoring errors and therefore much of the remaining chapter focuses on the prescribing error literature. Where there are studies exploring errors in monitoring, these are introduced in the relevant sections.

Rates of Prescribing and Monitoring Errors

One of the first large-scale attempts to record the number of actual and potential ADEs in medication ordering in the hospital setting was conducted in the United States (Bates et al. 1995a). Its finding that ADEs were common and often preventable highlighted the extent of this issue for patient safety. Preventable ADEs occurring at the ordering stage (i.e., during prescribing) were most common, representing 56% of preventable ADEs. Since 1995, many researchers in health care settings around the world have attempted to measure the exact scale and magnitude of prescribing errors (Lewis et al. 2009). The vast majority of these studies have been set in hospitals, with much less of a focus on general practice.

Despite this work, the true prevalence of prescribing errors in the hospital setting remains elusive, with wide variation in reported rates. Rates have been reported to be as low as 0.3% (Lesar et al. 1997a) and as high as 24.2% of medication orders (Webbe et al. 2007). A systematic review of prescribing error prevalence reported a median rate of 7% of medication orders (Lewis et al. 2009). The stark contrast between a study by Parke (2006), which recorded an error rate of 0.9% of medication orders, and the study by Webbe and colleagues (2007), which reported an error rate of 24.2% of medication orders, can be explained by the methods used. The former collected data via incident reports, and the latter collected data via pharmacists' routine checking of prescription charts.

The definition used can also partly explain variations in error rates as some studies' definitions will be broader than others. Variations in error rates can also be explained by the specialties in which the study was conducted. For instance, studies in high-dependency or intensive care may have higher rates of error due to the use of multiple medications (Cullen et al. 1997b) and the complexity of calculations used when giving medications intravenously (Kane-Gill et al. 2010), as discussed below.

There are far fewer studies on prescribing errors in general practice, yet errors occurring within this setting are an important cause of patient harm. Prescribing problems have been found to account for roughly one third of preventable drug-related admissions to hospital (Howard et al. 2007), some of which could be attributed to prescribing in general practice. The limited studies that do exist exhibit similar issues of heterogeneity as those in the hospital setting. A U.K. study reported an error rate of 7.46% of prescribed items after checking over 37,000 prescriptions written by doctors across three general practice surgeries (Shah et al. 2001). A U.S. study reported a very similar error rate of 7.6% after screening 1879 prescriptions (Gandhi et al. 2005). A more recent U.K. study screened patients' records from 15 general practices and found prescribing errors in 4.1% of prescribed items (Avery et al. 2013). However, other studies have reported error rates of 0.6% (Chen et al. 2005) and 40% (Claesson et al. 1995). This variation is likely to be attributable to the same factors as discussed in the hospital literature (i.e., differences in definitions, settings, and the type and rigor of data collection techniques).

Data on monitoring errors in general practice are scarce. A study reporting on monitoring errors found an error rate of 0.9% of prescribed items (Avery et al. 2013), yet when the analysis used only those items that specifically required monitoring as the denominator, this increased to 6.9%.

Stage of Hospital Stay

Interfaces of care are associated with high levels of prescribing errors. Much of the literature has reported that errors are most common at admission to a hospital (Dornan et al. 2009; Ryan et al. 2014). Prescribing on admission can be problematic as patients may be unable to provide a detailed history of their medication or may provide erroneous information. Particular difficulties can be experienced during weekends and evenings as general practices may be closed. In many settings, pharmacists will provide a medication reconciliation service that aims to provide a complete and accurate list of a patient's medications. Therefore, studies that record prescribing errors on admission sometimes include any *planned* omissions as prescribing errors. Errors are also common on discharge from a hospital (Abdel-Qader et al. 2010) or transfer to another care setting. The topic of errors at interfaces of care is discussed further in Chapter 5.

Drugs Associated with Error

Most studies do not record the numbers of medications prescribed per drug class, and it is clear that those medications that are most commonly prescribed will be associated with the most errors. Studies that report prescribing rates that take into account the denominators for individual classes of drugs may well tell a different story.

A systematic review of preventable ADEs in ambulatory care found that three categories of medication were responsible for over four-fifths of preventable ADEs—cardiovascular drugs, analgesics, and hypoglycemic agents (Thomsen et al. 2007). Preventable ADEs can arise from errors in any part of the medication process, but studies of hospital prescribing errors have also found that cardiovascular medications and analgesics are associated with high numbers of errors (Bobb et al. 2004; Lesar et al. 1997b). Anti-infectives are also commonly associated with prescribing errors (Bobb et al. 2004; Lesar et al. 1997b) and, in a systematic review of hospital prescribing errors, were the class most commonly associated with error from 22 included studies (Lewis et al. 2009). Other common associations reported were with cardiovascular, central nervous system, and gastrointestinal medications.

The classes of medications most commonly associated with high numbers of errors in general practice also reflect those drugs that are commonly prescribed; for example, musculoskeletal drugs (Gandhi et al. 2005), drugs used on the skin, eye, ear, nose, and throat (Avery et al. 2013), infection medications (Gandhi et al. 2005), and cardiovascular medications (Avery et al. 2013), but also drugs for malignant disease and infection (Avery et al. 2013). Cardiovascular drugs have been found to be most commonly associated with monitoring errors (Avery et al. 2013).

PRESCRIBING ERRORS IN SPECIALIST AREAS AND SPECIFIC PATIENT GROUPS

Rates and types of prescribing errors are likely to vary depending upon the specific setting of data collection and with particular subgroups of patients. Pertinent areas in which nuances of prescribing errors are evident include critical care, mental health, and extremes of age (i.e., children and the elderly). Errors can also occur that are unique to e-prescribing systems and these are covered in detail in Chapter 19.

Prescribing Errors in Critical Care

A study of 24 U.K. critical care units reported a prescribing error rate of 15% (Ridley et al. 2004)—twice the median error rate reported in a systematic review of all prescribing errors in hospitals (Lewis et al. 2009). However, the potential severity of prescribing errors appears similar to that reported in the overall hospital setting. High pharmacological intervention, complex cases (often with multi-organ failure), and high use of the intravenous route may contribute to a higher incidence of errors in the critical care setting (Cullen et al. 1997b). The nature of errors that occur in this setting reflect the clinical area, with more errors involving intravenous medications, particularly potassium chloride, and cardiovascular medications (Ridley et al. 2004).

However, these types of medications are also likely to be prescribed more frequently within this setting and do not *necessarily* represent a higher risk, as data on usage rates for route and drug class are often not reported.

Prescribing Errors in Mental Health

The relatively few studies conducted in this setting have found errors to be at least as common as in other settings (Keers et al. 2014; Paton and Gill-Banham 2003; Stubbs et al. 2004). A large study by Stubbs and colleagues (2004) in nine U.K. mental health units found an error rate of 2.4% of prescription items; a more recent U.K. study in three mental health hospitals found an error rate of 6.3% of prescription items. Grasso and colleagues (2003) used a review team to detect prescription errors and reported an error rate of 165 errors per 1000 patient days in one U.S. psychiatric hospital. Variations in error rates are likely to be at least partly due to the same issues as in other settings. For example, Grasso and colleagues (2003) used a review team to detect errors, whereas Stubbs and colleagues (2004) and Keers and colleagues (2014) asked pharmacists to detect errors during the course of their routine work.

However, there are also particular prescribing challenges in psychiatry that may alter the characteristics of prescribing errors in this setting. These include prescribing "off label," as many psychotropic medications are prescribed outside of their product license (Chen et al. 2006). Prescribing "off label" means that information regarding the dose, route, duration, and monitoring requirements is not available in standard reference texts, which may be a risk for error. Prescribing for patients who are unable to communicate or may not question their care is another challenge that will be encountered in this setting, leaving patients more susceptible to error. Patients with mental health needs will often have other co-morbidities requiring treatment, and the need to prescribe many non-psychotropic medications could generate problems for mental health prescribers, who may have less familiarity with such medications. Studies have found that errors are more likely to occur when prescribing non-psychotropics than psychotropics in a psychiatric setting (Haw and Stubbs 2003; Rothschild et al. 2007). There is also some evidence that more serious errors are more likely to involve non-psychotropics than psychotropics (Stubbs et al. 2006).

Overall, the types of prescribing errors reported in mental health are similar to those reported elsewhere, with errors in dosing commonly encountered (Paton and Gill-Banham 2003; Stubbs et al. 2006). Less common error types specific to mental health include prescribing psychotropic medications for a detained patient without the necessary legal paperwork to authorize the prescription (Stubbs et al. 2004) or prescribing a drug without registering the patient with the drug company—a specific requirement for clozapine (Stubbs et al. 2006).

Prescribing Errors in Children

The rate of potential ADEs is reported to be three-times higher in pediatric inpatients than in adults, with the majority (79%) occurring at the prescribing stage (Kaushal et al. 2001). A U.K. study of the incidence and nature of prescribing errors in pediatric inpatients identified 391 prescribing errors in 2955 medication orders across five hospitals, resulting in an error rate of 13.2% (Ghaleb et al. 2010). This is slightly higher than the 8.9% error rate reported in a study utilizing the same detection method and a similar (but not identical) error definition across all wards of 19 U.K. hospitals (Dornan et al. 2009). In general practice, a U.K. study of prescribing errors reported an 87% excess risk of error if the patient was younger than 15 years old (Avery et al. 2013).

Prescribing for children presents particular challenges. The wide range of bodyweights and weight-based dosing leads to wide variation in prescribed doses. Prescribers might then overlook erroneous dosages as they are not immediately obvious—what comprises a large overdose in a small child could be an appropriate dose for an older child. The added complexity of performing calculations based on individual patient parameters provides greater opportunities for error than is perhaps encountered when prescribing for adults. Young patients' inability to communicate also adds complexity to the prescribing scenario (Kaushal et al. 2010). Furthermore, the way that neonates and children handle drugs can be complicated by their immature renal and hepatic systems. Children are different from adults in the way that their bodies distribute and handle drugs (pharmacokinetics) and the effects that drugs have on their bodies (pharmacodynamics) in comparison to the adult population. For example, intravenous ciprofloxacin is often prescribed at 400 mg twice a day in adults but, because of increased hepatic extraction by the liver (due to the high volume-to-surface area ratio in children), children must be prescribed this drug three times daily (10 mg/kg up to a maximum dose of 400 mg three times a day). This may seem counterintuitive, but failure to increase the frequency could be fatal in severe infections.

Dosage errors are believed to be the most frequent type of medication error in children, as in adults (Wong et al. 2004). Tenfold dosing errors, due to misplacement of a decimal point or addition of a zero, are particularly problematic, often causing considerable harm if they reach the patient (Wong et al. 2004). The fact that medications are often given using formulations developed for use in adults (as pediatric formulations are often not available) probably goes some way to explaining why errors may reach the patient. Due to the substantial risk of dosing errors resulting from erroneous calculations, it is not surprising that e-prescribing is often promoted as a mechanism to reduce errors in settings such as pediatric units (Fontan et al. 2003). As with the adult population (Lewis et al. 2009), anti-infectives, analgesics, and sedatives are commonly associated with errors in children (Kaushal et al. 2001). Seriously ill children are more likely than less acutely ill children to experience a prescribing error (Kozer et al. 2002), which is reflected in the finding that the highest rates of prescribing errors occur in pediatric critical care.

Prescribing Errors in the Elderly

Within general practice, it has been reported that the elderly (as well as the young) are twice as likely to be given a prescription with an error (Avery et al. 2013). There are many reasons for this; for example, elderly patients will often have multiple comorbidities and are often subject to polypharmacy. The greater the number of prescribed items, the higher the risk of error (Ryan et al. 2014).

As with children, elderly patients are likely to experience harm from medications due to age-related differences in pharmacodynamics and pharmacokinetics. For example, the elderly are much more sensitive to the effects of benzodiazepines (Mangoni and Jackson 2004), leading to ADRs such as over-sedation and hallucinations, even with small overdoses—such neuropsychiatric events are a common type of preventable ADE (Gurwitz et al. 2005). Furthermore, cognitive impairment and dementia may make prescribing difficult due to limited dialogue with the patient and the reduced ability of the patient to act as a defense against prescribing errors. Indeed, a study by Nirodi and Mitchell (2002) found that errors in dosage and illegibility were more common in patients with dementia than for those with functional psychiatric illnesses. Furthermore, elderly patients will frequently move across different care settings, being exposed to errors at the interface.

A large U.K. study of the prevalence of medication errors in care homes (referred to as "long-term care facilities" in the United States or "aged-care facilities" in Australia) reported a prescribing error rate of 8.3% of prescriptions, with nearly four in ten care home residents experiencing an error (Barber et al. 2009), a similar rate to that found in the hospital setting. Yet a general practice study of errors, discussed previously, found that for patients over the age of 75 years, the prevalence of error was 41.9% of prescribed items, compared to 18.7% for patients of all ages (Avery et al. 2013), with the risk of error for elderly patients nearly twice that for 15–64-year-olds. A U.S. outcome-based study investigating the incidence of ADEs in long-term care facilities reported that 59% of 338 preventable ADEs were associated with prescribing errors and that 80% were associated with errors at the monitoring stage, most commonly inadequate monitoring (Gurwitz et al. 2005). Monitoring errors are a significant problem in the elderly, due to the types of medications that are more commonly prescribed in the elderly, such as warfarin, diuretics, and angiotensin-converting enzyme inhibitors. In a U.K. study, nearly one in five care home residents who were prescribed medications that required monitoring had an error in the monitoring of this medication (Barber et al. 2009).

THE CAUSES OF PRESCRIBING ERRORS

Although many studies have attempted to determine the prevalence of prescribing errors, far fewer studies have set out to uncover the causes. A systematic review identified only 16 empirical studies of the causes of and factors relating to prescribing errors (Tully et al. 2009). This section will explore some of the commonly cited causes of prescribing errors within the literature. Many studies of the causes of errors apply Reason's accident causation model (Reason 2000) or the Theoretical Domains Framework (Lawton et al. 2012) when analyzing their findings. These are covered in detail in Chapter 8 and therefore not repeated here.

It is important to note that much of the literature on causes of errors has focused on hospital prescribers, with very little conducted in primary care. There has also been a general focus on the prescribing of junior doctors, as they have been identified as being more likely to make errors and responsible for the majority of prescribing in the hospital setting (Dornan et al. 2009; Lesar et al. 1990; Ryan et al. 2014).

In line with the accident causation model (Reason 2000), the error-producing conditions at the time of prescribing include busyness, time of day, and complexity of the patient; latent conditions include wider management or design decisions around, for example, e-prescribing software. The various factors shown to be statistically associated with prescribing errors, as well as being suggested by prescribers themselves, are discussed below.

Individual Factors

Prescribers' lack of knowledge, skills, or experience is often cited as a cause of prescribing errors (Coombes et al. 2008b; Dean et al. 2002a; Leape et al. 1995). Knowledge of drug doses and consideration of potential drug interactions or contra-indications are particular issues highlighted in the literature (Coombes et al. 2008b; Dean et al. 2002a, b; Lewis et al. 2014). Doctors' ability to perform drug calculations is also problematic, particularly for children (Menon et al. 2006; Rowe et al. 1998) and has been found to be unrelated to experience and qualifications (Menon et al. 2006). There is some debate on the influence of medical education on prescribing errors, but there is no doubt that prescribing training and pharmacology education is a concern to many, including junior doctors themselves (Dornan et al. 2009). As such, prescribers have been the focus of various educational interventions, as discussed in Chapter 16.

Experience is a factor of interest to many researchers, which is unsurprising as junior doctors have been found to make more errors than their senior counterparts (Dornan et al. 2009). Therefore, some large studies have focused exclusively on this particular group. The EQUIP (Dornan et al. 2009) and PROTECT (Ryan et al. 2014) studies set out to investigate the reasons why U.K. junior doctors make prescribing errors. The results of these studies found many similar causes to those reported in the general literature and both identified these causes as being multi-factorial; lack of experience was not the single main cause.

The physical and mental well-being of prescribers is another commonly cited individual factor. Tiredness is a reported factor in errors made by more junior doctors, as these doctors may work long hours, including through the night (Coombes et al. 2008b). Other physical causes of errors discussed in the literature include hunger, thirst, low mood, and feeling unwell (Coombes et al. 2008b; Dean et al. 2002a). Doctors' attitudes towards prescribing is also a contributory factor in prescribing errors, with some attaching a low importance to prescribing, particularly re-prescribing, such as re-writing a patient's drug card or writing out a patient's medication for discharge without the need for any alteration of therapy (Barber et al. 2003; Coombes et al. 2008b; Dean et al. 2002a). The fostering of such attitudes may have its origins in the medical culture and, as such, can also be considered a latent failure of the system.

Team Factors

A lack of support from senior doctors is a common theme discussed by junior doctors when talking about prescribing errors in hospitals (Coombes et al. 2008b; Dornan et al. 2009). There are some differences between specialties, with doctors on surgical rotations reporting feeling being particularly isolated. More generally, doctors may perceive pharmacists as a safety net for prescribing errors, affording less vigilance when prescribing (Dornan et al. 2009).

Poor or absent communication is a commonly cited error-producing condition. Problems occur, for example, when instructions from seniors to prescribe medications lack detail and merely include the name of the drug (Nichols et al. 2008; Ross et al. 2012).

The nursing team has also been implicated as a contributory factor in doctors' prescribing errors, with some doctors reporting being unduly influenced by nursing requests and incorrectly assuming nurses will always flag up clinically relevant patient information when prescribing (Dornan et al. 2009; Lewis et al. 2014). Within general practice, nurses' requests for prescriptions were also believed to contribute to prescribing errors made by doctors (Slight et al. 2013).

Work Environment Factors

Low staffing levels and high workload are common issues in the health care setting, leading to an increased pressure on prescribers. Such an environment has been commonly linked with the emergence of prescribing errors, both in hospitals (Coombes et al. 2008b; Dornan et al. 2009; Ryan et al. 2014) and in the general practice setting.

Within hospitals, prescribing when on call during weekends and overnight has been associated with a higher rate of prescribing errors (Hendey et al. 2005), and these periods have been identified by prescribers as difficult times to prescribe due to increased busyness. Prescribing outside of normal working hours has sometimes been associated with a lack of support when prescribing, demonstrating the interlinking of error-producing conditions (here, team factors in addition to the working environment). Other working environment factors include a lack of facilities such as a desk (Dean et al. 2002a) or computer (Lederman and Parkes 2005).

Patient Factors

Prescribing for patients who are unfamiliar and for whom the prescriber has no prior medical knowledge of (as is often the case in specialties such as accident and emergency or while working out of hours) is thought to contribute to prescribing errors (Dornan et al. 2009), as is prescribing for patients with multiple or complex conditions or who are in need of acute care (Coombes et al. 2008b; Dean et al. 2002a). Difficulties communicating with patients who have dementia or who do not speak English have also been linked with errors (Coombes et al. 2008b). A study of errors in general practice reported that women were less likely than men to experience a prescribing or monitoring error (Avery et al. 2013) and that patients who were prescribed multiple medications were more commonly associated with prescribing errors (Avery et al. 2013). However, in their study of hospital prescribing errors, Fijn and colleagues (2002) did not find patient gender or the number of co-prescriptions predictive of prescribing error.

Latent Conditions

Latent conditions in prescribing errors include design features of medication charts (Coombes et al. 2008b; Dornan et al. 2009) and e-prescribing systems. Such errors can arise from a lack of familiarity with a particular format, a problem that is particularly encountered by junior doctors in the United Kingdom who move between hospitals. The layout of medication charts can also contribute to error with, for example, sections that need to be folded out and are easily missed.

Other latent conditions associated with prescribing errors include the steep hierarchal structure present in hospital practice, which can prevent more junior doctors from questioning seniors and seeking help (Coombes et al. 2008b; Lewis et al. 2014). Junior doctors' fear of appearing incompetent can also prevent them from asking for help (Dean et al. 2002a) or checking their prescribing decisions. Furthermore, less experienced doctors may blindly follow orders, assuming their seniors' decisions to be correct (Lewis et al. 2014).

Management decisions surrounding staffing and working hours have been linked to the occurrence of prescribing errors (Coombes et al. 2008b; Patterson et al. 2004). Such decisions have ramifications for workload, a major contributory factor in prescribing errors, as discussed previously.

CONCLUSION

It is clear that prescribing errors can be attributed to many factors. Importantly, studies reveal that these factors do not occur in isolation; the principal conclusion of this body of work is that prescribing errors generally occur due to several failures, rather than having a singular discrete cause.

There is considerable empirical evidence on the prevalence of prescribing errors in health care. Such studies are disparate in their methodologies, making comparisons difficult across studies, and therefore future work should consider the utilization of established and validated methods. Studies of the causes of errors have established some common thematic patterns and future work could now focus upon interventions to reduce error occurrence. It is generally agreed that interventions need to be both complex and multifaceted as the causes of errors are multifactorial. Section III includes several chapters that will address these issues in more detail.

3 Dispensing Medication

K. Lynette James

CONTENTS

KEY POINTS

- Dispensing errors can arise at any stage of the dispensing process, and most commonly involve labeling errors and supply of the wrong drug and the strength or dose of medication.
- Errors occur due to human fallibility, organizational deficiencies, and working conditions. Look-alike, sound-alike drugs, low staffing levels, high workloads, and interruptions have all been cited as causes.
- The reported rate of dispensing errors is influenced by the definitions and error detection methods used, making it challenging to compare studies.
- Understanding dispensing errors allows risk-reduction strategies to be implemented to ensure the quality and safety of patient care.

INTRODUCTION

Dispensing involves the preparation and supply of medication according to a prescription or medication order. Dispensing is a complex process that has evolved from the compounding of natural products to the supply of commercially manufactured and packaged medicines. Nevertheless, errors still occur.

THE EVOLUTION OF DISPENSING PRACTICE

The origins of dispensing can be traced back to ancient civilizations that investigated the medicinal properties of plants, animals, and minerals ("galenicals"). The resulting remedies and compounding techniques continued to be used by physicians, apothecaries, druggists, and pharmacists until after the Second World War. The post-war period was heralded as the golden age of drug discovery. Pharmaceutical companies began to develop a wide range of new chemical drugs formulated into various dosage forms and marketed as proprietary products. The need to compound medicines has diminished as the availability of commercial products has increased. By the 1980s, about 70% of medications were commercially available as solid dosage forms, mainly tablets and capsules (Anderson 2001), and dispensing mainly involved supplying commercially manufactured medications.

These days, pharmacists are only occasionally required to extemporaneously compound a medicinal product according to a formula. These extemporaneously prepared products are generally dispensed for an individual patient when a comparable commercial product is unavailable or unsuitable (American Society of Hospital Pharmacists 1994; Jackson and Lowey 2010). Extemporaneously dispensed medicines are not subject to the same regulatory safeguards as commercially manufactured products and the pharmacist is accountable for their quality, safety, and efficacy (American Society of Hospital Pharmacists 1994; Jackson and Lowey 2010). In 2000, a pharmacist and trainee were prosecuted for manslaughter in the United Kingdom after a baby died from a 20-fold overdose of chloroform, following administration of peppermint water prepared in a community pharmacy using the wrong strength of chloroform water (Anon. 2000). Owing to the risks associated with compounding, pharmacists in many countries prefer to outsource extemporaneous dispensing to specialist pharmaceutical compounding companies (Candlish et al. 2003).

In contrast, the increasing use of parenteral medication in hospitalized patients has led to greater involvement of hospital pharmacists in the preparation of intravenous (IV) admixtures. Aseptic or sterile manufacturing units are common in hospital pharmacies. In these units, specialist pharmacists assume responsibility for the preparation of IV admixtures, cytotoxic medications, and parenteral nutrition within a controlled environment, thereby minimizing microbiological contamination and maintaining the physicochemical stability of the prepared product.

Dispensing practice has therefore evolved considerably. Modern dispensing practices also adopt a holistic approach of pharmaceutical care centered on the supply of medicines to patients. This approach enables pharmacists to utilize their knowledge of pharmacology, drug disposition, formulation, and therapeutics to rationalize and optimize patients' medication, as well as dispensing the medication required.

THE DISPENSING PROCESS

Dispensing is a complex, multistage process encompassing all activities from receipt of the prescription to issue of the medication to the patient. Figure 3.1 illustrates the key stages involved in both community and hospital pharmacies (American Society of Hospital Pharmacists 1980; James et al. 2009; Spivey 2012; The Pharmacy Guild of Australia 2013), although details can vary between countries as a result of differences in regulatory frameworks and drug distribution systems (such as unit dose versus original pack dispensing). For example, in some settings, pharmacists might not clinically review prescriptions for appropriateness nor complete a final accuracy check of dispensed medication (Silva et al. 2008).

Extemporaneous compounding additionally involves completion of appropriate documentation. These documents typically specify the compounding formula, preparation procedure, details of all raw materials used (name, strength, quantity or volume, batch number, and expiration date), storage requirements, and expiration for the compounded product and the staff involved in compounding. Guidance on documentation and environmental requirements for extemporaneous and sterile compounding are available from professional and regulatory bodies (American Society of Health-System Pharmacists 2014a; Beaney 2005; Medicines and Healthcare Products Regulatory Agency 2014; Pharmaceutical Society of Australia 2010).

The dispensing of medication must be supervised by an appropriately qualified practitioner. In many countries, this role is undertaken by a pharmacist, who accepts the legal and professional responsibility for verifying the legitimacy of the prescription, clinically assessing the appropriateness of medication for an individual patient, and undertaking any remedial action required. In addition, pharmacists generally perform final accuracy checks on dispensed medications to ensure medications are accurately supplied in accordance with the prescription. Suitably trained support staff may assist the pharmacist in dispensing and compounding. For example, in Scandinavia, dispensary staff may be prescriptionists with a Bachelor's degree in pharmacy (Nordic Pharmacy Association 2008), compared to the five year Master's degree held by pharmacists. In contrast, pharmacy technicians in some

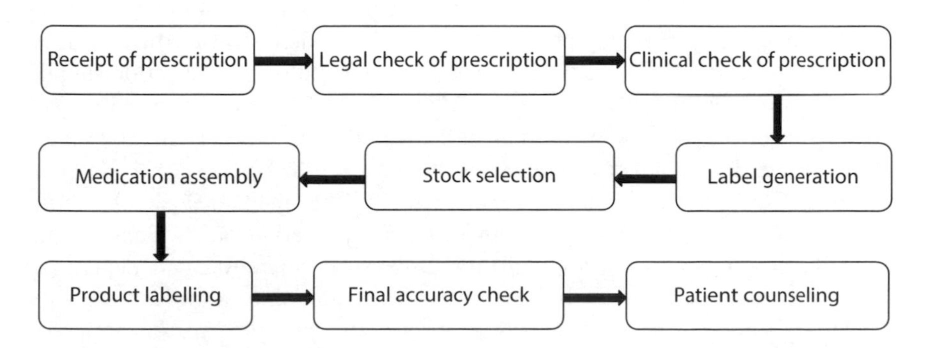

FIGURE 3.1 Typical steps involved in dispensing medication to individual patients in community and hospital pharmacies (inpatient and outpatient). In the inpatient setting, the legal and clinical check may take place on the ward.

other countries complete an accredited course that involves experiential learning and didactic teaching (American Society of Health-System Pharmacists 2014b; General Pharmaceutical Council 2014; The Pharmacy Guild of Australia 2014). In some countries, accredited checking technicians who have completed a competency-assessed training program can perform final accuracy checks on dispensed medications, provided the prescription has previously been scrutinized by a pharmacist for clinical appropriateness (General Pharmaceutical Council 2014).

While there is some variation in dispensing processes within community and outpatient settings, there is even wider variation in the hospital inpatient and care home settings. This next section will consider the different distribution systems used to dispense medications to individual hospital inpatients or care home residents, including various forms of unit dose and multi-dose dispensing.

Unit Dose Dispensing

Many hospitals within the United States, Canada, and other countries use unit dose dispensing. This involves the separate packaging of each individual dose unit (such as a tablet or capsule) in a protective sealed pack, labeled with the name, strength, and dose of the medicine, batch number, and expiry date (O'Leary et al. 2006). These unit dose packs are typically supplied to the wards via drug trolleys or cabinets; packs may also include the patient's name. During drug administration, nurses retrieve the required unit dose(s) and administer the medication to the patient. Ward-based automated dispensing systems, comprising an electronic storage cabinet and/or drug trolley, are used in many hospitals to aid the storage and supply of unit doses. These are discussed further in Chapter 20.

Multi-Dose Dispensing

Many terms are used to describe this dispensing system, including original pack dispensing, calendar pack dispensing, and multi-dose individual patient dispensing. It is used within community pharmacies and many hospitals in the United Kingdom and Australia, and involves a quantity of medication being supplied and labeled for an individual patient. Community pharmacies supply the quantity of medication specified on the prescription. However, in hospitals, the quantity of medication supplied will depend on local practice. Historically, U.K. hospital pharmacies generally supplied inpatients with seven days' supply of medication, dispensed from larger multi-dose containers in the pharmacy department. However, in January 1994, European Directive (92/27) came into force, which specified that medicines should be dispensed in an original pack complete with a patient information leaflet. Consequently, many U.K. hospitals adopted original pack dispensing, whereby inpatients are supplied with an original manufacturers' pack, often equating to a month's supply of medication. When labeled with directions for use, these original packs can then be used to fill discharge prescriptions on the ward, reducing patients' waiting times for medication and facilitating efficient hospital discharge.

Monitored Dosage Systems

Community and hospital pharmacies sometimes prepare medications in monitored dosage systems to aid patient adherence with medication regimens and to support the administration of medication to patients in care homes. There is a variety of monitored dosage systems available, both multi-dose and unit dose. Multi-dose systems comprise cassettes that contain all the different doses of medication to be administered to an individual at a particular time of day, for each day of the week. An information sheet outlining the details of each medication (name, strength, form, and directions) is usually attached. In contrast, unit dose systems are often supplied to care homes supported by registered nurses. These involve the preparation and use of a separate blister card for each medication that a patient is prescribed. Each blister card contains individually sealed unit doses of the medication to be administered each day of the month. The blister card is labeled with the patient's name, drug, dose, and administration time.

Monitored dosage systems can be filled manually by pharmacy staff who remove the medication from the original manufacturers' packs and place the doses in the appropriate compartment of the cassette or blister card prior to sealing the system, or filled using automated filling systems.

DISPENSING ERRORS

Dispensing errors can arise at any stage of the dispensing process. Technical errors may occur during the process of checking prescriptions for legal validity, product assembly or preparation, labeling, and completion of appropriate documentation or registers. Clinical judgment errors may also occur during the screening of prescriptions for clinical appropriateness or during patient counseling. However, in both practice and research, the term "dispensing error" generally refers only to technical errors arising during the process of medication preparation and supply, and excludes clinical judgment or counseling errors.

Dispensing errors have been defined as "deviations from a written prescription occurring during the dispensing process of selecting and assembling medication (drug/content errors), generating and affixing of dispensing labels (labeling errors) and issue of the dispensed products to patients (issue errors)" (James et al. 2013). Dispensing errors include deviations from local protocols, procedures, guidelines, and professional and regulatory references. They can be subdivided into

- External errors—dispensing errors detected and reported after medication has left the pharmacy, which may or may not lead to patient harm
- Internal errors ("near-misses")—dispensing errors detected during dispensing before the medication has been issued to the patient, ward, or clinical area

Error categories used to classify various types of dispensing error are outlined in Table 3.1, although few studies include documentation errors (Beso et al. 2005).

TABLE 3.1
Types of Dispensing Errors

Drug/Content Errors	Labeling Errors	Issue Errors	Documentation Errors
Wrong drug dispensed	Wrong drug on label	Failure to supply drug	Incorrect patient details recorded in register
Wrong strength dispensed	Wrong strength on label	Drug supplied that was cancelled/not prescribed	Incorrect prescriber's details recorded in register
Wrong formulation dispensed	Wrong formulation on label	Incorrect outer bag label	Incorrect drug name recorded in register
Wrong quantity dispensed	Wrong directions on label	Drugs incorrectly bagged	Incorrect drug strength recorded in register
Expired medication dispensed	Wrong cautionary label/warnings on label	Issued to the wrong patient/ward	Incorrect drug form recorded in register
Drug not dispensed	Wrong storage information on label	Failure to supply/inappropriate measuring device such as spoons or oral syringes	Incorrect drug quantity recorded in register
Errors specific to extemporaneous or aseptic preparation:	Wrong patient name on label	*Errors specific to extemporaneous or aseptic preparation:*	Incorrect date of supply/prescription date recorded in register
Wrong drug used in preparation	Missing/wrong expiry on label	Incorrect drug, diluent, and quantitative particulars for unofficial products	Other incorrect documentation error
Wrong strength of drug used in preparation	Completely incorrect label on product	Incorrect container/packaging of product	*Errors specific to extemporaneous or aseptic preparation:*
Wrong form of drug used in preparation	*Errors specific to extemporaneous or aseptic preparation:*	Missing/incorrect administration devices such as filters	Incorrect/missing patient information transcribed from prescription to worksheet
Wrong diluent used in preparation	Incorrect route of administration on label		Incorrect/missing drug details (name, form, and strength) on worksheet
Wrong strength of diluent used in preparation	Incorrect rate of administration on label		Incorrect working formula on worksheet
			Incorrect calculation of ingredients required to prepare product
			Incorrect method of preparation/compounding on worksheet

Methods of Investigating Dispensing Errors

Various techniques have been used to detect and investigate dispensing errors (Tables 3.2 and 3.3). These include self-report, incident reports, case note review, and observation, all of which are discussed in more detail in Chapter 7. Covert patients have also been used to investigate the incidence and types of dispensing errors in the community pharmacy setting. Allan and colleagues (1995) employed three covert patients to each present one prescription per pharmacy, for warfarin, carbamazepine, or theophylline. One hundred randomly selected pharmacies were visited by the covert patients and the dispensed medication checked by a researcher. Twenty-four errors were detected among 100 dispensed prescriptions, giving an error rate of 24% (Table 3.2). The use of covert patients potentially provides an accurate estimate of the incidence of dispensing errors. However, only a limited number of drugs, with higher risk of dispensing errors, were included in the study, potentially inflating the error rate. The use of covert patients to detect dispensing errors is also time-consuming and expensive, and potentially raises ethical issues.

Interviews (Beso et al. 2005; Gothard et al. 2004), surveys (Bond and Raehl 2001; Peterson et al. 1999), and focus groups (Chua et al. 2003) have been used to explore the causes of dispensing errors. Surveys enable a large amount of data to be collected from a wide range of pharmacy staff in a short period. However, survey response rates may be low. Focus groups and interviews allow in-depth explorations of the circumstances surrounding errors, but participants may not fully divulge details of the errors with which they were involved. Participant responses to surveys, interviews, and focus groups are also potentially subjective and rely on recall.

Incidence and Prevalence of Dispensing Errors

As for other types of medication error, quantifying the incidence of dispensing errors is complex. These issues are discussed in more detail in Chapter 7. The rate of dispensing errors is influenced by the operational definitions (error definition, classification of error types, and error rate) and error detection methods employed (James et al. 2009). Allan and Barker (1990) recommended that error rates be calculated as "the number of actual errors (incorrect in one or more ways) divided by the opportunities for error...multiplied by 100 to arrive at a percentage." Nevertheless, there is wide variation in the nature of the numerators and denominators used in the literature (Tables 3.2 and 3.3). Furthermore, studies using observation and covert patients to detect dispensing errors generally reveal higher dispensing error rates than those using self-reported incidents or case note review. Consequently, it is often not possible to compare studies in a meaningful way.

The most common types of both external and internal (near-miss) dispensing errors identified by both incident reporting and observation generally involve the wrong drug, wrong strength, wrong dosage form, wrong quantity, or wrong labeling of medication with incorrect directions (James et al. 2009).

TABLE 3.2

Examples of Studies Reporting the Incidence of External and Internal ("Near-Miss") Dispensing Errors in Community Pharmacies, Presented according to Country

Country	Study	Error Type	Number of Pharmacies	Data Collection Method	Key Findings
Denmark	Knudsen et al. (2007)	External	40	Incident reports	203 errors in 1,466,043 prescriptions (0.01%)
United Kingdom	Kayne (1996)	External	4	Incident reports	50 reports in 5004 prescriptions (0.99%)
	Chua et al. (2003)	External and internal	4	Incident reports	39 (0.08%) external errors and 247 (0.48%) internal errors in 51,357 dispensed items
	Ashcroft et al. (2005b)	External and internal	35	Incident reports	50 (0.04%) external errors and 280 (0.22%) internal errors in 125,359 dispensed items
	Franklin and O'Grady (2007)	External	11	Observation	95 errors in 2859 dispensed items (3.3%)
	Barber et al. (2009)	External	259 care home residents	Observation	187 errors in 1915 dispensed (or omitted) items (9.8%)
	Franklin et al. (2014b)	External	15	Observation	Non-electronically transmitted prescriptions: 608 labeling errors in 12,624 items dispensed (4.8%) and 170 content errors in 12,624 items dispensed (1.3%) Electronically transmitted prescriptions: 277 labeling errors in 3733 items dispensed (7.4%) and 52 content errors in 3733 items dispensed (1.4%)
United States	Allan et al. (1995)	External	100	Covert patients	24 of 100 prescriptions had one or more external error type (24%)
	Flynn et al. (2002)	External and internal	Unclear	Observation	91 of 5784 prescriptions had one or more external error type (1.6%) 74 of 5784 prescriptions had one or more internal error type (1.3%)
	Flynn et al. (2003)	External	50	Observation	77 of 4481 prescriptions had one or more error type (1.7%)
	Teagarden et al. (2005)	External	1 (mail order)	Observation	16 errors in 21,252 prescriptions (0.08%)
	Hoxsie et al. (2006)	External	18	Observation	5 errors in 950 observations (0.5%)
	Varadarajan et al. (2008)	External	1	Observation and incident reports	16 of 3337 prescriptions had one or more error type (0.48%)

TABLE 3.3
Examples of Studies Reporting the Incidence of External and Internal ("Near-Miss") Dispensing Errors in Hospital Pharmacies, Presented according to Country

Country	Study	Error Type	Number of Hospitals	Data Collection Method	Key Findings
Australia	Thornton et al. (1990)	External	1	Observation	Numerator and denominator not specified; error rate 0.08%
	De Clifford (1993)	External	1	Observation	6 errors in 964 items dispensed across three study periods (0.6%)
	Parke (2006)	External	1	Observation	0 errors in 24,174 dispensed items (0%)
Brazil	Anacleto et al. (2007)	External	1	Observation	345 drugs with one or more errors in 422 prescriptions (81.8%)
Spain	Gonzales et al. (2005)	External	1	Observation	20 errors in 2827 medication lines (0.7%)
United Kingdom	Bower (1990)	External	5	Incident forms	21 errors in an unspecified number of dispensed items (0.02%)
	Banning (1995)	Internal	1	Incident forms	180 errors in 38.846 dispensed items (0.4%)
	Spencer and Smith (1993)	External	19	Incident forms	181 errors in 1,002,095 dispensed items (0.02%)
	Wu (2000)	Internal	1	Incident forms	352 errors in 37.828 dispensed items (0.94%)
	Noott and Phipps (2003)	External and internal	1	Incident forms	52 external errors in 332,501 dispensed items (0.02%) pre-intervention; 16 external errors in 165,212 dispensed items (0.009%) post-intervention 351 internal errors in 332,501 dispensed items (0.11%) pre-intervention; 359 external errors in 165,212 dispensed items (0.22%) post-intervention
	Barker (2003)	External	1	Incident forms	Numerator and denominator not specified; error rate 0.02%
	Beso et al. (2005)	External and internal	1	Incident forms	32 external errors (0.02%) and 130 internal errors (2.7%) in 194,584 dispensed items
	Adedoye (2007)	Internal	1	Incident forms	89 of 3930 dispensed items contained one or more error type (2.26%)
	Franklin et al. (2008)	Internal	2	Observation	Site 1: 245 errors in 9161 items dispensed pre-automation (2.7%); 93 errors in 9289 items dispensed post-automation (1.0%) Site 2: 217 errors in 16,283 items dispensed pre-automation (1.3%), 46 errors in 7894 items dispensed post-automation (0.6%)
	James et al. (2007)	External	1	Incident forms	42 of 391,467 dispensed items containing one or more error type (0.01%) pre-automation; 32 of 429,911 (0.008%) post-automation
	James et al. (2008)	External	20	Incident forms	915 of 5,564,969 dispensed items containing one or more error type (0.02%)

(Continued)

TABLE 3.3 (Continued)

Examples of Studies Reporting the Incidence of External and Internal ("Near-Miss") Dispensing Errors in Hospital Pharmacies, Presented according to Country

Country	Study	Error Type	Number of Hospitals	Data Collection Method	Key Findings
	James et al. (2011)	External and internal	5	Incident forms	35 of 221,670 dispensed items containing one or more external error type (0.02%) 291 of 221,670 dispensed items containing one or more internal error type (0.13%)
	James et al. (2013)	Internal	1	Incident forms	235 of 36,719 dispensed items containing one or more error type pre-automation (0.64%); 147 of 52,808 dispensed items post-automation (0.28%)
United States	Guernsey et al. (1983)	External	1 outpatient pharmacy	Observation	1165 errors in 9394 prescriptions (12.4%)
	Buchanan et al. (1991)	External	1 outpatient pharmacy	Observation	369 of 10,888 prescriptions containing one or more error type (3.39%)
	Kistner et al. (1994)	External	1	Observation	1229 of 9849 prescriptions containing one or more error type (12.5%)
	Flynn et al. (1996)	External	1	Observation	164 of 5072 prescriptions containing one or more error type (3.23%)
	Cina et al. (2006)	External and internal	1	Observation	1059 external errors (0.75%) and 4016 internal errors in 140,755 doses (2.9%)
	Klein et al. (1994)	External	1	Observation	34 errors in 4029 doses with a manual system (0.84%); 25 in 3813 doses with an automated system (0.65%)
	Becker et al. (1978)	External	1 satellite pharmacy	Observation	45 errors in 2420 doses dispensed by a pharmacist (1.86%); 21 errors in 2403 doses dispensed by a technician (0.87%)
	Spooner and Emerson (1994)	External	1	Observation	34 errors in 3116 doses dispensed by a pharmacist (1.09%); 10 errors in 7571 doses dispensed by a technician (0.13%)
	Ness et al. (1994)	External	3	Observation	107 errors in 49,718 doses dispensed by a pharmacist (0.22%); 50 errors in 55,470 doses dispensed by a technician (0.09%)
	Seifert and Jacobitz (2002)	External	Regional poisons center	Case note review	40 errors in 6450 unintentional therapeutic exposures (0.62%)

Community Pharmacy

Table 3.2 summarizes the rates of external and internal (near-miss) errors reported in the community pharmacy setting. The majority of researchers have quantified both internal and external error rates using the numbers of prescriptions or items dispensed as a denominator. Others have used surveys to explore their occurrence. For example, Newgreen and colleagues (2005) identified that 162 dispensing errors were reported to the Pharmacy Board of Australia between July 1, 1998 and December 31, 2004. A postal survey of Tasmanian pharmacists revealed that 71% ($n = 134$) of respondents were aware of making a dispensing error in the previous six months, with a median of three errors made by each pharmacist (Peterson et al. 1999).

One study has specifically examined the incidence of dispensing errors associated with monitored dosage systems for care home residents. Barber and colleagues (2009) identified an external dispensing error in 9.8% of prescription items dispensed (or omitted) for 94 English care home residents. Most were labeling errors (7.3%), with drug or content errors (2.3%) and clinical errors (0.2%) being less common. Cassette-type monitored dosage systems had higher odds of a dispensing error compared to blister card systems (odds ratio = 2.88, 95% confidence interval = 1.5–5.6, $p = 0.0012$), due to an increase in labeling errors (Alldred et al. 2009).

Hospital Pharmacy

Table 3.3 summarizes the rates of external and internal (near-miss) dispensing errors in a variety of countries across the world. As might be expected, the rate of internal (near-miss) dispensing errors (U.K.: 0.28%–2.7%; U.S.: 2.9%) is higher than that for external dispensing errors (U.K.: 0.008%–0.02%; U.S.: 0.75%).

INCIDENCE AND TYPES OF EXTEMPORANEOUS AND ASEPTIC DISPENSING ERRORS

Very little is known about the incidence of aseptic and extemporaneous dispensing errors. In 2006, the U.S. Food and Drug Administration conducted a survey of extemporaneously compounded medicines produced by pharmacies (Food and Drug Administration 2006). A total of 125 active pharmaceutical ingredients and 73 compounded finished products were sampled and analyzed according to the techniques specified in the U.S. Pharmacopeia. All 125 samples of active pharmaceutical ingredients were found to contain the correct drug and quantity of active substance. Results of analytical tests could only be presented for 36 of the 73 compounded finished products sampled. Of these 36 products, 33% ($n = 12$) failed the assay for active ingredient and/or the test for content uniformity. The potency of compounded finished products ranged from 67.5% to 268.4% of the quantity specified on the product label. All compounded finished products sampled passed the identity test.

Various extemporaneous preparation errors have been reported in the literature (Gudeman et al. 2013). Variations in the potency of compounded medications have also been identified by the State Boards of Pharmacy in Missouri, Ohio and Texas (Gudeman et al. 2013). Analysis of nitroglycerin (glyceryl trinitrate) ointment prepared in accordance with 84,000 prescriptions revealed that 46% of preparations failed tests for potency and content uniformity (Azarnoff et al. 2007). Furthermore, Mahaguna and colleagues (2004) identified that only one of ten progesterone suppositories prepared by pharmacies met the specified potency. There was also great variation in the pH of the suppositories, and some were contaminated with microorganisms. Use of incorrect preparation techniques has also been reported, accounting for 6% of all medication errors reported to the U.S. Pharmacopeia between 1999 and 2000 (Cowley et al. 2001).

Table 3.4 summarizes the literature evaluating the incidence of external and internal (near-miss) errors in the preparation of injectable medicines in pharmacy sterile compounding units. There is considerable variation in the rate of injectable preparation errors, with research based on self-reported incident forms not surprisingly reporting lower error rates than observation. The most common types of errors occurring during the aseptic preparation of injectable medicines in the pharmacy environment are incorrect doses, labeling errors, incorrect diluents, and incorrect final containers (Bateman and Donyai 2010; Flynn et al. 1997; Limat et al. 2001).

CAUSES OF DISPENSING ERRORS

Human error is frequently implicated as a cause of medication errors. Reason's accident causation model (1990) has been used to investigate errors in health care, as discussed further in Chapter 8. Application of this theory to staff reports of dispensing errors given during interviews and responses to incident report forms reveals that internal dispensing errors (near-misses) are caused by a complex interweaving of active failures, error-producing conditions, and latent conditions (Beso et al. 2005; Gothard et al. 2004; James et al. 2008). Slips included selecting the wrong drug or strength of medication and computer selection errors; lapses included forgetting to remove inappropriate cautionary labels during label generation. Knowledge-based mistakes included assumptions that products were interchangeable, whereas rule-based mistakes involved dispensing from labels rather than prescriptions and supplying the wrong quantity of medication. Latent conditions included workforce planning, similar drug names, poor labeling or packaging of medicines, and unclear presentation of drug selection lists on the computer software used to generate labels (Beso et al. 2005; Gothard et al. 2004; James et al. 2008). Error-producing conditions cited as factors contributing to dispensing errors were work environment (high workload, low staffing, noise, lighting, interruptions, or distractions), the team (poor communication and lack of leadership), personal factors (lack of knowledge and skills, competence, ill health, and personality), the task (complex prescriptions and a lack of protocols or guidance), and patient factors (complex patients, language, personality, and social factors).

Workload

There is conflicting evidence regarding the impact of workload on dispensing errors. Research conducted by Guernsey and colleagues (1983) over 12 days in a U.S. hospital outpatient pharmacy involved observers inspecting dispensed medication to

TABLE 3.4

Examples of Studies of the Incidence of External and Internal ("Near-Miss") Errors in the Preparation of Injectable Medicines in Pharmacy Sterile Manufacturing Units, Presented according to Country (IV = Intravenous)

Country	Study	Error Type	Number of Hospitals	Product	Data Collection Method	Key Findings
France	Ranchon et al. (2011)	External	1	Cytotoxic	Incident reports	26 errors in 22,138 preparations (0.12%)
	Limat et al. (2001)	External	1	Cytotoxic	Incident reports	140 errors in 30,819 preparations (0.45%)
Spain	Serrano-Fabia et al. (2010)	External	1	Cytotoxic	Incident reports	58 errors in 16,473 preparations (0.35%)
	Escoms et al. (1996)	External	1	Cytotoxic	Incident reports	314 errors in 4734 preparations (6.6%) and 94,680 opportunities for error (0.3%)
United Kingdom	Bateman and Donyai (2010)	External and internal	Multiple	All IV admixtures including cytotoxic and parenteral nutrition	Incident reports	24 of 958,532 preparations containing one or more external error type (0.0025%) 4667 of 958,532 preparations containing one or more external error type (0.49%)
United States	Flynn et al. (1997)	External	5	All IV admixtures including cytotoxic and parenteral nutrition	Observation	145 of 1679 doses containing one or more error type (8.6%)
	Sacks et al. (2009)	External	1	Parenteral nutrition	Incident reports	18 errors in 4730 preparations (0.4%)

detect dispensing errors. The study reported a linear relationship between the number of prescriptions dispensed and potentially serious dispensing errors. However, a similar study by Kistner and colleagues (1994) in a different U.S. pharmacy found no correlation between the number of prescriptions dispensed per hour and the number of dispensing errors. In the United Kingdom, James and colleagues (2013) investigated the association between workload and internal dispensing errors in a hospital pharmacy. Over a six-week period before and after the installation of an automated dispensing system, data on dispensary workload were determined by a non-participant pharmacist observer. Pharmacy staff were also asked to self-report details of internal dispensing errors on standardized forms. Automation significantly increased throughput (median 9.2 items/person/hour pre-automation and 13.7 post-automation; $p < 0.001$) and decreased the rate of internal dispensing errors (0.64% of items dispensed pre-automation and 0.28% post-automation; $p < 0.0001$). A positive association was identified between throughput and the occurrence of internal dispensing errors both pre-automation ($p = 0.015$) and post-automation ($p < 0.001$). Pharmacy automated dispensing systems are discussed further in Chapter 21.

Lighting

Research suggests that lighting levels have a significant effect on dispensing errors. Buchanan and colleagues (1991) investigated the impact of three different lighting levels (484, 1098, and 1572 lux) on the incidence of dispensing errors in a high-volume U.S. military pharmacy. During the study period, 10,889 prescriptions were inspected by a pharmacist observer for dispensing errors. The study found that increasing the lighting level from 1098 to 1572 lux significantly reduced the dispensing error rate from 3.9% to 2.6% of dispensed items ($p < 0.01$). However, increasing illumination from 484 to 1098 lux had no impact on the dispensing error rate.

Interruptions, Distractions, and Noise

Flynn and colleagues (1996, 1999) investigated the impact of interruptions, distractions, and ambient sounds on dispensing errors in a U.S. ambulatory care pharmacy. Over 23 days, two video cameras recorded the pharmacy working environment and a researcher inspected dispensed medications for errors. Details of dispensing errors were compared to information on ambient noise, interruptions, and distractions obtained by reviewing the videotapes. Pharmacy staff were considered to have been interrupted if an external stimulus resulted in the cessation of the dispensing activity. In contrast, staff were considered distracted if they continued with the dispensing task. During the study period, 2022 interruptions occurred, affecting 1143 prescriptions, and 2457 distractions affected 1329 prescriptions. Prescriptions affected by interruptions and distractions had a higher dispensing error rate. Analysis of interruptions ($p = 0.004$) and distractions ($p = 0.012$) per half hour were both significantly associated with dispensing errors (Flynn et al. 1999). In contrast, it has been reported that unpredictable audible stimuli, controllable audible stimuli, and noise were statistically associated with a lower incidence of dispensing errors (Flynn et al. 1996). However, as the noise level increased, the dispensing error rate initially increased and then decreased. Various explanations were proposed for this finding, including that audible stimuli increased pharmacist concentration on the dispensing task. It was also suggested that pharmacists may have a mechanism that controls stress induced by noise, allowing them to dispense accurately (Flynn et al. 1996).

The same study evaluated the relationship between pharmacist distractibility and dispensing error rate. To determine distractibility, each pharmacist was asked to complete a "group embedded figures test" for field dependence. The study found that pharmacists who had a higher score (higher tolerance to distractions) made fewer dispensing errors (Flynn et al. 1999).

Look-Alike, Sound-Alike Drugs

Anto and colleagues (2011) investigated the relationship between similar drug names and drug selection errors (labeling and content errors). Details of medications involved in 911 dispensing errors reported by a U.K. hospital pharmacy between January 2005 and December 2008 were analyzed. The study found that dispensing frequency influenced the probability of drug selection errors occurring. Dispensing errors were more likely to occur if the prescribed medication was rarely supplied by the pharmacy. A large proportion (84%) of drug selection errors also involved pairs of medications that were orthographically similar (related to spelling), as quantified using the edit distance. The edit distance is defined as "the number of characters required to be deleted/inserted to transform one word to another" (Anto et al. 2011). The distribution of drug selection errors peaked at edit distances of two and seven, corresponding to the medication strength and dosage form, respectively. Furthermore, approximately 50% of the medication pairs involved in drug selection errors (prescribed and incorrectly supplied) appeared in close proximity to each other on the computer system menus.

CONCLUSION

Dispensing medication is complex and inherently risky. This chapter demonstrates that dispensing errors do occur, with internal ("near-miss") errors occurring more frequently than external errors. Dispensing errors most commonly involve labeling errors and the supply of the wrong drug and strength/dose of medication. They occur due to human fallibility, organizational deficiencies, and working conditions. Look-alike, sound-alike drugs, low staffing levels, high workloads, and interruptions have been cited as causes. Understanding dispensing errors allows risk-reduction strategies to be implemented to ensure the quality and safety of patient care.

4 Administration of Medication

Katja Taxis

CONTENTS

KEY POINTS

- Medication errors are common in the inpatient setting—a median medication administration error rate of about 10% of opportunities for error has been reported, excluding wrong time errors.
- Much higher error rates of about 50% have been reported for intravenous medication.
- Next to wrong time errors, omission of medication and dosing errors are consistently reported as the most common error types.
- Observation-based methods give a good picture of the quality of the process of drug administration, but other methods are needed to detect rare, serious incidents.

INTRODUCTION

As highlighted earlier, the use of medication is one of the most frequent interventions in today's medical practice. Exact figures are not available, but the majority of medication administrations involve patients taking their own medication at home. In other instances, medication taken at home is administered by formal or informal carers (for example, spouses, parents, children, nurses, or care assistants). The remaining administrations take place in the hospital and other institutional settings, such as nursing or residential homes, predominantly by nurses or care assistants.

Most of what we know about medication administration errors (MAEs) is based on studies in the hospital inpatient setting. One of the earliest studies investigated MAEs in anesthesia (Beecher and Todd 1954). Subsequently, much of the early work was carried out by researchers such as Barker and his group at Auburn University (1962), who wrote some fundamental papers on observation-based medication error research (for example, Allan and Barker 1990). Such studies of MAEs triggered the development of solutions such as the unit dose system, the role of the clinical pharmacist in safe medication management, and various technological developments to support medication use in the hospital setting, many of which are discussed in more detail in Section III. Much less is known about MAEs in medication administered by nurses or carers in patients' own homes, and few studies have been done in other organizational settings, such as nursing homes. Separately, there is a large body of literature on patients' adherence with their medications. Some forms of non-adherence (mainly unintentional non-adherence) can be considered as MAEs (Barber et al. 2005; Furniss et al. 2014), but non-adherence or other types of administration errors by the patient themselves or their informal carers are outside the scope of this chapter.

The main focus of this chapter is MAEs in the hospital setting. We start by describing the process of drug administration and by giving a brief overview of specific methodological issues in the study of MAEs. The prevalence of MAEs in the inpatient setting, their nature, and their common causes are then discussed. This is followed by a summary of what is known about MAEs in nursing homes and patients' own homes.

THE PROCESS OF MEDICATION ADMINISTRATION

The process of medication administration can be broadly divided into a preparation step (for example, dissolving soluble tablets or drawing up a solution with a syringe) and the actual drug administration step. MAEs can occur at either step. Compared to other types of medication errors, such as prescribing errors, administration errors are less likely to be intercepted before they reach the patient (Leape et al. 1995).

METHODOLOGICAL ISSUES IN MEDICATION ADMINISTRATION ERROR RESEARCH

General issues relating to the measurement of medication error rates are discussed in more detail in Chapter 7; here, some specific issues relevant to MAEs are highlighted.

Definition of a Medication Administration Error

A wide range of definitions are used for MAEs, both in research and in clinical practice. A frequently used definition is "A MAE is the administration of a dose of medication that deviates from the prescription, as written on the patient

TABLE 4.1
Subcategories of Medication Administration Errors

Omission	Wrong dose preparation/preparation technique
Wrong dose/improper dose	Wrong patient
Wrong dosage form	Wrong rate of administration
Deteriorated drug	Drug incompatibility
Extra dose/unordered dose	Wrong time
Wrong drug	Wrong administration technique
Unordered/non-prescribed drug	Fast intravenous bolus
Wrong route	Wrong diluent

Source: Adapted from McLeod, M.C., Barber, N., and Franklin, B.D. 2013. *BMJ Qual Saf* 22: 278–289.

medication chart, or from standard hospital policy and procedures. This includes errors in the preparation and administration of intravenous medicines on the ward" (Ghaleb et al. 2010). Prescriptions are assumed to be correct. Some studies also include transcribing errors as administration errors (Taxis et al. 1999). In general, deviations from hospital procedure, such as not checking patients' identity bands or not labeling intravenous (IV) infusions, are not considered to be errors if they do not result in a MAE. There are some studies that report such procedural deviations separately, including failures to follow standard aseptic techniques during the preparation and administration of IV medications (Taxis et al. 2004; Westbrook et al. 2010b). Importantly, there are often inconsistencies between studies in terms of inclusion or exclusion criteria, even if they use the same definition of a MAE. Examples include the inclusion or exclusion of wrong time errors and doses left at the patient's bedside without the nurse witnessing their consumption (McLeod et al. 2013). Such variation can have a major impact on MAE rates. Table 4.1 gives an overview of the commonly used subcategories of MAEs (from McLeod et al. 2013).

Calculating Medication Error Rates

To calculate a MAE rate, a suitable numerator and denominator are required. The numerator can be the number of doses with one or more medication errors, so that a dose is either correct (associated with no errors) or incorrect (associated with one or more errors). Alternatively, some studies use the total number of errors as the numerator. In this case, the error rate can theoretically exceed 100% (Franklin et al. 2009). The majority of studies use the notion of "total opportunities for error" as the denominator, defined as the sum of the total number of doses ordered plus any unordered doses given (or the equivalent: the number of doses given plus any doses omitted). Some studies instead use the total number of observed doses as the denominator, which therefore excludes any doses omitted. Studies are not always clear regarding whether or not omissions are included (McLeod et al. 2013).

Data Collection Methods

The gold standard method for detecting MAEs is the observation of actual practice: a researcher observes drug preparation and administration and identifies any errors. The purpose of the observation can be "disguised" (an alternative explanation is given so that those observed are not aware of the true purpose of the study) or undisguised. Few studies have investigated the effect of the observer's presence, but those that have explored this issue suggest there is little impact (Dean and Barber 2001). A disadvantage of observation is that it is time-consuming for data collectors and it is potentially intrusive for those observed. Therefore, in most studies, observation data are only available for a few days or weeks. As a consequence, rare events may not be detected using this method, and so other techniques, such as incident reporting systems, are valuable for gathering examples of such errors. Methods less commonly used include the analysis of reports to poison centers, focusing on serious and rare events mainly outside the hospital setting (for example, Lavon et al. 2014). There are also some studies investigating MAEs using simulation techniques, including a recent example in ambulatory chemotherapy (Prakash et al. 2014). Other methods for studying medication errors are discussed in Chapter 7, together with methods for assessing their severity.

MEDICATION ADMINISTRATION ERRORS IN THE HOSPITAL SETTING

Incidence and Prevalence

In recent international reviews, the median MAE rate was around 10% of the total opportunities for error (Berdot et al. 2013; Keers et al. 2013b), excluding wrong time errors. However, a formal meta-analysis has not been conducted due to the heterogeneity in denominators and error classifications (Berdot et al. 2013). A much higher rate of errors of around 50% (excluding wrong time errors) has been reported in studies of IV medication (for example, Taxis and Barber 2003b), as highlighted in an international review (Keers et al. 2013b). Similarly, in an analysis of observational studies in the United Kingdom, errors were five-times more likely in IV doses than in doses given via other administration routes (McLeod et al. 2013). In contrast, Berdot and colleagues (2013) found a median MAE rate of about 10% of the total opportunities for error, including four studies on injectable medicines. These differences are largely due to different study inclusion criteria and reflect the lack of standardization in this field.

The incidence of error may also vary according to clinical discipline and type of ward studied. There is some suggestion of higher error rates in pediatrics than adults, due to the need to calculate doses individually, based on the patient's age, weight, body surface area and/or their clinical condition, as well as the high proportion of unlicensed and off-label medication used in this setting (Ghaleb et al. 2010; Kaushal et al. 2001). However, other data suggest similar error rates in the two settings (Keers et al. 2013b). A slightly higher error rate has also been reported in the critical care setting (Kiekkas et al. 2011), but this may be at least partly due to the high percentage of IV medications administered in this clinical area. The majority of evidence originates from Europe, the United States, and Australia, while the few studies carried

TABLE 4.2
Examples of Observation-Based Studies of Medication Administration Errors in the Hospital Inpatient Setting

Reference	Participants and Setting	Error Rates	Most Common Types of Errors	Comments
Calabrese et al. (2001)	5 intensive care wards in different hospitals, United States	3.3% of 5744 observations	Wrong infusion rate, not specified errors, dose omissions	Only specified high-risk medications included. Every time data collection took place for an individual patient, this was considered an observation, so each observation could be related to one or more doses of medication.
Ghaleb et al. (2010)	10 pediatric wards across 5 hospitals, United Kingdom	19.1% of 429 opportunities for error	Preparation errors, wrong rate of IV administration, wrong time	A dose with a preparation step (such as dissolving the medication) was counted as having two opportunities for error; a dose not involving preparation procedures was counted as one opportunity of error.
Westbrook et al. (2010b)	6 wards in 2 teaching hospitals, Australia	25.0% of 4271 drug administrations	Wrong time, wrong IV administration rate, wrong dose	Error rate calculated using doses having at least one clinical error as a numerator and drug administrations as a denominator.
Nguyen et al. (2014)	An ICU and a PSU in an urban hospital, Vietnam	ICU: 48.9% of 407 opportunities for error; PSU: 64.1% of 281 opportunities for error	Wrong administration technique, wrong preparation technique, wrong dose	Error rate defined as the number of doses with clinically relevant errors divided by the total opportunities for error, with one opportunity for error per dose administration.

Note: ICU = intensive care unit; IV = intravenous; PSU = post-surgical ward.

out in resource-restricted settings suggest similar or higher MAE rates (Alsulami et al. 2013; Nguyen et al. 2014; Romero et al. 2013). Table 4.2 summarizes the results of some studies of MAE rates in hospitals around the world.

Types and Nature of Errors

Wrong time errors are among the most frequent types of errors identified in observation-based studies. Studies using a strict definition of doses given within 30 minutes of the prescribed time report a median wrong time error rate of 27%; studies using a value of within 60 minutes report median error rates of less than 10% (Berdot et al. 2013). In general, wrong time errors are regarded as clinically minor; some studies do not even include this type of error at all (Keers et al. 2013b). However, there are clearly exceptions, such as for insulin, where timing can be critical (Cousins et al. 2011). Next to wrong time errors, omission of medication and dosing errors are consistently reported as being the most common. In studies of IV doses, errors were common in reconstituting the drug as well as in the actual administration to the patient, such as involving the wrong administration rate (McDowell et al. 2010). In pediatrics, ten-fold administration errors are a potentially very serious error type (Doherty and McDonnell 2012).

Few studies have focused on MAEs associated with the use of infusion devices. A study in a U.S. hospital by Husch and colleagues (2005) found one or more errors in 67% of 426 infusions given using an IV pump, where the definition of an error also included procedural issues, such as not having the infusion rate on the medication label or a patient not wearing an identification bracelet. No rate on the label, unauthorized medications, patient identification errors, and rate deviations were the most commonly identified. Pump programming mistakes were relatively infrequent, but had potential to cause harm.

Studies that report error rates by medication category cite errors as being commonly associated with the following: nutrition and blood, gastrointestinal system, cardiovascular system, infections, and central nervous system. However, such an analysis does not always take into account the frequency of use of these medicines, so it remains unknown whether certain drug classes have a much higher risk of error (Keers et al. 2013b).

Studies that include the potential severity of MAEs report that the majority are likely to have little clinical impact (Berdot et al. 2013; Keers et al. 2013b). In some instances, health care professionals may prevent harm by intervening (Bates et al. 1995a); in other cases, potentially harmful errors are not actually harmful in an individual patient. However, given the large number of drug administrations taking place every day in every hospital, even a small proportion of clinically important errors can cause considerable patient harm.

As highlighted above, methods such as spontaneous reporting may be used to ascertain information on rare and serious MAEs. The following are examples of serious MAEs based on case reports, although such sources do not provide meaningful data on their prevalence or incidence (Anon. 2005; Cousins et al. 2011; Dart and Rumack 2012; Gilbar 2014; Medicines and Healthcare Products Regulatory Agency 2010; Noble and Donaldson 2010; Reeve and Allinson 2005):

- Ten-fold dosing errors of paracetamol (acetaminophen) infusions in neonates and babies leading to serious liver toxicity, due to confusion between the dose volume in milliliters and the dose in milligrams.
- Antineoplastic agents, such as vincristine, which are intended for IV administration, but inadvertently given intrathecally, leading to profound toxicity and death, due to IV and intrathecal drugs not being adequately separated in time and/or place.
- Inadvertent fast injections of potassium chloride leading to fatal arrhythmias, due to the resemblance of potassium chloride ampoules with solutions such as sodium chloride.

- Ten-fold insulin dosing errors leading to serious hypoglycemia, seizures, coma or death, due to the abbreviation "u" being used for "units" and misinterpreted as a zero.

Such data, in combination with observation data, have led to a number of medications being classified as "high-alert medications," defined as drugs that bear a heightened risk of causing significant patient harm when used in error, including MAEs. The U.S.-based Institute of Safe Medication Practices (ISMP) periodically updates a list of potential high-alert medications based on error reports submitted to the ISMP National Medication Errors Reporting Program (Institute for Safe Medication Practices 2014c), the published literature, and input from practitioners and safety experts (Institute for Safe Medication Practices 2014b). Examples of medications on the list are antithrombotics, insulins, narcotics, and concentrated potassium chloride for injection, as well as epidural and intrathecal medications.

Causes

Reason's accident causation model (2000), discussed further in Chapter 8, has been used to inform investigations of the causes of MAEs. Slips and lapses are among the most commonly identified failures, followed by violations and knowledge-based mistakes (Keers et al. 2013a). Factors contributing to errors and violations can be grouped as being related to patients, staff, medications, or working conditions (Table 4.3), although little is known about the relative contributions of these factors. According to Reason, there are also further organizational (higher-level)

TABLE 4.3
Factors Contributing to Medication Administration Errors

Factor	Examples
Patient	Lack of, difficulty with, or delays in waiting for intravenous access, absent/sleeping patient
Staff	Health and personality: fatigue, sickness/discomfort, stress, nervousness
	Inexperience, inadequate training, lack of knowledge
Working environment	Absence of policies or complicated/unsuitable policies
	Insufficient equipment such as computers
	Communication failures such as illegibly written information, failure to pass on information
	Lack of supervision of junior nurses
	Heavy staff workload (end of shift/patient transfer pressure), lack of qualified staff
	Distractions and interruptions
	Lack of medicine supply (logistics)
Medication	Look-alike medicines, sound-alike medicine names

Source: Adapted from Keers, R.N. et al. 2013a. *Drug Saf* 36: 1045–1067; McDowell, S.E. et al. 2010. *Qual Saf Health Care* 19: 341–345.

contributing factors. For MAEs, these would include decisions at a hospital level, such as whether to have pharmacy-based drug preparation or supply ready-to-use products. Other higher-level factors concern the organization and content of training and education, such as the extent to which the knowledge and skills of nurses relating to medication administration are formally assessed during their basic training (Gonzales 2012).

Causes of MAEs have been identified in some observational studies alongside establishing their frequency. In some cases, observation has been complemented with informal conversation with nurses (Taxis et al. 2004). Other studies have used formal interviews (for example, Gladstone 1995), questionnaires, or self-report methods to identify MAE causes (for example, Orser et al. 2001). Alternatively, formal prospective or retrospective risk assessment methods have been used (Guchelaar et al. 2005). Some studies have tested the impact of specific causes, such as distractions, on MAEs rates (Westbrook et al. 2010b) or investigated personality characteristics related to risk-taking and error-prone behavior (Gonzales 2015). Many other causes have been suggested based on incident reporting, mainly from large incident reporting systems, as described later in Chapter 15.

MEDICATION ADMINISTRATION ERRORS IN THE NURSING HOME SETTING

Fewer studies of MAEs have been carried out in nursing homes than in hospitals. People admitted to nursing or care homes may have a greater risk of MAEs. There are several potential reasons for this. In general, such patients receive a large number of medicines. A considerable proportion of patients may also have swallowing difficulties, which may result in inappropriately manipulating medication. Many residents are cognitively impaired, which makes them less able to be involved in the medication administration process themselves or to alert staff about potential problems. Finally, because of age-related pharmacokinetic and pharmacodynamic changes and frailty, they are probably more vulnerable to experiencing adverse drug events once an error has occurred (Avorn and Gurwitz 1995). However, a large U.S. study comparing error rates between hospitals and skilled nursing facilities (nursing homes) did not find significant differences in error rates (Barker et al. 2002). Table 4.4 summarizes some recent studies in nursing and care home settings.

Certain types of MAEs seem to be more frequent in this setting. These include the inappropriate crushing of medications that should not be crushed, such as slow-release formulations (van den Bemt et al. 2009a) and the splitting of tablets that should not be split (Verrue et al. 2010). Such errors may be more likely in the nursing home population because of a higher prevalence of swallowing problems. In the U.K. care home setting, Alldred and colleagues (2011) found significantly higher error rates for inhalers and liquid medicines than for solid oral doses. About half of all inhalations were given incorrectly, commonly involving not shaking the device before use, incorrect inhalation techniques, and the wrong number of inhalations given. Such problems are likely to reflect the number of administration steps typically required, in the right sequence, to ensure successful administration.

TABLE 4.4
Examples of Medication Administration Error Studies in the Nursing or Care Home Setting

Reference	Setting	Error Rate	Common Types of Errors
Barber et al. (2009)	55 care homes (residential and nursing), United Kingdom	8.4% of 1380 opportunities for error	Omission, wrong dose
Barker et al. (2002)	12 skilled nursing facilities, United States (part of a larger study in a total of 36 health care facilities)	22% of 1451 opportunities for error	Wrong time, omission, wrong dose
Verrue et al. (2010)	2 nursing homes, Belgium	2.0% of 817 opportunities for error in home 1; 1.7% of 2650 opportunities for error in home 2	Omission, inappropriate tablet splitting, wrong dose
van den Bemt et al. (2009a)	3 nursing homes, the Netherlands	21.2% of 2025 opportunities for error	Wrong administration technique (such as incorrect crushing of tablets), wrong time, omission

Few studies have investigated the causes of MAEs in this setting. Barber and colleagues (2009) reported patient-related factors such as patients' unawareness of their medications, and patient behavior as contributing factors. Other factors were similar to those identified in hospitals, including lack of staff and lack of knowledge about drug administration issues, such as inhaler technique. Other factors were related to failures in communication and poor working environment. Lack of responsibility was reported as an organizational factor contributing to errors.

MEDICATION ADMINISTRATION ERRORS IN PATIENTS' HOMES

Home health care is a system of care provided by skilled practitioners to patients in their homes under the direction of a physician. This might include administration by trained carers in a patient's own home or in assisted living facilities. Little information is available on MAEs in these settings. Small-scale studies in a small number of countries would suggest there being high error rates (Meyer-Massetti et al. 2012; Young et al. 2013). More work is needed to investigate these problems, especially since there is a trend towards increased home care for the elderly and other patient groups. There is also a trend towards medication-related tasks being increasingly delegated to unskilled home care assistants (Craftman et al. 2013).

CONCLUSION

MAEs are common in the inpatient setting. Observation-based methods give a good picture of the quality of the process of drug administration, but other methods are needed to detect rare serious incidents. Although observation-based studies suggest that few errors have the potential to cause harm, the large number of drug administrations means that this remains an important area for improvement. Much of the research has been carried out in Europe and the United States, but smaller studies from the rest of the world suggest similar medication error rates. Relatively high error rates have also been identified in the nursing or care home setting, but little is known about MAEs that occur as a part of health care delivered in patients' own homes or assisted living facilities.

5 Professional Communication and Medication Safety

Patricia M.L.A. van den Bemt,
Cor J. Kalkman, and Cordula Wagner

CONTENTS

KEY POINTS

- Medication discrepancies are frequent and occur at all transition points in health care, including hospital admission, transfer between wards, and at discharge.
- This leads to patients experiencing clinically relevant medication discrepancies and potential or actual harm.
- Factors contributing to medication transfer errors can be divided into factors related to medication, to the patient, to information technology, and to health care professionals.

INTRODUCTION

Professional communication is essential in health care to ensure that all health care professionals who are caring for the same patient share the same information, and that any potential problems associated with their care are resolved. In relation to

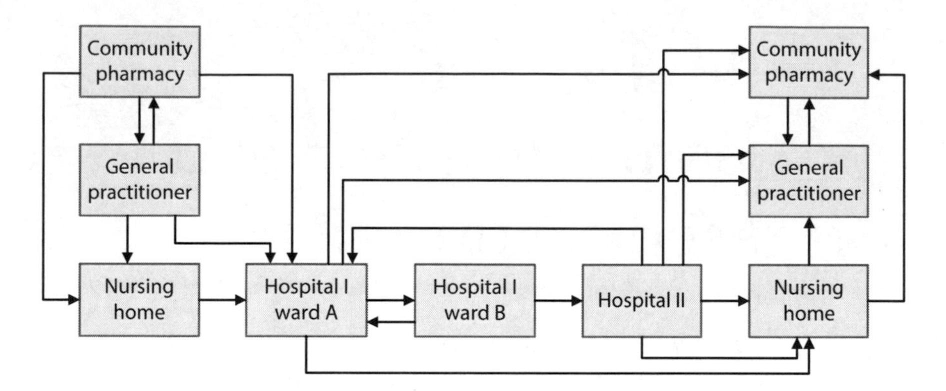

FIGURE 5.1 Overview of frequent transitions in health care, illustrating points of transition.

medication, this means that information on all medicines (both prescribed and over the counter) used by the patient should be available, including dosages, formulations, indications, contra-indications, and allergies. Correct information on medication is vital when prescribing new medications (in order to avoid duplication, contra-indicated medication, or drug–drug interactions) and when diagnosing diseases (symptoms may be caused by adverse drug reactions [ADRs] rather than an underlying disease).

Insufficient transfer of medication information may lead to problems regarding new prescriptions or appropriate diagnosis and management. Additionally, it may lead to medicines being stopped or started inadvertently, being continued unnecessarily, or being dosed incorrectly. Medication transfer problems may arise at any transition within the health care system, but are most important between hospital care and primary or long-term care, both at admission and discharge, and between different hospital departments (such as between intensive care units [ICUs] and other wards; see Figure 5.1).

In this chapter, we will identify various common errors that occur during care transitions and describe the extent of the problem, both in terms of the frequency and type of medication discrepancies and in terms of the clinical relevance of these discrepancies. In addition, we will present several example cases to illustrate the types of problems encountered (Boxes 5.1 through 5.3). Finally, potential causes for such medication transfer problems will be presented, with a special focus on interprofessional cooperation between pharmacists and doctors, and attitudes and barriers towards such cooperation.

SCOPE OF THE PROBLEM

Medication Discrepancies on Hospital Admission

When patients are admitted to a hospital, it is essential to obtain the home medication list, including not only the medication the patient is prescribed, but also information on "over-the-counter" (i.e., non-prescription) medication, contra-indicated medication (for example, due to a prior ADR), and drug allergies. Box 5.1 outlines some examples

BOX 5.1 EXAMPLES OF CASES WITH PROBLEMS IN COMMUNICATION

CASE 1: A MISSED BETA BLOCKER

Judith was admitted for planned surgery on her shoulder. In the electronic patient record, the admitting surgical resident found the following notes from the cardiologist, who was involved in the pre-operative screening period: "Unclear whether patient uses metoprolol slow release 200 mg once daily. This should be checked with GP and if she does not use it, metoprolol 25 mg twice daily should be started, together with digoxin." The resident misinterpreted this and prescribed the following as admission medication: digoxin 0.125 mg once daily and metoprolol 25 mg twice daily. During her hospital stay, Judith developed symptoms of withdrawal of beta blockade (tachycardia and symptoms of angina pectoris). After checking the community pharmacy record, it appeared that she had been previously taking metoprolol slow release 200 mg once daily. By inadvertently changing this medication regime, the dosage of metoprolol was decreased to a quarter of the dose that Judith was used to. This may have explained her symptoms.

CASE 2: A MISSED CONTRA-INDICATION

James, who had been diagnosed earlier with psoriasis, was prescribed a beta blocker by his GP. Subsequently, his psoriasis was exacerbated and the GP changed the therapy to another blood pressure-lowering drug. He added a note to the medical record, which was not transferred with the medication list when James was later admitted to a hospital. In the hospital, a beta blocker was started, again leading to an exacerbation of his psoriasis.

CASE 3: A MISSED ALLERGY

Amy was sent to the emergency department of a hospital by her GP with a diagnosis of exacerbation of her pulmonary obstructive disease. Without being asked about allergies, she was administered amoxicillin and clavulanic acid by intravenous infusion. Amy immediately developed symptoms of a serious allergic reaction, for which she had to be admitted to the ICU of the hospital.

where missing medication, a missing contra-indication, and missing drug allergy information have occurred as a result of poor communication about medication.

When a patient is admitted to a hospital from primary care, the community pharmacist and general practitioner (GP) collectively have most of the relevant information on the medication use of the patient. However, the records from community pharmacists only reflect dispensing histories, and recent therapy changes may be missing. For example, when a patient speaks to their GP about a complaint that may indicate an ADR and is told to stop the drug or to lower the dosage, this new information typically fails to reach the community pharmacist. Furthermore, a patient may visit multiple community pharmacies. The GP records may also be incomplete because of multiple prescribers: this information could be available to the community pharmacist, but may not reach the GP. Ideally, the patients themselves would be the best source of information, but this is also not always the case (for example, in cognitive impairment or critical illness). The best way to obtain information on the actual medications used before hospital admission would be to combine as many sources as possible. This is not an easy task, as sources may conflict. This also explains why many medication discrepancies, defined as differences between the in-hospital admission medication list (i.e., the list of medications composed after hospitalization but before new prescriptions are added) and the home medication list (i.e., the list of medications the patient actually uses in the pre-hospital setting) can occur at the point of hospital admission.

Tam and colleagues (2005) published a systematic review on the frequency of medication discrepancies (referred to as "medication history errors" by the authors) on hospital admission. They included 22 studies with 3755 patients. Studies were mainly performed on medical wards, with others on mixed medical and surgical wards, surgical wards, an emergency department, and a psychiatric hospital; four studies did not mention the actual setting. The percentage of patients with one or more discrepancies in their prescribed medication was 10%–67% across the studies. This increased to 27%–83% when non-prescription drugs were included and to 34%–95% when information on drug allergies or prior ADRs was also considered.

These wide ranges may be at least partly explained by different definitions of medication discrepancies in the individual studies: some studies only included omission errors, while others also included dosage errors and commission errors (such as erroneously adding medications not used before admission). Three of the included studies that evaluated all types of errors showed that 42%–59% of all errors involved omissions and 30%–42% were dose or frequency errors. One study suggested that commission errors occurred less frequently, representing only 11% of all errors (Cornish et al. 2005). However, the literature overall suggests that only including omission errors would lead to a substantial underestimation of the overall frequency of errors. In addition, many of the studies included both unintentional and intentional medication discrepancies, but only unintentional discrepancies should be included as a true discrepancy. Intentional discrepancies are the result of treatment changes carried out by the physician and are, therefore, less important in terms of patient safety (provided the reasons for the treatment changes are documented and appropriate).

The studies reporting the percentage of unintentional discrepancies found that 19%–75% of all discrepancies were unintentional. This wide range may be explained by the methodology of acquiring the medication lists: not all studies used as many sources of information as possible. Therefore, the results of this systematic review are difficult to interpret, but a conservative estimate would be that 20% of medication discrepancies are unintentional. This would result in 7%–19% of patients having one or more unintentional medication discrepancy (including non-prescription drugs, allergies, and ADRs).

There are a number of important aspects to consider when designing a study aimed at identifying medication discrepancies at hospital admission. The "best possible medication history list" should be created by using as many sources as possible, including information from the GP, the community pharmacy, the patient, and by checking the labels on the patient's medication containers. This should be compared to the medication list obtained immediately after admission, in order to focus on unintentional discrepancies rather than therapeutic changes (i.e., intentional discrepancies), and all types of medication discrepancies (omissions, commissions and formulation, dosage, and frequency changes) should be sought. In addition, "over-the-counter" medications and information on drug allergies and contra-indications (including prior ADRs) should be included, but reported separately.

Several studies have been published since the review by Tam and colleagues (2005) that comply with these recommendations on study methodology. For example, in one study in a Swedish hospital that included 670 patients, 47% of patients had at least one unintentional medication discrepancy on hospital admission (Hellstrom et al. 2012). Within the World Health Organization's (WHO) High 5s project, in the Netherlands, the percentage of patients with one or more unintentional discrepancies was 62% in the pre-intervention period. In this project, aimed at estimating the effect of pharmacy-led medication reconciliation, over 1500 patients from 12 hospitals were included (van den Bemt et al. 2013). A recent study in France, which included 256 patients from one hospital department, identified that 33% of patients had one or more unintentional medication discrepancy (Qulennec et al. 2013).

Taking these more recent studies into account, which only looked at unintentional medication discrepancies, it appears that the estimation from the above review may have been *too* conservative. A more realistic estimate would be that 30%–60% of patients admitted to a hospital will have one or more unintentional medication discrepancy.

Given the high frequency of medication discrepancies, it is appropriate to ask the question of whether all of these discrepancies are clinically relevant. Tam and colleagues (2005) included several studies that estimated the proportion of medication discrepancies considered to be clinically relevant, defined as "having the potential to cause moderate or severe patient discomfort or deterioration in the patient's condition." Often, this was estimated by panels of pharmacists and/or doctors, who had to reach a consensus on their estimates. The review concluded that 11%–59% of all medication discrepancies were clinically relevant, although this was based on only a few studies. Prospective studies to identify actual outcomes such as harm due to medication discrepancies are difficult to perform. When one discovers a discrepancy, it is not ethical to wait and see what will happen; one should correct the discrepancy and thus no adverse drug events (ADEs) should occur. While retrospective studies avoid these ethical issues, such studies suffer from the fact that ADEs are rarely documented accurately in medical records. If they are, the establishment of a causal relationship with medication discrepancies is difficult. Studies on clinically relevant outcome parameters have generally only been performed when investigating the effects of health care improvement interventions. Most of these intervention studies looked at health care use after hospital discharge as an outcome (for example, emergency department visits or readmissions). Intervention studies are beyond the scope of this chapter and will be dealt with in Chapter 17.

Medication Discrepancies Associated with Transfers within a Hospital

Even within the same hospital, patients can be transferred from one ward to the other. Common transfers are between the emergency department and inpatient wards, between general wards and ICUs, and between operating theaters and inpatient wards. In many hospitals, these different clinical areas use different medication systems. For example, when computerized prescriber order entry (CPOE) is implemented, emergency departments, ICUs, and theaters are often exempt from implementation or use their own tailor-made systems that better accommodate their specific patient processes. This means that transfer between these departments and other wards can be particularly prone to medication discrepancies. A Canadian study of 129 patients showed that 62% of all internally transferred patients had one or more unintentional medication discrepancy due to medication reconciliation failures (Lee et al. 2010). More than a third of the patients had at least one unintentional medication discrepancy with the potential to cause harm.

A large cohort study using administrative patient records evaluated the rates of potentially unintentional discontinuation of medication following hospital or ICU admission (Bell et al. 2011). The study did not assess whether or not the discontinuation was actually unintentional, but it did include the discontinuation of chronic medication generally intended for long-term use. The study showed that patients being admitted to a hospital were more likely to discontinue chronic medication than were patients who had not been hospitalized. This effect was stronger when a patient was admitted to an ICU while hospitalized, compared to being hospitalized on a general inpatient ward. Thus, an ICU stay added to the discrepancies already occurring while in the hospital.

Medication Discrepancies on Hospital Discharge

At the time of hospital discharge, a current medication list should be prepared, ideally including (and differentiating) the following:

- Discontinued hospital medications (for example, hypnotics)
- Discontinued long-term medications, together with the reasons for discontinuation (for example, due to ADRs)
- Details of newly started long-term medications that will need to be continued
- Ongoing long-term medications that are to continue unchanged
- Specified durations or stop dates for short-term medications (for example, antimicrobials, anti-emetics, and analgesics for acute pain)
- Details of allergies, compliance aids, and future monitoring requirements

Subsequently, this current medication list needs to be communicated to the next health care professional and to the patient, preferably before discharge (but, as a minimum, during the discharge process). The health care professional should then update the patient's medication record with the new information. As with hospital admission, this is a complicated process in which multiple opportunities for error may arise, as seen in Box 5.2.

BOX 5.2 EXAMPLE OF COMMUNICATION PROBLEMS AT HOSPITAL DISCHARGE

CASE 4: A PATIENT NOT STOPPING TYPICAL HOSPITAL MEDICATION

A male elderly patient, Joseph, was hospitalized for a broken leg. Since his discharge from the hospital, Joseph continued using high dosages of metoclopramide (an anti-emetic). He began to develop symptoms of Parkinson's disease. Metoclopramide has prokinetic properties on the gastrointestinal tract by acting as an antagonist on dopamine receptors. Dopamine antagonists can cause Parkinson's-like symptoms as a side effect.

Joseph's son, being a doctor himself, asked the GP why the metoclopramide was prescribed. The GP did not know, but he stated that "as the drug was prescribed in hospital, it was surely indicated," and he was reluctant to stop the metoclopramide. The son subsequently asked the hospital doctor caring for his father why metoclopramide had been prescribed: the doctor's response was for hiccups, which is an off-label indication for metoclopramide. It should have been stopped prior to hospital discharge, because the hiccups had been gone for weeks. With this information, the GP felt secure in stopping the metoclopramide, and within a week, the Parkinson's-like symptoms resolved. The symptoms had severely hampered Joseph's daily routine for months, which could have been prevented by preparing an adequate discharge medication list, including stopping short-term hospital medication.

An important source of medication discrepancies on hospital discharge are discrepancies that occurred earlier, on admission to the hospital (Climente-Marti et al. 2010) or during internal hospital transfer. If not discovered in time, these discrepancies will remain unresolved and the patient will be discharged with the wrong medication information. In most studies on discharge medication discrepancies, this fact is acknowledged, but a clear distinction between discrepancies occurring earlier versus at the time of discharge is not always made. Cornu and colleagues (2012) have estimated that unresolved discrepancies at admission may be associated with at least half of the discrepancies at discharge. Pippins and colleagues (2008) assessed this, and found that 72% of discharge errors originally occurred at admission. However, most studies just report the frequency of medication discrepancies on hospital discharge, irrespective of their origin. The recent review by Kwan and colleagues (2013) can be used as a source of information on the frequency of medication discrepancies on hospital discharge: three studies from this review report the proportion of patients with one or more discrepancy at discharge. In their study of 150 patients discharged from a tertiary care hospital, Wong and colleagues (2008) reported that 41% of patients had at least one unintentional medication discrepancy at discharge. The same frequency was identified in an earlier and smaller study in 56 patients (Vira et al. 2006). Kripalani and colleagues (2012) performed a large intervention study in 851 patients that showed no effect of the intervention, implying that the combined results of the control and intervention group can be used as a measure of medication discrepancies in usual care. They showed that 51% of patients had one or more clinically relevant medication error (defined as preventable or ameliorable ADEs and potential ADEs due to medication discrepancies or non-adherence) after discharge. Non-adherence accounted for approximately half of all errors. Combining these studies gives an estimate of 25%–40% of patients with one or more medication discrepancy on hospital discharge. Some 50%–75% of these discrepancies may have arisen earlier, on hospital admission, leaving a frequency of "true" discharge problems of 6%–20%.

From the patient's point of view, it is irrelevant, however, where these discrepancies arose: when he or she is discharged with unintentional discrepancies, there is a problem. The types of medication discrepancies occurring on hospital discharge are comparable to the types identified on hospital admission. In the study by Vira and colleagues (2006), omission errors occur most frequently, followed by incomplete information on dosages, while in the study by Wong and colleagues (2008), this was the other way round. The study by Kripalani and colleagues (2012) was the only study to also identify commission errors at discharge, which occurred less frequently than both omission and dosage errors.

As mentioned above, an adequate current medication list on discharge should also contain the reasons for stopping a medication (for example, ADRs). Failure to do so can have serious consequences, as the patient vignette in Box 5.3 clearly shows. In fact, being aware that medication discrepancies occur during hospitalization makes primary health care workers prone to restarting a medication when they do not know the reason it was stopped. They may assume that the medication was stopped unintentionally due to a medication reconciliation error and thus restart the drug.

The patient in Case 5 (Box 5.3) is certainly not the only one to experience the inadvertent restart of an intentionally stopped medication. Van der Linden and colleagues (2006) studied this subject and found that 27% of medications intentionally stopped during hospitalization because of an ADR were re-prescribed within six months of discharge. Only 51% of all ADRs were mentioned in the discharge

BOX 5.3 EXAMPLE OF A CASE WITH SERIOUS CONSEQUENCES DUE TO COMMUNICATION PROBLEMS

CASE 5: A CASE OF PANCREATITIS

In the Dutch multicenter HARM study (Leendertse et al. 2008), a male patient was included who was admitted because of pancreatitis due to simvastatin use. Because of the seriousness of this reaction, simvastatin was stopped, but the reason for stopping it was not adequately communicated to primary care. Consequently, the simvastatin was restarted by the GP and pancreatitis reoccurred, necessitating readmission to the hospital. Again, the medicine was stopped, but for the second time simvastatin was restarted in primary care when the patient was discharged. It was only after the third admission for pancreatitis that the simvastatin was stopped permanently, because finally the reason for stopping was adequately communicated.

letter to the GP and, even for serious reactions, this was still only 62%. Furthermore, the community pharmacist did not receive any information on ADRs. This lack of communication may contribute to the inadvertent restart of intentionally stopped medication, but it is not the only reason. In order to prevent such erroneous restart of medication, the ADR needs to be properly documented within the next health care professional's information system. Van der Linden and colleagues (2006) have also shown that this occurs only for 22% of communicated ADRs.

The studies looking at the frequency of medication discrepancies at discharge have the advantage over many studies of hospital admission, in that only data on clinically relevant discrepancies were collected and hence reported. However, the concerns about methods for assessing clinical relevance remain the same as for the hospital admission studies. This hampers the ability to draw conclusions on the actual clinical relevance of medication discrepancies on hospital discharge.

POTENTIAL CAUSES OF MEDICATION TRANSFER PROBLEMS

The first part of this chapter has dealt with the scope of the problem. The information presented clearly illustrates that medication transfer problems can arise at every transition point in health care, even within the same hospital. In this second part, we will discuss potential causes for medication transfer problems.

Medication-Related Causes

A well-known risk factor for medication errors in general is the use of multiple medications: the more medicines are used by a patient, the higher the chance that something goes wrong. This also applies to medication transfer errors.

Specific categories of medication are more prone to error, especially medications with irregular dosage regimes. The mistakes with methotrexate for rheumatoid arthritis are notorious: the correct dosage of once weekly can be inadvertently transcribed as once daily, with (sometimes lethal) detrimental effects for the patient. Errors occur more often in sound-alike or look-alike medication names; for example, the unintentional switch of Losec 40 mg (omeprazole, a gastric acid-inhibiting drug) to Lasix 40 mg (furosemide, a diuretic).

Patient-Related Causes

The patient plays a key role in providing adequate information about his or her medication use, as illustrated by a study evaluating medication reconciliation at discharge with and without the help of the patient (Karapinar-Çarkıt et al. 2009). In this study, 5.3 interventions per patient were performed after patient counseling, compared to 2.7 interventions per patient before patient counseling. This means that the patient provided additional information that was necessary to construct the most accurate medication discharge list (requiring more interventions). However, due to several factors, the patient may not be able to fulfill this role: cognitive impairment, critical illness, and language barriers may all hamper communication and thus lead to medication transfer problems.

Furthermore, the patient may not always know what information is important to share; for example, "over-the-counter" medication is not always perceived as "real" medication and thus this information will not always be shared with health care professionals. Patients may also not consider certain formulations, such as transdermal patches or ear or eye drops, to be "medication," or may not wish to share details of herbal or similar medicines with conventional medical practitioners. Finally, patients may assume that health care professionals automatically know which medicines they are taking and not realize the importance of keeping a list themselves.

Information Technology-Related Causes

Although the introduction of CPOE systems together with clinical decision support systems can lead to safer prescribing of medications (Ammenwerth et al. 2008), these systems may also contribute to (new) medication errors, as described in Chapter 19.

Often, individual groups of health care professionals run their own electronic medication records: the GPs have their system, the community pharmacists have their own system, hospitals run other systems, and even within hospitals, separate departments may operate on different systems. Due to the lack of an integrated health care information system, communication gaps arise that necessitate the application of alternative (non-information technology-based) communication routes. For example, when patients are admitted to the hospital, the medication record from the community pharmacy may be printed and then faxed to the hospital. Subsequently, the fax is used for medication reconciliation, when all medicines need to be re-entered into the medication system of the hospital. This is, therefore, an error-prone process with multiple transcription steps. Additionally, because information is not being transferred electronically, additional labor-intensive steps are needed to document the transferred information in the right location in the hospital or GP system. As mentioned earlier, Van der Linden and colleagues (2006) have shown that this rarely occurs, even for vital information on ADRs.

When attempts are made to link systems electronically, other errors may arise. One of these problems is the correct translation of dosing codes from one system to another. This problem may give rise to dosing errors after transfer from one health care professional to another (Cheung et al. 2014).

A final problem arises because of the original purpose of community pharmacy records: they are mainly dispensing records, providing an overview on all dispensed medicines. Whether this medication is actually used by the patient and in what dosage is not recorded in these community pharmacy systems. Yet often these records are interpreted as records of actual use, which may lead to errors. For example, a patient may have been dispensed morphine slow-release tablets 10 mg twice daily, but because of increasing pain, he has spoken to his physician, who told him to double the dosage (i.e., 20 mg twice daily). As the patient still has enough tablets at home, he can increase the dosage without informing the community pharmacy so, in their system, the incorrect dosage of 10 mg twice daily will persist, until the dispensing of a new prescription with updated dose instructions.

Health Care Professional-Related Causes

In order to transfer information on medications, adequate professional communication is essential. This applies to doctor–doctor communication, but also to pharmacist–pharmacist and pharmacist–doctor communication. Kelly and colleagues (2013) explored pharmacist and physician views on collaboration using a survey. Of 518 invited pharmacists, 407 (79%) responded, while for doctors the response rate was only 33 of 462 (7%). With respect to the preferred methods of communication, pharmacists preferred telephone or face-to-face communication, while doctors preferred telephone or fax communication. Of the pharmacists, 57% stated that electronic transfer of information should be explored. Physicians and pharmacists agreed on the barriers to collaboration: they both regarded lack of time and financial compensation and the need to collaborate with multiple physicians/pharmacists as the most significant barriers. Successful communication will involve respect for each other's preferred mode of communication and attention to the barriers for collaboration.

The perceptions of hospital health care professionals (physicians, nurses, and pharmacists) regarding medication reconciliation were studied using focus groups in three U.S. hospitals (Vogelsmeier et al. 2013). Results of the focus group discussions were that clearer definitions of medication reconciliation are necessary and that responsibilities also needed to be defined. Alignment of disciplines regarding these definitions and responsibilities is a prerequisite for adequate communication. When establishing interprofessional communication networks aimed at medication reconciliation, this should be borne in mind.

CONCLUSION

This chapter has provided an overview of the scope of the problem of medication transfer problems and potential causes for these problems, illustrated by specific patient case reports. This provides a basis for the improvement of medication information transfer practices, as considered in Chapter 17.

Section II

Approaches to Understanding and Resolving the Problems

6 Introduction

Mary P. Tully and Bryony Dean Franklin

Section I has provided a detailed description of the various aspects of the medication use process and how errors at each stage can cause actual or potential patient harm. Section II next comprises seven chapters that cover a variety of approaches to understanding why those errors occur and how we can design ways to resolve them. The approaches vary from the theory based, such as psychological and educational theories in Chapters 8 and 9, to the practical, such as the aspects of measurement discussed in Chapter 7. Both types of strategy are needed to ensure rigorous and robust approaches to both research and the development and evaluation of interventions in clinical practice.

It is essential to have theory that underpins research and development in medication safety. Theory allows a much deeper understanding of the entire context of medication use and facilitates greater understanding of the complexities involved. Without it, researchers and practitioners tend to focus on narrow technical solutions, often within limited professional boundaries. These chapters serve as primers to the area for everyone working in medication safety, whether as a researcher investigating an aspect of the process or a health care professional studying the impact of a new service.

There are multiple theoretical approaches within the broad research paradigms described in this section. There are also other paradigms, such as sociology and anthropology, which we have not had the space to include, but which may also provide researchers with theories to investigate particular research questions about the medication use process. Consideration of the focus that each theoretical approach affords can help the researcher decide which theories to consider. This may include deciding whether the individual, the local team, or the broader organization is the focus of interest. Chapter 8, for example, describes psychological theories that focus on the individual. Other approaches, such as the systems approach in Chapter 11, include the individual, but also expand outwards to eventually consider the whole organization. Chapter 10 addresses the complex interactions between people and objects (i.e., technology in its widest sense, whether it is a vast interconnected electronic health record or a single preprinted sheet of paper used to write prescriptions). That chapter also highlights the two-way nature of these interactions. Because technology is woven into the social context by the people who use it, if the latter reject the technology as not fitting into the existing social norms they want to retain, the technology may fail to deliver the outcomes initially anticipated.

Each theory alone can cast light on a particular area; used together, it is possible that they can illuminate even more by giving both a deeper and a broader understanding of the area to which they are applied. This is a theme that runs through several of these chapters. For example, the Theoretical Domains Framework described in Chapter 8 is a combination of psychological theories, taking the core components of each to create a new whole. Chapter 9 describes the possibility of combining different educational theories and particularly advocates the application of complexity theory alongside others. Uncertainty and ambiguity are inherent within a complex adaptive social system, such as the real-life environment of health care practice. It is not difficult to see how important it is that we remain cognizant of the whole medication use process, within which we may be investigating one specific aspect.

Improving medication safety requires the introduction of interventions into the health care system, and rigorous measurement is needed to ensure that any intervention has the desired impact. Chapter 7 describes approaches to defining errors and highlights the importance of producing operational definitions, so that all data collectors are clear on how to recognize and classify errors when they are encountered. The quantitative measures described in Chapter 7 can then be subject to statistical analysis of their likely impacts. Other methods of data collection (quantitative and qualitative) are described throughout the section, especially in Chapters 11 and 12.

The approach to measurement in Chapter 7 is not the only way to show the impact of interventions. Chapter 12 describes the use of measurement in improvement science, where smaller samples of data are collected than would be traditional in other study designs, but they are collected more frequently. These data are not analyzed by inferential statistics, but by statistical run charts or process control charts, which show changes over time. Such charts can be posted on an office notice board, for example, to show the relevant health care professionals what has been found and to encourage the change to new ways of working. Thus, in this circumstance, measurement is not a neutral act. The process of collecting and reporting the data may well change what is being measured—and this concept is at the very heart of improvement science.

Several chapters in this section also highlight the use of qualitative methods, which can be used to conduct an in-depth analysis of how and why errors occur and whether interventions can change behavior as desired to improve patient safety. Again, combining qualitative and quantitative methods in a mixed-methods approach can be invaluable when conducting a rigorous examination of such a complex system as the safe use of medicines. The chapters here introduce many of the approaches that can be used and cite other sources that can provide the novice with a useful introduction to this vast area.

7 Measuring Medication Errors

Monsey McLeod

CONTENTS

KEY POINTS

- Medication errors are defined and classified in many ways: operational definitions with clear inclusion and exclusion criteria are essential whenever planning a study, collecting or analyzing data, or interpreting findings.
- There are different methods for detecting medication errors; more than one approach may be required in quantitative studies to determine an accurate medication error rate.
- The setting can affect the types of medication errors that occur; knowing the setting is important for understanding and interpreting published studies.
- Severity assessment tools are usually designed to assess either actual or potential harm; validated tools are available and should be used where appropriate.

INTRODUCTION

Measurement is essential to understanding the problems and to monitoring the effects of interventions. There are many studies reporting medication error rates. Yet it remains unclear whether medication errors have reduced over the years or, more importantly, whether patients are any safer (Vincent et al. 2008). Attempts to analyze reported medication error rates across studies are limited by differences in settings and methods, and insufficient reporting (Ferner 2009; Franklin et al. 2005, 2009; McLeod et al. 2013). This chapter aims to help clinicians and researchers involved in planning, conducting, and interpreting quantitative studies of medication error rates. It also describes some possible data collection methods for qualitative studies that investigate the causes of error.

Medication errors pose a threat to three key dimensions of quality care: patient safety, clinical effectiveness, and patient experience (Department of Health 2008), and reducing medication errors is an important priority worldwide. Fortunately, the majority of medication errors do not cause actual patient harm. Where such harm does arise, this is generally referred to as an adverse drug event (ADE).

THE RELATIONSHIP BETWEEN MEDICATION ERRORS AND ADVERSE DRUG EVENTS

An ADE is defined as an injury or patient harm resulting from medication use (Bates et al. 1995a); this includes harm due to both appropriate and erroneous medication use. This is an important concept, as the concepts of ADEs and medication errors overlap, but are not interchangeable (Figure 7.1). Morimoto and colleagues (2004) suggest that incidents that are ADEs can also be classified according to whether or not the event that led to harm was preventable or was ameliorable. As depicted in Figure 7.1, not all medication errors contribute to patient harm; those that lead to potential harm are classified as potential ADEs and those that lead to actual harm are actual ADEs. Furthermore, not all ADEs are preventable; an example of a non-preventable ADE would be a patient who experiences an allergic reaction to a new medication. This latter scenario is an example of an adverse drug reaction. An ameliorable ADE is "an injury of which the severity or duration could have been substantially reduced if different actions had been taken" (Morimoto et al. 2004).

DEFINITION AND CLASSIFICATION OF MEDICATION ERRORS

What is a medication error? Early studies in the 1960s were pivotal in highlighting the problem of medication errors as a threat to patient safety. A seminal paper defined a medication error as "the administration of the wrong medication or dose of medication, drug, diagnostic agent, chemical, or treatment requiring the use of such agents, to the wrong patient or at the wrong time, or the failure to administer such agents at the specified time or in the manner prescribed or normally considered as accepted practice" (Barker and McConnell 1962). Sharp-eyed readers may have spotted that this definition focuses on errors at the medication administration stage. Yet the use of medicines comprises a range of stages: these can be broadly

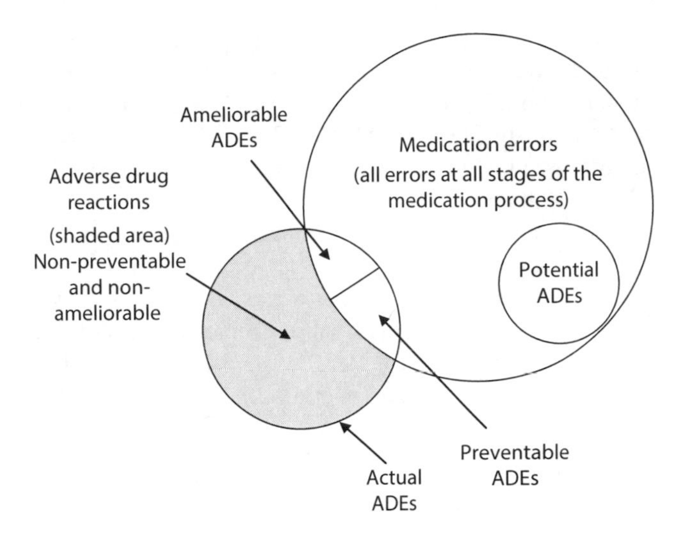

FIGURE 7.1 The relationships between medication errors, actual adverse drug events (ADEs), and potential ADEs. (Adapted from Bates, B.W. et al. 1995a. *J Gen Intern Med* 10: 199–205; Morimoto, T. et al. 2004. *Qual Saf Health Care* 13: 306–314.)

considered as prescribing, dispensing, administering, and monitoring (Chapters 2 through 4). In some medication systems, there may also be a transcription stage. Errors can occur at any of these stages, and a more encompassing definition of a medication error is

> …any preventable event that may cause or lead to inappropriate medication use or patient harm while the medication is in the control of the health care professional, patient, or consumer. Such events may be related to professional practice, health care products, procedures, and systems, including prescribing; order communication; product labeling, packaging and nomenclature; compounding; dispensing; distribution; administration; education; monitoring; and use (National Coordinating Council for Medication Error Reporting and Prevention 2011).

One of the common pitfalls in searching for and interpreting medication error studies is that the term "medication error" is sometimes used to refer to all types of medication error, and sometimes to a specific type, such as prescribing errors or medication administration errors (MAEs). It is only in recent years that the distinction between different types of medication errors has become increasingly explicit in the literature. In this chapter, the term "medication error" is used as a generic term that encompasses all types of medication error. The terms "prescribing error," "MAE," "dispensing error," or "monitoring error" are used when information specific to each of these is reported. In some studies, errors occurring during nurses' preparation of medications prior to administration have been described as dispensing errors. However, the term is more often used to refer to pharmacy dispensing errors, which is the convention used here.

Literature reviews have identified a myriad of definitions: 28 definitions for a prescribing error including some types of monitoring error (Lewis et al. 2009), 21 for a dispensing error (James et al. 2009), and 30 for a MAE (Keers et al. 2013b). To further complicate matters, the same definition has sometimes been associated with multiple functional meanings, due to variation in inclusion and exclusion criteria and the use of different error subcategories (McLeod et al. 2013). Specific issues in defining prescribing and monitoring errors, dispensing errors, and MAEs are discussed in Chapters 2, 3, and 4, respectively.

Definition is not synonymous with classification; a decision first needs to be made as to whether an event meets the definition of a medication error before it can be classified according to relevant subcategories, such as type of error (wrong drug, wrong dose, and so on) and/or potential causes. Subcategories of specific types of medication error were discussed in Chapters 2, 3, and 4; further reading is also available elsewhere (James et al. 2009; Keers et al. 2013b; Lewis et al. 2009; McLeod et al. 2013).

Effect of Health Care Setting on Medication Errors

When reading the literature on medication errors, it is important to consider the health care system, settings, and countries in which the studies have been carried out. Variations in systems, processes, and staff responsibilities are likely to bring about different risks and types of errors and therefore affect the extent to which study findings can be extrapolated to other contexts. For example, in countries such as the United Kingdom and many parts of Europe, medications for use on hospital inpatient wards are typically supplied as a mixture of ward-based or "floor" stock and multi-dose supplies labeled for individual patients. Individual doses are then prepared on the ward by nursing staff during regularly scheduled drug rounds, conducted several times each day; this includes intravenous doses, which are known to be at high risk for error (McLeod et al. 2013). By contrast, in countries using a unit dose distribution system, pharmacy staff are responsible for preparing individual doses for nurses to administer. Thus, MAEs in countries using a floor stock and/or multi-dose individual patient dispensing systems will include any errors made during the preparation of doses on the ward (as well as during the administration stage). Such errors are less likely with unit dose systems. In other countries, patients are supplied with manufacturers' original packs of medication without a dispensing label specifying the patient's details or individualized instructions for use. The likely effects of such international variations on medication errors has been explored to some extent (Dean et al. 1995; Taxis et al. 1999; Wirtz et al. 2003), but the true extent of variation in medication systems and the corresponding effect on medication errors is unknown.

Even within the same country, whether a study was conducted in the community, nursing home, or hospital has important implications for the interpretation of the findings. As highlighted in Section I of this book, medication error rates, subtypes, severity, and causes vary across different health care settings. An important consideration, both for interpreting and reporting studies, is to identify factors associated with the setting that could affect medication errors. This includes recognizing that even the same process within the same hospital may subtly differ according to the ward, patient cohort, and skill mix of the health care team.

APPROACHES TO DETECTING MEDICATION ERRORS

Having established the type(s) of medication errors of interest, it is necessary to consider the method of error detection. Each type of medication error may be identified using a range of methods, and the decision as to which is the most appropriate will require a careful balance between achieving the study objectives and recognizing the limitations of the chosen method(s). The following are common methods used for detecting medication errors and their associated strengths and limitations.

Document Review

This is sometimes called retrospective review or chart review, and involves one or more medication-related sources of information (paper drug charts, electronic medication administration records, medical notes, laboratory results, and so on) being reviewed by a researcher (Dean and Barber 2001; Franklin et al. 2009; Ghaleb et al. 2010; Seifert and Jacobitz 2002). Strengths include that it: (1) allows review of sequential errors, which may not be possible with methods such as observation unless consecutive shifts are observed; (2) allows identification of ADEs as well as some errors that do not result in harm; and (3) allows actual data collection to take place at relatively flexible times. Limitations include that the method: (1) does not capture errors that were not documented or are not apparent from document review; (2) is dependent on accurate documentation that reflects the true nature of what happened; (3) is relatively time-consuming; and (4) as a retrospective method, it may require consideration of changes in practice over time that could influence how and what information is documented, such as changes in recommended clinical practice or legislation.

Trigger Tools

Trigger tools refer to a specific form of document review based on identifying specific events that may indicate that harm has occurred (Cavell 2009; Franklin et al. 2010a; Resar et al. 2003). Examples include identifying the use of specific antidotes, such as vitamin K, which may indicate a warfarin overdose, and laboratory results, such as a high aminoglycoside level. The presence of these events "triggers" a more detailed document review. As such, trigger tool assessment can be carried out via manual document review or as an automated process and is less time-consuming than unstructured document review. It also focuses on harm. However, there are a number of limitations associated with the use of trigger tools to identify medication errors: (1) they typically focus on detecting specific types of medication error associated with certain drugs; and (2) they have relatively low sensitivity and specificity. Confirmation of a medication error is required following a positive trigger, which usually involves additional manual document review.

Review of Incident Reports

Incident reports can provide useful information on some aspects of medication safety within the organization; this approach has the added advantage of being relatively low cost, as many organizations are routinely collecting and reporting on these data (Ferner 2009). However, the level of under-reporting makes it the least accurate method for quantifying medication error rates. It has been estimated that only about 1 in 100 prescribing errors and even fewer MAEs are reported as incident reports (Franklin et al. 2007b, 2009; Hall et al. 1985; McNally and Sunderland 1998). In a study of 3337 prescriptions, observation detected 16 dispensing errors, while incident reports detected none (Varadarajan et al. 2008). Some of the reasons for under-reporting include: (1) reporting relies on someone being aware that an error has occurred and knowing how to report it; (2) reporting may be hindered by an actual and/or perceived lack of time to report; and (3) there may be psychological barriers to reporting, such as fear of disciplinary action and/or perceived waste of time.

Self-Report

Self-report during routine screening or double-checking differs from incident reporting. Self-report involves asking staff to document medication errors they detect during their routine tasks, such as pharmacists who double-check dispensed medications for accuracy (Franklin et al. 2009; Ghaleb et al. 2010). Incident reporting, on the other hand, is usually encouraged for any kind of incident (not just those picked up during routine double-checking), but is often limited to incidents that staff perceive to be important to report and/or associated with actual patient harm. The strengths of self-reporting include: (1) it is a useful and practical method for detecting prescribing, dispensing, medication administration, and monitoring errors in health care settings where medications are routinely screened or double-checked prior to the next stage of the medication use process; and (2) it can be a relatively efficient method. Its limitations include: (1) it potentially increases staff workloads if additional documentation of detected medication errors is required; and (2) there is risk of inter-staff variation in the identification and documentation of medication errors (for example, due to differences in experience).

Surveys, Interviews, and Focus Groups

These methods are generally used to explore perceptions and causes of medication errors rather than provide data on their incidence, although some also ask participants to estimate the frequency with which they experience certain types of events. A strength of the survey approach is that it can be used to reach a large number of potential participants, while interviews and focus groups can provide rich data for exploring the causes of medication errors. Their limitations include: (1) there may be psychological barriers to discussing errors, such as fear of disciplinary action; (2) these methods rely on recall; and (3) responses can be highly subjective.

Observation

This involves a researcher who is trained to detect medication errors shadowing participants as they carry out their routine tasks (Allan and Barker 1990; Barker and McConnell 1962; Dean and Barber 2001). The method has most often been used for MAEs and dispensing errors, but may also be useful for detecting other types of medication errors. The strengths of the observational approach include: (1) it is a validated method for detecting MAEs (Dean and Barber 2001), which are difficult to identify using other methods; (2) errors can be detected in real time, allowing the observer to prevent patient harm; and (3) the observer may also be able to identify some potential contributing factors. The limitations of observation include: (1) it is relatively time consuming and costly; (2) training is required; (3) there is potential for the observer to affect those observed, which could theoretically increase or decrease medication error rates, although evidence suggests that this is not significant provided that the observer is non-judgmental and any interventions are made tactfully (Dean and Barber 2001); (4) if the observer intervenes to prevent errors before those involved have every opportunity to identify and correct them, perhaps during a final check, this could overestimate error rates; (5) there is a risk of observer fatigue influencing detected errors; (6) rare events are unlikely to be observed; and (7) it is generally considered impractical for regular monitoring.

Simulation

Here, participants are invited to carry out a task and/or respond to a simulated real-life scenario in an artificial environment. The strengths of simulation are (1) medicines prescribed, dispensed, administered, or monitored during simulation can be checked for accuracy by a researcher or trainer to potentially increase the reliability of error detection; and (2) different experimental conditions can be tested in a controlled manner without affecting patient care. Its limitations include: (1) the environment may not incorporate all the real-life human and system factors that could contribute to medication errors; (2) additional time is required by staff as it requires taking time out of their routine work in order to participate; and (3) simulations can be time consuming and costly to set up.

Covert Patients

Analogous to the "mystery shopper" approach used in market research, in this approach, participants continue working in their normal environment and are blinded as to the true motive of the patient they are treating. The strengths of the covert patient approach include: (1) medicines prescribed, dispensed, or administered to covert patients are likely to reflect real-life practice; and (2) it allows for the standardization of scenarios for comparison between health care professionals or facilities. Its limitations include: (1) it requires an individual to be trained to consistently portray a particular patient case, as well as to assess the health care professional according to pre-specified criteria; (2) it is impractical for measuring medication error rates in hospital inpatient settings, as a very large sample of interactions would be required; and (3) it raises various ethical issues due to both the use of deception and the use of health care resources by someone who is not a genuine patient.

To decide which method of detecting medication errors is most suitable, it may be useful to consider whether the main objective is to identify the frequency with which medication errors occur or to identify a sample of errors to allow more detailed

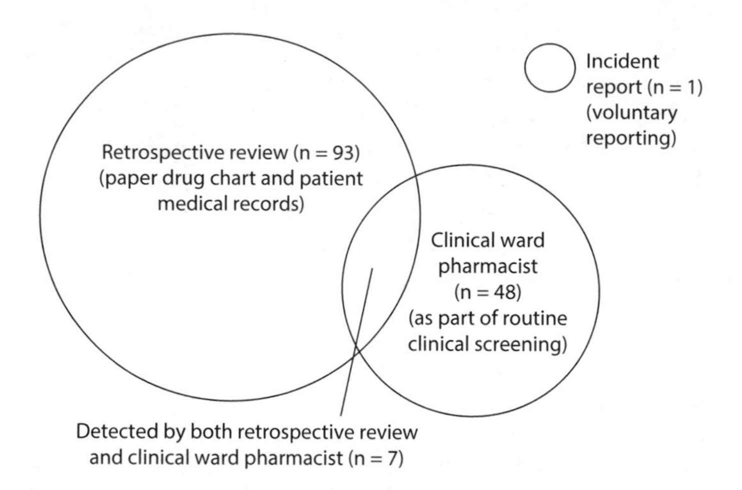

FIGURE 7.2 Differences in prescribing errors detected using three methods in a U.K. hospital: retrospective review, detection by a clinical ward pharmacist, and incident reports. (Adapted from Franklin, B.D. et al. 2009. *Pharmacoepidemiol Drug Saf* 18: 992–999.)

subsequent exploration of the contributory factors. To achieve the former, one or more methods that capture errors accurately and comprehensively are required, while for the latter, a less comprehensive but more in-depth method of identifying the errors may be more appropriate. For instance, identifying the prescribing error rate for hospital inpatients may be best achieved via a combination of data collection by pharmacists involved in the routine screening of medication orders and retrospective review of patient medication orders and notes by a researcher (Figure 7.2 and Franklin et al. 2009). By contrast, the use of interviews and focus groups may be appropriate to exploring their causes. Ultimately, irrespective of the method, the underlying assumption in medication error detection is that the person(s) collecting data have adequate knowledge of what the correct medication process should be; this may be obvious when planning a study, but the relevant skills or training of the data collector is a detail that is often omitted in the literature.

MEASUREMENT OF MEDICATION ERROR RATES

Variation in medication error definitions (the numerator) and their functional meanings will clearly have a significant impact on reported medication error rates. The denominator plays an equally vital role: "The denominator is a measure of the exposure to the hazard" (Ferner 2009). Denominators could include medication orders, prescriptions, items dispensed, opportunities for error, patient days, or patient admissions. The choice broadly depends on considerations such as the purpose of describing the medication error rate and the importance of comparison with other studies. The type of medication error is an indirect consideration; there is no accepted rule that states, for example, that prescribing error rates have to be calculated as a percentage of medication orders, or dispensing error rates as a percentage of prescriptions or items dispensed. However, to facilitate interpretation and comparison of the findings in published research studies, it is recommended that: (1) established denominators are used; and (2) more than one medication error rate (calculated using different denominators) can, and probably should, be reported in quantitative studies (McLeod et al. 2013).

In establishing the denominator, the next important consideration is its definition and functional meaning. Similar to the challenge with medication error definitions, the same denominator may have different functional meanings in different studies (Lewis et al. 2009; McLeod et al. 2013). Several literature reviews of different types of medication error have helpfully summarized the range of denominator terms, definitions, and inclusion/exclusion criteria used (Franklin et al. 2010b; Ghaleb et al. 2006; James et al. 2009; Keers et al. 2013b; Lewis et al. 2009; McLeod et al. 2013). Other key considerations for determining and interpreting medication error rates are summarized below.

Patient Population

The patient population should be clearly defined to facilitate interpretation of the findings and allow potential comparisons to be made with other studies; the profile of errors detected between different populations is likely to vary.

Method of Data Collection

Occasionally, an inappropriate method is used to collect medication error data, such as using incident reports to determine medication error rates or using a setting-specific trigger tool that was not designed for use in other settings. For studies that involve multiple data collectors, it is important that inter-rater reliability is also assessed.

Definitions

Both the numerator and the denominator used to calculate the medication error rate should be defined, so that it is clear whether or not actions such as procedural violations are included. A comprehensive set of inclusion and exclusion criteria is essential to facilitate data collection and the interpretation of findings, providing specific examples of errors and cases not considered to be errors where possible. During data collection and analysis, there may also be borderline cases that do not completely fit within the definitions used. To ensure consistency and clarity, borderline cases should be reviewed, previous categorizations amended accordingly, and such examples reported. Many studies have developed a cumulative list of "case law" to help ensure a consistent approach to error definition by multiple researchers over the course of a study (Avery et al. 2013; Barber et al. 2009). Where appropriate, consider using and/or adapting existing medication error and denominator definitions. Furthermore, consider how local policies affect the definition of an error: this context information is important for others.

Denominator

There are many potential problems associated with the denominator in quantitative medication error studies. Medication error rates are sometimes presented as a percentage without specifying the denominator; however, the numerical value and unit of the denominator should both be clearly stated. In some studies, percentage medication error rates can sometimes exceed 100. It is important to specify how many errors are possible per denominator unit—for example, can one dose be associated with more than one error? The number of doses or patients (or other relevant denominators) with at least one error should also be calculated if more than one error per denominator unit is possible. Calculating a denominator can be particularly problematic for determining prescribing error rates (Franklin et al. 2010b; Lewis et al. 2009). Errors of prescribing omission are sometimes included and sometimes excluded from the denominator; it is important to clearly state whether omissions are counted as errors and whether or not they are also included in the denominator.

ASSESSING SEVERITY OF HARM

As highlighted earlier, the majority of medication errors do not cause patient harm. Unfortunately, those that do can sometimes lead to devastating consequences. Assessment of the error severity is important for determining clinical relevance and facilitating prioritization and better management of follow-up activities (American Society of Hospital Pharmacists 1993). As such, the inclusion of severity assessment is usually recommended when studying medication errors both in research and in practice. The severity of medication errors has been assessed in a variety of different ways. One systematic review identified 40 tools (including adaptations of other tools) that were used for assessing the severity of prescribing errors (Garfield et al. 2013). Another identified ten tools (also including adaptations of other tools) used for assessing the severity of MAEs (Keers et al. 2013b).

A key issue in assessing error severity (also sometimes referred to as clinical significance or clinical importance) is whether *actual* or *potential* harm is assessed. This is an important point as tools validated for assessing actual harm are not necessarily appropriate for potential harm, and vice versa. The obvious advantage of identifying actual harm is its high validity and relevance; patients either did or did not suffer harm, and the level of actual harm is less subjective than assessing the potential harm. However, assessing actual harm is not always practical, particularly where there might be a considerable time delay between the occurrence of the error and any harmful effect, or where someone intervenes to prevent an error from actually reaching the patient. It would be unethical, for example, to stop pharmacists from routinely correcting erroneous prescriptions in order to measure actual harm. Moreover, the same medication error may lead to minor harm in one patient and severe harm in another; this makes it difficult to generalize the likely outcomes of medication errors for informing broader management. By contrast, assessing potential harm does not require knowledge of actual patient outcomes and offers a more practical approach that can be used in research to generate more comparable findings. Nonetheless, the risk of bias due to inter-rater subjectivity with potential harm assessment tools should be considered; a validated and reliable tool should be used where appropriate.

Despite the range of tools reported in the literature, only a few have been validated and/or are deemed to be reliable (Dean and Barber 1999; Forrey et al. 2007; Garfield et al. 2013; Taxis et al. 2002). Table 7.1 summarizes the main attributes of three commonly used severity assessment tools. The National Coordinating Council for

TABLE 7.1
Common Severity Assessment Tools

	NCC MERP (2011)	Dean and Barber (1999)	Folli et al. (1987)
Description	Clinical severity of an error is classified according to nine categories (A–I), where "A" indicates an event that had the capacity to cause an error but an actual error did not occur, and "I" indicates that there was an error that led to patient death.	Clinical severity of an error is scored on a visual-analogue scale from 0–10 by a number of health care professional judges, where 0 represents no potential effect and 10 represents an incident that would result in patient death.	Clinical severity of an error is classified into one of three categories: potentially lethal, serious, or significant.
Judges recommended by the tool	Not specified	United Kingdom (medication administration errors): minimum 4 judges from any of doctors, nurses, or pharmacists; Germany (medication administration errors): researchers used the tool from Dean and Barber (1999) in the context of a German health care system and suggested a minimum of 3 judges, with one each of a doctor, nurse, and pharmacist being required (Taxis et al. 2002).	Folli et al. (1987) used a member of the pediatric faculty or attending physician and two pediatric pharmacists.
Harm assessed	Intended for assessing actual harm	Intended for assessing potential harm	Intended for assessing potential harm
Comments	Has been used in published studies to assess *potential* harm (Garfield et al. 2013). Forrey et al. (2007) collapsed the 9 categories into 6 and found the adapted tool to be valid and reliable. The tool includes a "non-error" category, so it can also be used for recording whether or not an error has occurred.	Validated for use in the United Kingdom and Germany. The mean severity score for medication administration errors was lower in Germany than in the United Kingdom for the same cases; the reasons for this are unclear, but may be related to differences in culture and systems between the two countries.	Lowest harm rating is "significant"; some studies have expanded on the categories of Folli et al. (1987). No data are available on the validity of the tool.

Note: NCC MERP = national coordinating council for mediation error reporting and prevention.

Mediation Error Reporting and Prevention's index was developed to classify actual harm, but has been adapted for use in assessing the potential harm associated with prescribing errors (Bobb et al. 2004; Forrey et al. 2007; van den Bemt et al. 2002).

CONCLUSION

The measurement of medication errors is an essential step for understanding, monitoring, and developing effective interventions to increase patient safety. This chapter has presented an overview of the main challenges and approaches associated with measuring medication errors, together with relevant recommendations.

8 Psychological Theories

Rebecca Lawton and Gerry Armitage

CONTENTS

KEY POINTS

- Psychological theories, particularly around cognition and behavior, are the most commonly used theories in medication safety research.
- Theories from cognitive psychology help researchers to understand the reasons why errors occur.
- Theories from understanding and predicting behavior help researchers to understand why health care professionals act in less-than-ideal ways.
- Used in combination, these theories can help researchers and health care professionals to design interventions that reduce errors and violations and hence increase safety in medication use.

INTRODUCTION

Psychological theories are among the most widely used in patient safety research, which is perhaps not surprising given that, ultimately, safety depends largely on the performance and behavior of those people (doctors, nurses, midwives, pharmacists, and other health professionals) who provide care. This chapter will review two main psychological paradigms and how these apply to the study and prevention of medication error. To do this, we will go back in time and look at how contemporary thinking has developed from these two paradigms and the benefits of uniting these to develop error-reduction strategies. First, we look at the discipline of cognitive psychology and how this has helped us better understand the reasons why we human beings do not always get it right. We will then explore the literature on understanding and predicting behavior, which offers an insight into why human beings do not always behave as required. In conclusion, we propose a case for combining these two paradigms and considering their mutual benefits in better understanding and promoting medication safety.

THE HISTORY OF INDIVIDUAL BLAME

The earliest studies of the human contribution to accidents focused on variations in the performance of individuals in the workplace and their so-called "accident proneness." This approach, supported by the statistical work of Greenwood and Woods (1919), held that humans make errors because they are incompetent or lack moral integrity. At the time of the First World War, factory managers were keen to understand why rates of accidents seemed to be higher in some cases. Greenwood and Woods (1919) submitted a report to the British government that identified some workers as more "accident prone" than others and suggested that this was due to cognitive and personality flaws. This finding was comforting for factory owners, who could then justify ignoring the ever-increasing pressures on workers and the repressive high-risk conditions to which people were exposed. Instead, the research findings conveniently supported a potent strategy for many factory managers: dismiss those workers who were thought to be inherently flawed, while ignoring any responsibility for providing safe working conditions. Similar theorizing continued in other statistical studies between the two world wars, when it was consistently believed that accident-prone individuals suffered accidents both in work and at home. Arguments that personality characteristics were associated with higher levels of reporting, rather than higher levels of accidents, or that those experiencing more accidents were exposed to higher levels of risk, were largely ignored (Lawton and Parker 1998). Findings of this kind cemented the foundations for focusing blame on those proximal to errors, at what is known as the "sharp end" of the workforce (Reason 2000).

COGNITIVE PSYCHOLOGY AND THE UNDERSTANDING OF ERROR

In the 1970s and 1980s, cognitive psychologists began to study mental processes in order to try to understand how human beings became skilled performers, how we solved complex problems, and what attention and memory (information processing) resources were required for complex versus routine tasks. Much was learnt about "normal" performance from the study of breakdowns in that performance. Increasingly, it became clear that certain types of human errors (slips and lapses) were by-products of becoming skilled performers. Perhaps the most significant contribution to this field was that of James Reason (1990), who proposed in his book "*Human Error*" that each credit carries a related debit. For example, automaticity (a cognitive process that takes place largely independently of conscious control and attention) allows something called skilled performance (Rasmussen 1982), often at high speeds. However, if that schema of events becomes disrupted or a different sequence is required in a particular instance, then automaticity will very likely produce errors of execution (slips and lapses). In other words, the side-effect of learning to perform a task so that you can do it quickly without thinking about it is that, if that task changes, we become very susceptible to error.

Some examples will help to explain this. You drive to work the same way every day. On Saturday, you set off to the supermarket, which follows the same route as your work route until the third exit. You really need to concentrate on going the correct way at this exit, because your automatic response is to carry on and take the fourth exit towards work. If you are planning a shopping list, or talking to children in the back seat, or if your attention is otherwise distracted at this critical third exit, it is easy to take the route to work. In the health care context, a nurse will perform certain tasks in this automatic mode, and if interrupted or distracted, he or she may miss steps in the sequence of tasks. For example, when working through the various steps of preparing an intravenous injection, if a patient then simultaneously asks about their pain relief, the routine expiration date check for the medication can easily be omitted.

At the other extreme, novel tasks that we rarely perform require a great deal of our attention and access to previously encoded memories for information and events. Here, we are not operating automatically, and we default to what Rasmussen (1982) has called a knowledge-based level of performance—although this is labor intensive, it is potentially very effective. However, we are wired for efficiency, so when making decisions, we may rely on previously learnt rules: if X is apparent, do Y. For example, a junior doctor might learn "if a drug name ends with "cillin," then it contains penicillin." This type of rule-based performance (Rasmussen 1982) is helpful most of the time, but this doctor might then prescribe Augmentin (the brand name of a penicillin-containing compound antimicrobial) to a patient who is allergic to penicillin because they have over-applied the rule and assumed that all penicillin-containing antibiotics end in "cillin," which is not the case. The junior doctor has also failed to take into account a second rule: that brand names (for example, Augmentin) and compound names do not generally end in "cillin," even if they do contain penicillin.

When we are performing a task that is completely unfamiliar to us, we may not be able to draw on previously stored rules to help us, but we still remain subject to biases that impact on our decision-making, resulting in mistakes of planning. These are described in the literature (for example, Croskerry 2003) and include the availability bias—the tendency to judge things as being more likely if they come to mind easily. This might be related to the recency of experiencing a similar situation. We also tend to attribute blame for negative outcomes to personal failures (for example, the patient has no willpower and so is obese) rather than to situational factors (the patient has difficulty walking due to joint pain and, therefore, spends long periods of time in sedentary activities), a bias known as the fundamental attribution error. The study of mental processes (cognitive psychology) has led to the understanding that "to err is human." If we want to be able to walk and talk at the same time, or type quickly (as we are doing now), we have to accept the errors that come with it.

FROM COGNITIVE PSYCHOLOGY TO A MODEL OF ORGANIZATIONAL ACCIDENTS

James Reason and other psychologists such as Jens Rasmussen have spent many years studying human susceptibility to error, and Reason, in particular, has investigated the nature of human error in the context of organizational accidents. From a detailed analysis of large-scale accidents, he concluded that for managing errors in organizations, attention should be shifted from individual culpability to the conditions that are more likely to give rise to errors. Reason draws on a useful metaphor of reducing the prevalence of malaria. Killing mosquitoes after they have hatched (i.e., eliminating errors) is a short-term and largely ineffective strategy in the long term. Draining the swamps where mosquitoes breed (i.e., changing the conditions that give rise to error) and making mosquito nets available (i.e., improving defenses) are more effective ways of reducing risk (Reason 2000). We will draw on Reason's "organizational accidents" model (1995), upon which the London Protocol for accident analysis is based (Vincent et al. 2000), as the conceptual framework for analyzing those cognitive processes that can give rise to error. The accident causation model (as it is commonly called in health care) is probably the most commonly cited model for understanding error in the context of patient safety and has been widely used as a means of analyzing the causation of prescribing and medication administration errors (Dean et al. 2002a; Lewis et al. 2014; Sanghera et al. 2007; Slight et al. 2013; Taxis and Barber 2003a).

Following the U.K. Department of Health report *An Organisation with a Memory* (Chief Medical Officer 2000) and the U.S. Institute of Medicine's report (2000), many clinicians have actively embraced the accident causation model. This model recognizes that errors almost always stem from the interaction between individuals and systems and have psychosocial as well as technical components. Reason argues that accidents (and errors) are caused by active failures and latent failures. The former include slips, fumbles, lapses, and errors of planning known as mistakes; the latter often reside deep within an organization's operating mechanisms and range from problematic protocols to poorly designed information technology. While a clinician may make a slip, forget to do something, and incorrectly decide against a particular action, Reason maintains that, despite being directly linked to the clinician's behavior, these are the consequences of certain pre-existing conditions, rooted in the systems that surround clinicians. For example, a slip is more likely to occur when interruptions are frequent, and interruptions are more likely to occur when the pace of work is high. In turn, the pace of work may be influenced by the number of available nursing staff, and this can be determined by an organization's policies concerning guidance on an acceptable nurse-to-patient ratio. Latent failures are often omnipresent, but may not be obvious until an incident is investigated.

Charles Vincent and colleagues have used Reason's work to develop a model for the analysis and investigation of clinical incidents. This model acknowledges the role of latent conditions and also the local conditions surrounding a clinician (for example, access to clinical supervision) and any barriers to error. The resulting London Protocol (Taylor-Adams and Vincent 2004) encourages investigators to look at the processes underpinning clinicians' behavior, to stand back from an individual incident, and to consider whether the contributory factors identified in one incident have relevance to a clinical unit's overall safety, thereby facilitating a prospective analysis (i.e., looking at future risks to safety). This systems approach to patient safety has become well established in health care since the publication of the report *To Err is Human* (2000) and subsequent policy documents in the United Kingdom (Chief Medical Officer 2000, 2001). A number of other frameworks for studying latent failures have also been proposed that are based on the ideas contained in these reports;

for example, the Australian Incident Monitoring System (Runciman et al. 1993), the Eindhoven Classification (van Vuuren et al. 1997; World Health Organization 2009b) and the Veterans Affairs Root Cause Analysis System (Bagian et al. 2002). However, these frameworks are somewhat limited by a lack of empirical evidence and a reliance on latent failure classification systems from non-health care settings that are very different to the structures and processes of health care.

More recently, however, these ideas have been used to develop classifications of latent and active failures specifically for a health care context. For example, an interview study with U.K. nurses and nurse managers about the causes of medication administration errors (Lawton et al. 2012) identified ten themes. The primary theme was ward climate. This reflected the ward ethos, attitudes toward reporting of errors, professional regard, willingness to challenge and be challenged, etc. The more immediate precursors to medication error, and prevalent themes in the interviews, were human resource issues (particularly, too few qualified staff), workload (amount of and planning of work), and the local working environment (for example, noise, distractions, ward design, and equipment availability). Other important influences on staff behavior and performance were routine procedures (for example, admissions), bed management, team communication (written or verbal), and written policies and procedures, supervision, and leadership and training. There is a notable overlap between these data and those from another hospital-based study of contributory factors in medication errors based on an analysis of 40 interviews from a stratified sample of doctors, nurses, and pharmacists (Armitage et al. 2010). Here, as above, poor written and verbal team communication, high workload, and distractions were perceived as the dominant causes of medication errors. Indeed, these factors were also apparent in prospective studies of the causes of prescribing errors based on interviews with medical staff who had made errors in a U.K. hospital (Dean et al. 2002a; Lewis et al. 2014).

This growing body of knowledge about the causes of errors in hospitals has meant that it has been possible to develop an evidence-based framework of those factors contributing to patient safety incidents in hospitals. Based on a systematic review of 85 studies, Lawton and colleagues (2012) have done just that. Their framework (shown in Figure 8.1) identifies the proximal factors (for example, task, patient, team, and individual) and those that represent systems failures (for example, training and education, physical environment, and design of equipment) within a hospital context. This framework can be used as the basis for the investigation of medication incidents and can be built into routine incident reporting systems to capture important information about causation and so facilitate learning and action for patient safety. However, the lack of evidence within primary care about the factors contributing to errors means that a comparable framework is not currently available.

By as early as 1960, the tools and principles of human factors and ergonomics had been proposed as a means of addressing human limitations. Analyzing and redesigning work environments and workflows so as to compensate for human limitations is at the heart of this approach. For example, Chapanis and Safrin (as far back as 1960) called for changes to the medication administration task, recommending double-checking and changes to the design of the environment, in particular the nurses' station and medication preparation area. Furthermore, such analyses of work environments often led to the conclusion that changes to behavior are also required (for example, making handwriting more legible). It could be argued that, beyond education and training, the human factors and ergonomics approach is less able to address problems that involve people deliberately choosing to "do the wrong thing." Therefore, these behaviors must also be considered.

SOCIAL PSYCHOLOGY AND THE UNDERSTANDING OF BEHAVIOR

We turn next to a different field of psychology, which attempts to understand our deliberative behavior with reference to our attitudes. Allport (1935) described attitudes as "a mental and neural state of readiness, organized through experience, exerting a directive or dynamic influence upon the individual's response to all objects and situations with which it is related." Today, the attitude construct remains central to the study of social psychology, offering us, it could be argued, salvation from a purely behaviorist (stimulus–response) approach to understanding what drives behavior. While classical and operant conditioning explain behaviors in terms of automatic and learnt responses to stimuli, respectively, the attitude concept was offered as a means of understanding the mental processes (the "black box") that might intervene between the stimulus and behavioral response. For further reading on classical and operant conditioning, see Schwartz and colleagues (2001).

Psychology had proved itself as a science by focusing on the experimental study of observable concepts, stimuli, and responses. It was now to move into the

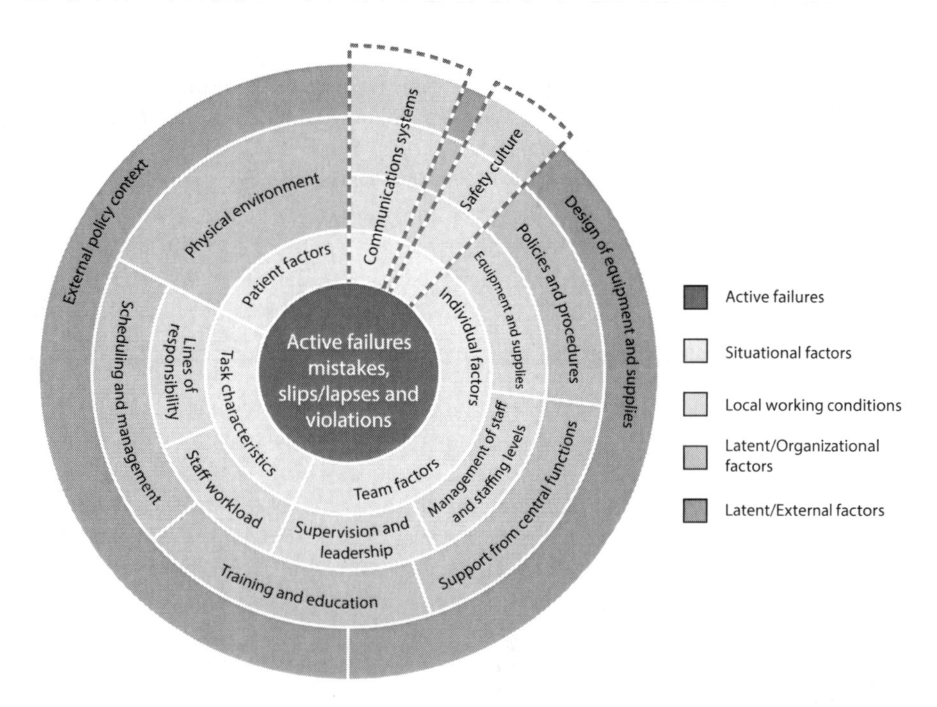

FIGURE 8.1 The Yorkshire contributory factors framework definitions. (Adapted from Lawton, R. et al. 2012. *BMJ Qual Saf* 21: 369–380.)

Factor domain	Definition
Active failures	Any failure in performance or behavior (e.g., error, mistake, violation) of the person at the "sharp-end" (the health professional).
Communication systems	Effectiveness of the processes and systems in place for the exchange and sharing of information between staff, patients, groups, departments, and services. This includes both written (e.g., documentation) and verbal (e.g., handover) communication systems.
Equipment and supplies	Availability and functioning of equipment and supplies.
External policy context	Nationally driven policies/directives that impact on the level and quality of resources available to hospitals.
Design of equipment and supplies	The design of equipment and supplies to overcome physical and performance limitations.
Individual factors	Characteristics of the person delivering care that may contribute in some way to active failures. Examples of such factors include inexperience, stress, personality, attitudes.
Lines of responsibility	Existence of clear lines of responsibility clarifying accountability of staff members and delineating the job role.
Management of staff and staffing levels	The appropriate management and allocation of staff to ensure adequate skill mix and staffing levels for the volume of work.
Patient factors	Those features of the patient that make caring for them more difficult and therefore more prone to error. These might include abnormal physiology, language difficulties, personality characteristics (e.g., aggressive attitude).
Physical environment	Features of the physical environment that help or hinder safe practice. This refers to the layout of the unit, the fixtures and fittings, and the level of noise, lighting, temperature, etc.
Policy and procedures	The existence of formal and written guidance for the appropriate conduct of work tasks and processes. This can also include situations where procedures are available but contradictory, incomprehensible, or of otherwise poor quality.
Safety culture	Organizational values, beliefs, and practices surrounding the management of safety and learning from error.
Scheduling and bed management	Adequate scheduling to manage patient throughput minimizing delays and excessive workload.
Staff workload	Level of activity and pressures on time during a shift.
Supervision and leadership	The availability and quality of direct and local supervision and leadership.
Support from central functions	Availability and adequacy of central services in support of the functioning of wards/units. This might include support from information technology, human resources, portering services, estates, or clinically related services such as radiology, phlebotomy, pharmacy.
Task characteristics	Factors related to specific patient-related tasks that may make individuals vulnerable to error.
Team factors	Any factor related to the working of different professionals within a group that they may be able to change to improve patient safety.
Training and education	Access to correct, timely, and appropriate training both specific (e.g., task related) and general (e.g., organization related).

FIGURE 8.1 (Continued) The Yorkshire contributory factors framework definitions. (Adapted from Lawton, R. et al. 2012. *BMJ Qual Saf* 21: 369–380.)

uncharted territory of non-observable stimuli that could only be accessed by asking people to introspect and self-report—methods largely viewed as unscientific. This felt risky to psychologists and, when researchers were only able to demonstrate weak relationships between attitudes and behaviors, the study of attitudes became less popular in the 1960s and 1970s. With the 1980s and the rise of cognitive psychology, interest turned again to the attitude constructs and, since that time, there has been a consistent drive to generate a scientific basis for understanding attitudes and their relationships to other aspects of the human condition. Theories were developed in the 1980s, of which the Theory of Planned Behavior (Ajzen 1991) is perhaps the most well-known. This theory attempted to explain the relationship between attitudes and behaviors in a more sophisticated way. Early work on the relationship between attitude and behavior identified only weak relationships between attitudes measured by a questionnaire item such as "eating healthily is good for my health" and self-reported or observed consumption of fruit and vegetables. What these new models did, however, was to emphasize the need for specificity and comparability of measures. So, if you want to know what someone will do based on their attitudes, you have to ask specific questions about the behavior in question. In the example above, we would need to ask specific questions about attitudes towards eating fruit and vegetables, rather than eating healthily, which might encompass a whole range of other eating habits (for example, eating wholemeal bread or reducing fat consumption).

The other important aspect of these models is that they recognized that attitudes about the outcomes of performing a particular behavior are only part of the puzzle. These new social cognition models, as they are known, also described other important constructs that were associated with behavior. So, attitudes, the Theory of Planned Behavior model argues, are based on different beliefs about the specific advantages and disadvantages of performing a behavior (if I drive fast, then either I will get to work on time or I am likely to get a speeding ticket). Beliefs about what other people think and do (a construct known as "social norms") and beliefs about whether or not I have control over my behavior (a construct known as "perceived behavior control") are also central to understanding whether or not I will perform the behavior in question. This is because all of these beliefs influence our levels of motivation and motivation is the most important driver of our behavior. Over the past 30 years, a range of social–cognitive theories, such as the Theory of Planned Behavior (Ajzen 1991), the Health Belief Model (Becker 1974), and the Protection Motivation Theory (Rogers 1975), have been developed, tested, and used to understand why people behave or respond in the ways that they do (for an overview of these models, see Conner and Norman 2005). Underpinning this work is the view that if we know which beliefs and feelings predict behaviors, then interventions that target these constructs might be used to encourage more of the desired behavior. The vast majority of these models have been used to understand, explain, and promote socially desirable behaviors (for example, eating healthily, recycling, driving safely, or medication adherence) and reduce undesirable behaviors (for example, smoking, drinking, unsafe sex, speeding, or prejudice).

FROM SOCIAL PSYCHOLOGY TO A THEORY OF BEHAVIOR CHANGE

We turn next to how these models can help in understanding the behavior of health professionals, including behavior relating to medication safety. With the evidence-based medicine (EBM) movement and the introduction of the U.K. National Institute for Clinical Excellence in 1999, there was a new vision for the National Health Service (NHS); research evidence, rather than clinical expertise and know-how, should guide clinical practice. A new science developed (variously known as implementation science, improvement science, or knowledge translation). The main focus of this new science was on how to change the behavior of health care professionals so that it corresponded to the research evidence. The starting point for the EBM movement was to produce guidelines that summarized the research evidence and defined the behaviors required of clinicians (for example, to administer aspirin to all pre-hospital patients following an acute myocardial infarction). It soon became clear, however, that simply improving the knowledge—or the knowledge and skills combined—of clinicians had only limited success in changing behavior (Grimshaw et al. 2004). Increasingly, there was a call for guideline implementation interventions to move beyond education and to be informed by, or even based on, theory.

The next question the scientific community had to address was which theory to deploy. At the same time, health and social psychologists were beginning to apply theories of behavior and behavior change to the study of health care professionals' behavior, particularly where this was prescribed by rules or guidelines. For example, Robertson and colleagues (1996) proposed a psychological framework for changing the behavior of doctors, while Walker and colleagues (2003) were attempting to identify the most appropriate psychological models for interventions to change clinical practice. Others followed and the focus now moved to which *psychological* theory was the best for understanding professional behavior change from an array of models developed by psychologists to understand and predict behavior in a variety of contexts. In 2003, a group of psychologists in the United Kingdom (from different disciplinary backgrounds) got together with the explicit aim of developing a framework (based on the many theories of behavior and behavior change) that would be of relevance in the context of understanding and changing health professional behaviors. This framework acknowledged the diversity of health behavior theory, but also the confusing array of similar constructs that might represent similar concepts but differed in name because of the psychological tradition from which the construct emerged. The group, led by Susan Michie, therefore attempted to assimilate and simplify the field so as to develop a framework that could be used by intervention developers (including non-psychologists) to identify the "determinants" of behavior change for their specific issue. The resulting Theoretical Domains Framework (TDF; Michie et al. 2005) includes 11 determinants of behavior change. As an interesting aside, a similar behavior change framework was also developed by Fishbein and colleagues (2000) in the United States. This framework focused on human immunodeficiency virus prevention behavior, but is comfortingly similar to the British TDF.

The TDF determinants represent targets for intervention and include: knowledge, skills, beliefs about capability, emotions, social influences, beliefs about consequences, social influences, motivations and goals, and behavioral regulation. Of specific relevance to working as a health professional is the determinant "professional identity." Two determinants of particular note, but somewhat less developed than the other factors, are "memory, attention and decision-making" (the realms of cognitive psychology) and "environment and resources." The latter refers to the structural and organizational characteristics (for example, staffing levels, equipment availability, or training resources) that make the behavior more difficult to perform. Those organizational factors that specifically impact on the ease with which quality improvement initiatives are implemented have also been explored in the literature (for example, Yano 2008). It is in these last two determinants in particular that we begin to see overlap with our previous models of organizational accidents (more on this later).

USING BEHAVIOR CHANGE THEORY TO IMPROVE MEDICATION SAFETY

To date, only a few studies have applied the TDF to understanding or changing any of the types of medication errors introduced in this book: prescribing, dispensing, or administration. We will take a look at these now. First, Hrisos and colleagues (2008) used the TDF as a basis for their intervention to promote the management of upper respiratory tract infection without the (inappropriate) prescribing of antibiotics. They targeted two constructs that were amenable to change and that predicted behavior—these were self-efficacy and beliefs about consequences. In the first instance, they used a graded task (a task that becomes increasingly more challenging over time) to help general practitioners become more confident in managing a consultation without prescribing antibiotics. In the case of beliefs about consequences, techniques of persuasive communication were used in which the consequences of different doctors' prescribing behavior were made salient to their practice.

More recently, Duncan and colleagues (2012) used the TDF to explore prescribing errors among junior doctors. The authors recognized that such studies are usually based on human error theory, and research conducted within this paradigm has identified that the majority of prescribing errors among junior doctors are mistakes caused primarily by stress, fatigue, high workload, inexperience, and gaps in training (supporting previous work by Dean et al. 2002a). Problems with a lack of feedback and poor communication have also been identified. However, Duncan and colleagues (2012) argued that for the purposes of developing interventions to target prescribing errors, a better understanding of the processes of behavior change is needed. To do this, they used the TDF to structure interviews with 22 junior doctors. This study served to identify specific determinants of prescribing errors and, based on this understanding and the wider literature, the authors were able to make specific suggestions for appropriate behavior-change techniques. These included modeling and the demonstration of good prescribing behavior by senior colleagues, feedback about the prescribing errors made and their outcomes, and the use of prompts or cues. The study also identified explicit links between determinants. Of particular interest was that junior doctors frequently referred to the environmental factors associated with the determinant "memory, attention and decision-making." These included distractions and interruptions arising from the busy clinical environment and the design of prescription charts that did not support decision-making.

There is growing recognition that this framework might provide a very sound basis for developing interventions to change behavior. We have used the TDF to co-design five different patient safety interventions with health care teams (Taylor et al. 2013). Two of these interventions have focused on promoting the safe use of medications (gentamicin and midazolam) in response to U.K. National Patient Safety Guidelines. Not only has this approach resulted in behavior change, but teams have responded positively to the process of change (Taylor et al. 2013), commenting on the impact of the intervention on teamwork and culture.

Earlier in this chapter, we referred to the research on health behavior change that influenced the development of social cognition models. In this field of research, there is an assumption that behavior is planned (i.e., subject to some kind of intentional decision-making process). There is recognition, of course, that with repetition, the enactment of a behavior requires very little attention and that past behavior becomes a very good predictor of future behavior. What these models are less capable of predicting are those behaviors that are not planned, or even desired by the individual, but sometimes represent automatic responses to environmental cues. It is in this domain that cognitive psychology offers us models that help in understanding exactly these breakdowns in performance; they are the by-products of an information-processing system that trades off complete accuracy for an ostensibly more efficient system, which has allowed us to adapt to work within complex systems and under time constraints, where automaticity equals speed.

It is also interesting to consider that while the models of Reason and fellow researchers began with an attempt to understand cognitive errors in performance, in later years, they strove to move beyond the cognitive system, recognizing that the social environment and the rules and norms of society also led to behaviors that were unsafe (Reason et al. 1998). Reason coined these behaviors that deviated from socially or formally prescribed norms and rules as "violations," and referred to cognitive errors and violations collectively as active failures. Others have also recognized the importance of this distinction. Parker and colleagues (1995a) demonstrated statistically (using factor analytic techniques) that it was useful to distinguish between errors (slips and lapses) and violations while driving a car. Moreover, they discovered that these two types of active failures showed different demographic profiles. While errors were more common among older and female drivers, violations were more common among young males. People who reported more violations on the road (i.e., those with a greater propensity to take risks [for example, deliberately exceed the speed limit]) also had more accidents (Parker et al. 1995b), whereas, in their studies, errors were unrelated to accident likelihood. These authors went on to study the predictors of road traffic violations further and, not surprisingly, they chose the Theory of Planned Behavior for this (Parker et al. 1992).

In another domain entirely—medication adherence—the same distinction has been made (although the terminology used is different). Barber and colleagues (2005), in their aptly named paper "*Can human error theory explain non-adherence*," address the issue of patient non-adherence to medication regimens. They conclude that while human error theory does a good job of understanding the causes of unintentional non-adherence (forgetting to take medication or taking the wrong number of pills), it did less well in explaining intentional non-adherence (for example, a patient's decision not to take a medication because they believe it is not helping them or they do not like to take medication). In a later study, Clifford and colleagues (2008) went on to use the Beliefs about Medicines questionnaire to better understand the differences between those patients who actually took their medications for chronic conditions and those who were intentional non-adherers (violators) and those who were unintentional non-adherers (error-makers). They found that the beliefs of the violators were significantly more negative than the beliefs of the other two groups (which did not differ). Collectively, what this evidence points to is that errors and violations might require different remediation strategies, because they appear to have a different cognitive basis.

It appears, then, that to fully understand and address medication errors, recognition of both of these two models that have grown out of a specific psychological paradigm and therefore are constrained by the principles of that paradigm is necessary. It is, we argue, important to consider both *performance* errors and *behavioral* violations in order to generate appropriate solutions. Take the example of speeding at 40 miles per hour in a 30 mile per hour zone. This could be a behavioral violation—the driver is late and needs to get to work on time, so deliberately drives faster than the rules allow. On the other hand, the driver might not see the signs that indicate that the speed limit on the road has changed. The latter is an example of a performance error. An intervention designed to introduce better signage or road markings might do very little to address the speeding of the first driver, but may well have the desired effect on the second driver. Therefore, if we want to improve road safety, we would need to understand whether, for the majority of drivers, speeding is intentional or not. If intentional, a different set of interventions would be required, which target beliefs or provide disincentives to speeding, perhaps through persuasive techniques and more speed cameras, respectively.

In the context of medication errors, the same principles apply. For example, the nurse who deliberately does not check patients' wristbands prior to administering medications may, therefore, give one patient's drug to another patient on the medication round. Alternatively, this same outcome could arise from a nurse misreading the details on a patient's wristband, or simply missing out this step due to an interruption. Once again, attempts to eliminate incorrect drug administration would need to account for both of these types of event. In the first instance, one might focus on promoting a general understanding among all staff about the consequences of not checking wristbands, and more senior staff might model the correct behavior. Supervisors could provide praise for checking wristbands correctly and provide clear instructions and explanations for how to perform the task safely. In the second case, however, the approach might be to redesign the conditions in which a medication round is conducted by asking patients and staff not to interrupt the administering nurse, by ensuring that wristbands are easy to read, or by using technology such as barcode verification to remove the need for humans to interpret the name on the wristband.

These different approaches to managing risks are a function of the different paradigms we have explored above. Those studying errors and mistakes have adopted a human factors and ergonomics approach to remediation, analyzing and redesigning work environments and workflows to compensate for human limitations and training people to be more aware of the biases that might impact on their decision-making. On the other hand, changing behaviors, and so reducing violations, involves working with and understanding people and their skills, motivations, social norms, feelings of control and emotions, and identifying and developing strategies that will affect these constructs and so bring about change. Both models originated from an attempt to understand individual performance and behavior. Both have culminated in models that account for contextual factors (for example, organizational systems and processes) and their impacts on individual performance.

What is different about the two models, however, arises perhaps from the extent to which the behaviors being studied are considered to be a function of free will, individual control and, hence, accountability. Slips and lapses are generally regarded as outside the individual's control, while violations are often deemed to be deliberate behaviors. This simplifies what is, in fact, a complex picture, with violations often becoming the routine over time and therefore associated with little planning (or free will), and some errors being associated with decision-making that is deliberative; however, the distinction remains and affects the focus of interventions. Thus, while the human factors approach tends to rely on interventions that target systems, teams, equipment, and tasks (Holden et al. 2013a), the individual is the starting point for the behavior change approach, which attempts to understand the thoughts and feelings of the people concerned, recognizing that, ultimately, these will influence our behavior. In this sense, the systems, teams, equipment, and tasks are important insomuch as they influence the thoughts and feelings of individuals and vice versa.

We would argue, therefore, that there is synergy between these two models and that, if they are considered together, this might enable a fuller analysis of the emotional, social, and cognitive mechanisms that intervene between the organizational conditions/local conditions and active failures.

COMBINING THE ACCIDENT CAUSATION AND BEHAVIOR CHANGE MODELS IN PROMOTING MEDICATION SAFETY

Let us think about a nurse who is under pressure on her duty roster due to another registered nurse being sick, plus there being a period of increased patient dependency. She is required to undertake a medication round and is aware that protocol demands she checks one particular intravenous medication with another registered nurse. She makes a conscious decision not to call on the second nurse, as she knows this will take time and, indeed, draw the second nurse away from a vital task. At the time of checking the volume of intravenous medication, she is interrupted and makes a dosing error, giving too much to the patient. As a consequence, the patient experiences some temporary nausea. The nurse reports the error and is subsequently disciplined. At her local hearing, she informs her manager of the local conditions and the manager responds by focusing on the nurse's deliberate flouting of the double-checking protocol, explaining that the outcome might have been avoided if she followed procedure.

However, an application of both our models could suggest a different approach, which could facilitate a better understanding of the error and hence a more appropriate intervention. First, it would be helpful if the manager were to ask the nurse to tell her about the day as it unfolded, providing an opportunity to uncover both the perceived contributory factors and the determinants of behavior. Questions might include: what caused the interruptions to her work (contributory factor)? How does the ward routinely respond to the unplanned loss of a trained nurse (contributory factor)? What are the potential advantages and disadvantages of deciding not to double-check an intravenous medication (behavioral determinant)? What do other nurses do in a similar situation (behavioral determinant)?

In summary, the manager should first elucidate the contributory factors, discriminating between organizational and local conditions that impact on safe practice, before probing the motivations behind her decision that day, and so determine whether the decision was affected by her degree of control, any local social norms and, of course, her emotions. In fact, in a study of double-checking as a contributory factor in medication errors, staff admitted that double-checking was often abandoned (with priority being given to the completion of the entire drug round at the expense of the double-checking safety net) when the pace of work was increased by factors beyond the control of those at the sharp end (Armitage 2008). The strength of our recommended approach, however, is in then synthesizing the behavioral perspective (behavior change theory) with a clear acknowledgement of the error-causing factors (organizational accident model). This established the impact of both highly dependent patients and the skill mix of staff on the likelihood of interruptions and distractions that can give rise to slips and lapses. We argue that this approach provides a more searching analysis, with lessons for the whole team. In addition, it is perhaps less likely to alienate staff and instead promote learning than is the response of disciplining the nurse, as described earlier.

As we set out above, to understand medication errors, it is necessary to recognize that the behavior of clinicians is a function of two processes: the first is automatic and intuitive, relying on well-learned associations and patterns of responding and uses little cognitive resource, being largely effortless (sometimes known as "system 1"). The second process is effortful and resource intensive, involving conscious thought and motivations to act (known as "system 2"). These dual processes (system 1 and system 2) are evident in models that have developed in various forms over the last 30 years to explain our responses to persuasive messages and the role of feelings and biases in decision-making, social judgments, and reasoning (for a review, see Evans 2008; Smith and DeCoster 2000). Such models can also be applied to help us better understand the prescribing, dispensing, and administration of medication.

It is important to acknowledge that when we perform a task, both of these processes (we will call them system 1 and system 2 to mirror existing research) will be operating. The default mode for human behavior is to minimize cognitive effort, meaning that what we do is very often a response to external and internal cues (for example, environmental prompts and feelings). Thus, when we are familiar with a task (for example, dispensing or administering penicillin as part of the drug round), it is likely that we will operate largely in this system 1 mode. However, there will be occasions within the performance of the task where reflection may be necessary and we may draw deliberately on our memories of similar situations and consider the outcome (a safe and timely drug round) we are aiming for and whether our current progress will allow us to achieve this outcome. When doing these checks, we will utilize system 2.

On the other hand, if we are engaged in a more complex or novel task for which a decision is required between a number of possible options (for example, prescribing an anticoagulant to an elderly patient), we will operate mostly within system 2, deliberately considering the options, drawing if possible on previous experiences and stored knowledge and seeking information from the environment (for example, asking senior colleagues or looking at a suitable reference source) to inform the decision.

By adopting this model, we can see that errors (slips, lapses, and mistakes) and violations are a function of the interplay of the two processes. Slips are more likely to occur when we are operating in system 1 and we fail to deploy the system 2 checking mechanism (we do not stop to check the allergy status of the patient), and mistakes happen when we are operating largely in system 2, but the reliance on automatic recall of a recent event (system 1; for example, the elderly lady that fell and had a nasty bleed after being prescribed warfarin) leads to less consideration of the possible alternatives (the stroke that might have been avoided because warfarin was prescribed). In the same way, violations of rules and procedures can be the result of automatic system 1 processing, i.e., the violation has become routine (Reason et al. 1998), perhaps as a result of repeated positive feedback (saved time and praise from the supervisor for tasks completed). Alternatively, a violation might involve the operation of system 2 processing; in other words, a conscious decision to deviate from rules (perhaps because there is disagreement with the rule or because the rule is impossible to apply on this occasion). This dual process model has only recently been applied to the study of clinician behaviors. In one such study of six primary care behaviors, including prescribing for blood pressure and glycemic control, the authors (Presseau et al. 2014) argued that the repeated nature of the behavior of health care professionals within similar contexts meant that reflective processing was unable to offer a full account of four of the six behaviors (including both prescribing behaviors). Thus, both processes must be considered.

CONCLUSION

This chapter has outlined two models that have been used to investigate the processes underlying the behaviors of health care professionals. We conclude that both processes (i.e., system 1 and system 2) must be considered in the further development of theoretical models to explain the behavior of health care professionals operating in an organizational context and in developing interventions to reduce errors and violations and so improve health care quality.

9 Educational Theories

Lucy McLellan, Anique de Bruin, Mary P. Tully, and Tim Dornan

CONTENTS

KEY POINTS

- Educational theories help to inform our understanding of the nature of learning and how we can improve education, practice and, ultimately, safe medication use.
- Three of the most relevant theoretical paradigms are cognitive psychology, social learning theories, and complexity thinking.
- Within medication safety, where education and practice are delivered together in a complex social world, these theoretical perspectives should be considered to be complementary, rather than oppositional.

INTRODUCTION

As highlighted in the first section of this book, safe practice is a core professional responsibility for those who prescribe, monitor, dispense, or administer medications. It is widely accepted that education plays a crucial role in fostering and sustaining the expertise required to do this. When we use the term "education," we refer not only to formal pedagogical activities, but also to learning that takes place during the course of everyday activities within the multiple workplaces where health care professionals perform their duties. Exploration of both these learning processes in terms of their virtues and deficiencies offers insights into how professionals' approaches to medication use could be influenced.

Educational theories provide perspectives on the nature of learning, which can then guide pedagogical research and practice. Educational theories provide sets of beliefs and assumptions, which suggest where we should focus our attention as we try to improve education and subsequently improve practice. For example, in order to reduce medication errors, should our attention be directed towards individuals' cognitive processes, the physical settings in which individuals prescribe or use medications, the social networks that exist within these settings, or the interactions between all of these?

This chapter contains three main parts. First, we explore the nature of professional knowledge and expertise, and how these may be attained. The second part provides a necessarily brief overview of some of the principles of three important paradigms in education: cognitive psychology (predominantly focusing on metacognition), social learning theories, and complexity thinking. We discuss the core values of each and explore their implications for research and practice in medication use. In the third part, we discuss the landscape of different theoretical perspectives and conclude that, when considering learning within a complexity world view, different theoretical paradigms can be considered complementary rather than oppositional. Throughout the chapter, we do not differentiate between "learners," "students," "teachers," and "professionals," as we view all who act in workplaces as having the potential to learn and contribute to learning, regardless of their role and title.

THE NATURE OF PROFESSIONAL KNOWLEDGE

Traditionally, the focus of health care professionals' education has been the acquisition of knowledge, skills, and attitudes. More recently, however, educators and researchers have acknowledged the importance of integrating these three components of competence, applying them in different contexts and acting in a variety of social environments. This moves us away from time-honored models of education, which focus on individual learners and their cognitive development, towards conceptualizations of learning as both an individual and a social process. This shift requires us to reconsider what we mean by "knowledge" in the context of professional learning.

The different stages of using medication safely, outlined in the first section of this book, illustrate how various types of knowledge are applicable to the overall process. For example, when prescribing a drug, a health care professional may draw on some or all of their codified knowledge (objective, explicit information, such as drug names and dosages), social or cultural knowledge (such as knowing where to find the prescription chart or how to access microbiology results), and personal knowledge. The latter is defined by Eraut (2010) as "what individual persons bring to situations that enables them to think, interact and perform," such as practical experience, memories of similar cases, and personal attitudes towards prescribing. Professional knowledge, then, refers to knowledge *in* practice, as well as knowledge *for* practice and knowledge *of* practice. In other words, within social work environments, knowing how to act and interact within different contexts and situations is a part of professional knowledge.

Knowledge and expertise are closely linked concepts. Professional expertise involves the integration and application of different types of knowledge in varied contexts, with the ability to monitor and adapt to situations. Billett (2001) proposed that vocational expertise is dynamic, negotiated and situated, and that the development of expertise occurs through participation in practice. He expressed clearly that expertise is a product of both individual activity and social practice and we must, therefore, avoid conceptualizing learning as either an exclusively individual or social process. In our own exploration of expertise development for prescribing (McLellan et al. 2012), we also concluded that individual cognition is situated within, and inseparable from, social contexts. Furthermore, responding appropriately to feedback from social environments is a fundamental part of expertise, particularly when it comes to transferring professional knowledge to new situations. Eraut (2004) suggests that knowledge transfer involves five inter-related stages:

1. The extraction of potentially relevant knowledge from the context(s) of its acquisition and previous use.
2. Understanding the new situation, a process that often depends on informal social learning.
3. Recognizing what knowledge and skills are relevant.
4. Transforming them to fit the new situation.
5. Integrating them with other knowledge and skills in order to think/act/communicate in the new situation.

Eraut (2004) acknowledges that these stages, which include individual (cognitive) and social components, do not occur as a simple sequence. Instead, they occur in response to complex situations and are interacting variables within the overall process. The question of how people learn and develop this expertise remains to be explored and can be viewed from a variety of perspectives, informed by a variety of educational theories. The next part of this chapter provides a brief insight into the cognitive, sociocultural, and complexity paradigms within educational theory, followed by a discussion of how these paradigms relate to one another and how individual and social perspectives can be reconciled in order to enhance our overall view of health care professionals' education.

EDUCATIONAL PARADIGMS

The Cognitive Paradigm

The process of using medications, outlined earlier in this book, is complex. It places high cognitive demands on those learning or performing the tasks of prescribing, monitoring, dispensing, or administering medication. In addition, it is a socially situated task, which involves communication within and across disciplines. A complex cognitive task can be defined as one requiring the coordination of different cognitive demands and the transfer of knowledge, skills, and attitudes to a variety of (social) contexts (van Merriënboer and Kirschner 2007). Even a seemingly simple task, such as prescribing paracetamol (acetaminophen), requires the health care professional to recognize the need to prescribe analgesia, select the appropriate drug, dose and formulation, understand the pharmacology, and know how to use the prescription chart or electronic prescribing system. Furthermore, it may involve social interaction with the patient, relatives, and colleagues. van Merriënboer and Kirschner (2007) write that complex learning should be addressed by holistic approaches to education: accounting for the interconnectedness of the separate parts of a task, rather than reducing the task to a set of disconnected fragments. They argue that education must integrate declarative (relating to codified knowledge), procedural (relating to practical skills), and affective (relating to feelings or emotions) learning by providing learners with whole-task, contextualized experience. For medication use, this is likely to mean at least some learning in clinical workplaces. Education frequently fails to achieve this, thereby limiting learners' development of expertise. For safe medication use, a holistic approach to the task would include both cognitive and social components and a feedback loop between the two, which connects intrapersonal experiences to social and physical contexts (Figure 9.1). Using prescribing as an example, a holistic approach would provide learners with opportunities to meet patients, make decisions about drug therapy, complete prescriptions, receive feedback on what they had done, and act upon it in future situations.

An important part of the feedback and self-regulation component of expertise, shown in Figure 9.1, involves metacognitive knowledge and skills. Metacognition refers to a person's knowledge about themselves and their relationship to a task, as well as their ability to adapt their behavior according to that knowledge (Flavell 1979). Put within the context of this book, a health care professional or student who

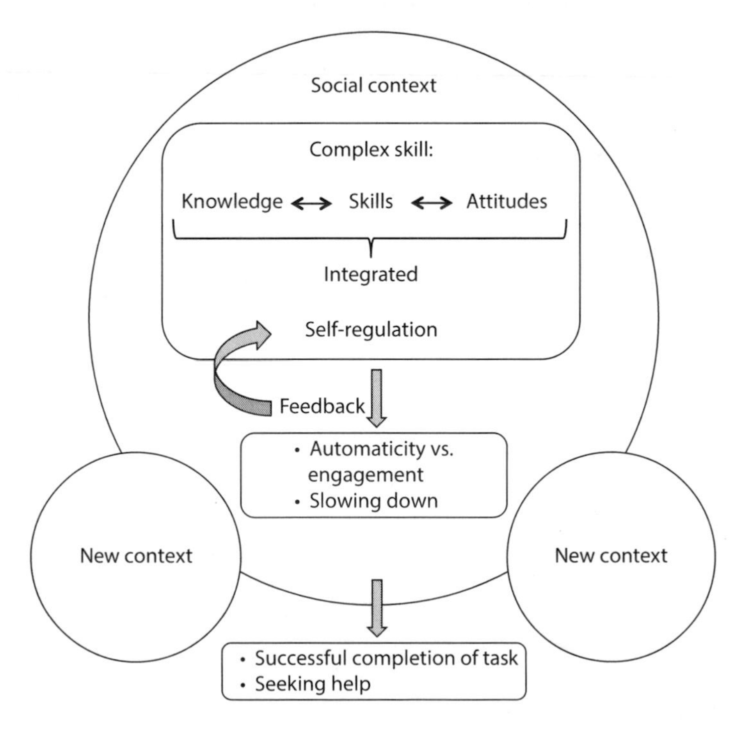

FIGURE 9.1 Model of expertise for prescribing. (From McLellan, L., Tully, M.P., and Dornan, T: How could undergraduate education prepare new graduates to be safer prescribers? *Br J Clin Pharmacol.* 2012. 74. 605–613. Copyright Wiley-VCH Verlag GmbH & Co. KGaA. Reproduced with permission.)

is able to recognize their ability (or inability) to complete a medication-related task without error and, furthermore, knows how and where to seek help when required is more likely to deliver safe patient care.

The importance of metacognition as a self-regulatory process is well recognized; however, research consistently demonstrates that people struggle to judge their learning and performance accurately (Dunning et al. 2003; Eva and Regehr 2005). Some suggested methods of improving individuals' metacognitive abilities include facilitated reflective practice, "think aloud" techniques, and the use of checklists and interactive assessments (Mattheos et al. 2004).

Metacognition research has traditionally been based on the assumption that there is a simple, linear relationship between monitoring, self-regulation, and performance; a perspective that has limited applicability in the "real world" of health care. Although metacognition is an internal process, it is influenced by external feedback and social interactions, as shown in Figure 9.1. For example, even if a health care professional can accurately assess their lack of capability or confidence regarding a task, other factors may limit the extent to which that judgment influences their behavior and actions. The following prescribing scenario, described by a junior doctor, illustrates how the desire to seek advice does not necessarily translate into action, because of the social context:

> You don't know them [the senior doctor] well enough to feel comfortable asking... they might think that's a stupid question so you, kind of, try and do a bit more yourself. Whereas...on my old ward I know I probably would've said to my [registrar], "Oh, what's the dose?" (Dornan et al. 2009).

Likewise, social and cultural aspects of workplaces may limit the extent to which individuals can engage in metacognitive activity. A different junior doctor explains how the nature of a ward round impairs the application of cognitive strategies for safe prescribing:

> The consultant had said on the ward round, you know, "Prescribe this," and you have, you're trying to hold the notes and hold the drug chart and hold everything and try and write ten things at once,...I mean, normally I would check the allergies before I prescribe, but I think it was just really, it gets really hectic on a ward round. (Dornan et al. 2009).

Consideration of metacognition as part of a bigger conceptualization of learning could provide important insights into how to improve practice. When viewed within the wider context of expertise, metacognition acts as a crucial link between the cognitive and social worlds, bridging the gap between individuals' cognitive experiences and their behavior, interaction, and engagement in social environments.

Social Learning Theories

In contrast with psychological perspectives, which focus on the individual acquisition and reproduction of knowledge, social learning theories emphasize the contexts and processes of knowledge production (as opposed to reproduction). Social theorists view learning as situated and dynamic, so the focus of their research is on the social construction and communication of meaning (Bleakley 2010). Individualism has been a prevailing perspective in medical research and practice for many years. Here, we provide a brief history of how social learning theories have challenged this individualistic view of education and explore the value of a non-individualistic—or communal—perspective on safe medication use.

Originating in post-revolutionary Russia in the 1920s and 1930s, sociocultural theories stem from Marxist roots. As the Soviet Union underwent major transformational change, its scholars laid the foundations for an important shift in conceptualizations of learning and development. In stark contrast to the leading behaviorist psychology theories of the time, which sought to develop simple explanations of the individual human mind, social learning theories emphasize that internal cognitive functioning cannot be separated from the external milieu. Vygotsky—one of the founders of social learning theories—explained that to study learning by artificially dividing the individual and social dimensions of the human mind was akin to exploring the fire extinguishing properties of water by studying the independent properties of oxygen and hydrogen:

> This man will discover, to his chagrin, that hydrogen burns and oxygen sustains combustion. He will never succeed in explaining the characteristics of the whole by analysing the characteristics of its elements. (Quoted in Moll 1992).

Much of Vygotsky's work focused on childhood development, but the underlying principles of his work influenced the social learning theories we discuss in this chapter, which are highly relevant to professional education. Wertsch (1985) identifies three major themes in Vygotsky's writing: (a) development is best understood by focusing on dynamic processes, rather than the static end products; (b) individual development originates from social sources; and (c) development is mediated by tools and signs, including language. Vygotsky's work was highly influential and formed the basis for subsequent sociocultural theories, including activity theory and communities of practice. These both share the assumption that learners' development does not occur in isolation from their social and cultural environment.

Activity Theory

Activity theory was developed by Leontiev, who shared Vygotsky's Marxist philosophy. This conceptual framework takes human activity as the unit of analysis, where "activity" refers to a socially situated phenomenon involving connections and relations between subjects, objects, and the environment, as the subject(s) pursue a desired outcome (Leontiev 1981). Engeström developed a model of activity theory that depicts how human action and social artifacts operate together. While remaining true to its philosophical roots, Engeström's model of activity theory has evolved over time, such that his second- and third-generation models are the principal ones in current use (Figures 9.2 and 9.3; Engeström and Sannino 2010).

Educating health care professionals to use medications safely could be viewed as an activity system in which the subject is a learner (either a student or professional), the object is achieving or maintaining expertise, and the outcome is patient safety.

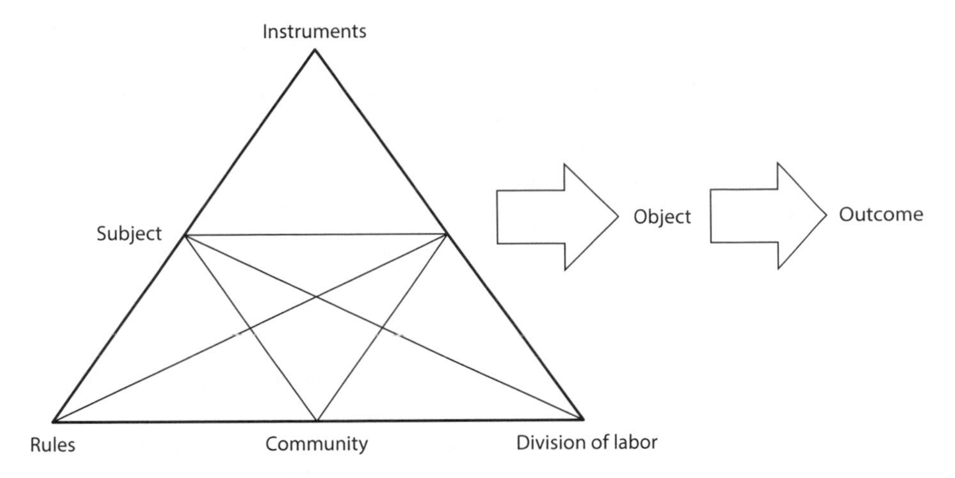

FIGURE 9.2 Second-generation activity theory. (Reprinted from *Educ Res Rev*, 5, Engeström, Y., and Sannino, A., Studies of expansive learning: Foundations, findings and future challenges, 1–24, Copyright 2010, with permission from Elsevier.)

Tools such as prescription charts and guidelines mediate the process. The subject is also part of a community, which may include colleagues, teachers, and patients, and the division of labor is the separate contributions of members of the community to the overall system. Importantly, the whole activity system is influenced by the culture and history of the individuals involved and their workplaces. Activity systems often include sources of conflict or contradiction that, when resolved, result in transformations, which Engeström terms "expansive learning."

Engeström's third-generation activity theory puts forward the idea of interacting activity systems that share, or partially share, an object (Figure 9.3; Engeström and Sannino 2010). De Feijter and colleagues (2011) illustrated the explanatory power of this model in their investigation of medical students' perceptions of patient safety. Viewing students as being part of two activity systems—learning to be a doctor and

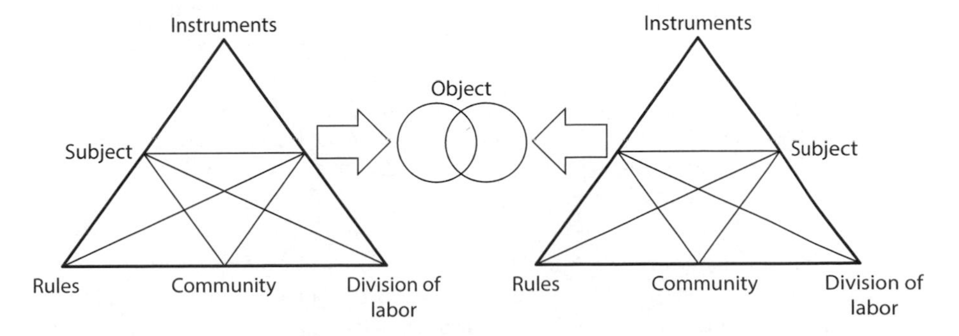

FIGURE 9.3 Third-generation activity theory. (Reprinted from *Educ Res Rev*, 5, Engeström, Y., and Sannino, A., Studies of expansive learning: Foundations, findings and future challenges, 1–24, Copyright 2010, with permission from Elsevier.)

providing safe patient care—revealed contradictions in roles and rules, which led to unsafe situations.

When thinking about safe medication use from a third-generation activity theory perspective, we could take account of multiple activity systems, such as learning to prescribe and performing the task of prescribing, while viewing patient safety as a shared object. Those readers who wish to know more about the application of this theoretical perspective as an intervention tool may find it helpful to refer to Kerosuo and colleagues' explanation of the Change Laboratory (2010), which enables the redesigning of activity systems through a social process of innovation.

Communities of Practice

The concept of communities of practice, initially developed by Lave and Wenger (1991) and then extended by Wenger (1998), focuses on learners as social participants in communities, whose identities are shaped as they interpret their experiences of practice. A core principle of this theory is "legitimate peripheral participation," which describes how novices become members of a community through different forms of involvement. For example, clinical medical students learn by participating in the overlapping practices of patient care and medical education (Steven et al. 2014). Their development is shaped by the forms of participation they have access to, which is largely dependent on their interactions with doctors. Wenger stressed the importance of relationships between learning, meaning, and identity by explaining that participation is not, in itself, sufficient for learning to occur; events must take on personal meaning in order for participants to (re)construct their identities because of these events. Bleakley (2011) described this view of learning as "a mutually transformative relationship between work and identity...focused on issues of existence and relationships such as 'being' and 'becoming'." Recently, Wenger (2010) has developed the theory to strengthen the emphasis on forming identity across multiple communities and systems of practice. This approach acknowledges the importance of knowledgeability across a much broader "landscape of practice," as well as competence within specific communities of practice. Knowledgeability is described as "a process of modulating identification across multiple locations of accountability... [which] involves a constant interplay between practices and identities." In the following quotation, a junior doctor reflects on making a prescribing error and illustrates that it is possible to hold the identity of a doctor without being all-competent or all-knowledgeable:

> I knew I should've looked it up cos I didn't really know it, but I, I think I just convinced myself I knew because I felt it was something that I should've known...because it is very easy to get caught up in, in being, you know, "Oh I'm a doctor now, I know stuff." (Dornan et al. 2009).

Exploring medication errors and patient safety from a communities or landscapes of practice perspective encourages researchers to consider the interplay between identity, learning, practice, competence, and knowledgeability, which can provide important insights into how social interactions influence behavior.

To summarize this part of the chapter on social learning theories, activity theory and communities of practice have many principles in common and are both highly applicable to medical education and patient safety research. Activity theory illustrates interactions that occur in pursuit of an object and focuses on the activities that occur within practice, while communities of practice explores how people become socialized into communities and how identities develop within and across systems. The concluding part of this chapter offers a more detailed discussion of how both of these perspectives take account of complexity and uncertainty within health care settings.

The Complexity Paradigm

Complexity theory, while originating in the mathematical domain, has been applied to a wide range of disciplines. Due to this transdisciplinarity, the paradigm is broad and evolving, so it eludes any single definition. In this chapter, we avoid discussing the intricacies of competing interpretations of complexity theory and, instead, go back to basics in order to answer the following questions: what does "complexity" mean and how are its insights relevant to education research and practice in relation to medication safety?

The term "complex" is derived from the Latin "*complexus*," meaning plaited together or interwoven. Awareness of this etymology leads us to understand that complex phenomena are made up of interconnected components that must be studied as systemic wholes, rather than reduced to their constituent parts (Alhadeff-Jones 2008). Simple phenomena, in contrast, exhibit straightforward relationships between parts, such that the overall whole is predictable by considering its component parts; system behavior is reproducible. Simplicity assumptions have traditionally formed the basis of most scientific enquiry, promoting linear cause–effect models as ways of explaining events or entities. Awareness and acceptance of complexity principles challenges these assumptions, calling for tolerance of ambiguity and the recognition that, in complex systems, reducing effects to their perceived cause(s) can generate false conclusions (Doll and Trueit 2010).

Although there appear to be no simple definitions when it comes to complexity, there are several characteristics that are demonstrated by complex systems. The interwoven elements of a system interact dynamically with one another, such that changes to one component alter others. Furthermore, relationships are nonlinear, so small changes have the potential to bring about unpredictable effects. Complex systems exhibit adaptive behavior in response to changes, evolving and learning in a process known as self-organization, which is closely related to feedback and self-regulation (Bousquet and Curtis 2011). Due to the way in which these systems transform over time, they are often termed "complex *adaptive* systems." Systems are sensitive to history and culture, so the past shapes the present and the future. Although systems have rules and boundaries, these are flexible. Overall, the behavior of a system emerges as a result of dynamic, complex interactions and, just as what emerges cannot be predicted by knowing the initial conditions of the system, the patterns of output cannot be fully understood by analytic methods that seek to reduce them to their component parts. The observed output of a complex system is, in any case, only representative of a fraction of myriad alternative and possible future outcomes (The Health Foundation 2010).

There are many examples of complex systems in the natural and human worlds, such as ant colonies, weather systems, jazz bands, families, business organizations, and governments. Most, if not all, social endeavors involve nonlinear, dynamic interactions between different people, their past experiences and their current environment, and can therefore be conceptualized as complex systems. These are sometimes referred to as "complex adaptive *social* systems." Health care professionals' education is one such example and, in fact, brings two systems together: education and practice. These may be thought of as nested complex systems, which operate in relation to each other and evolve together. A systems approach draws attention to relationships between education and practice at various levels and shows patterns and trends of learning. Learning, then, is not confined to individuals or social collectives, but also applies to the whole system as it adapts in response to emergent phenomena (Bleakley 2010).

Returning to the topic of medication use, this is a process that involves multiple layers of complexity for both education and practice. The system involves individuals who prescribe, administer, monitor or receive medications, all of whom can influence each other and are molded by past experience and personal history. These individuals interact with their environments, which requires them to work with tools or artifacts (such as prescription charts and computer systems) and exposes them to the cultures of workplace and organizations. Individuals exist as parts of professional communities and social groups; for example, a junior pharmacist may be a part of the pharmacy department team, the ward team where they provide clinical services, a community of other junior pharmacists on a postgraduate training course, and a wider social and professional network. Medication use by individuals and communities is also shaped by local and national policies, which are affected by global health issues, such as antibiotic resistance. These are just some examples of the interacting factors that come into play when it comes to safe medication use. A complexity approach acknowledges all these levels of activity and regards learning as an evolutionary process (Mennin 2010) involving all system levels. The focus of study is not individuals, communities, or societies, but the patterns of interactions and relationships within the system and the ways in which these adapt over time.

Controversially, this means that the complexity paradigm is able to accommodate a range of educational discourses, including those from cognitive and social paradigms, even though they are traditionally considered contradictory. This concept will be considered further in the discussion and conclusion at the end of this chapter, as we discuss how the theories we have explored here can complement or oppose each other (Davis and Sumara 2009; Mason 2008).

To bring our summary of the complexity paradigm to a close, we explore the advantages of bringing this approach to bear on education research and practice. Complexity principles are not just explanatory; they have immediate practical relevance and offer opportunities for transformation. Any interpretation of how a complex system functions can only partly explain what is going on. This means that, instead of changing systems by the careful analysis of events followed by the rigorous design of interventions, it is more appropriate to monitor patterns of emergent behavior and explore how those patterns change in response to trial-and-error interventions. These interventions can then be amplified or reduced according to whether they affect the system positively or negatively. Ultimately, the recognition that education is complex is not an obstacle to successful learning, research, and practice, but an opportunity for creative development and innovation. Murray Gell-Mann, Nobel Prize winner and co-founder of the Santa Fe Institute for Complexity Studies, sums up the importance of considering complex systems holistically by stating that "someone should be studying the whole system, however crudely that has to be done, because no gluing together of partial studies of a complex nonlinear system can give a good idea of the behavior of the whole" (Sturmberg and Martin 2013). The discussion that follows this part of the chapter explores how we might respond to this request, using an example from our own empirical work on how medical students learn to prescribe.

THE LANDSCAPE OF EDUCATIONAL THEORIES

To conclude our chapter on educational theories and safe medication use, we will explore the relationship between the aforementioned paradigms. Bleakley and colleagues (2013) suggested imagining "medicine, medical education and medical education research as overlapping fields that share a common concern—patient care and patient safety." Our discussion of educational theories shares this view that it is impossible to separate these domains. We start from an assumption that to practice, learn, or research the safe use of medications, the inherently complex nature of the world must be considered. In turn, the landscape of educational theories is a complex one; new perspectives emerge as a result of conversation, exploration, and open-mindedness at the level of individuals, activities, social groups, and organizations. We suggest that an overarching complexity worldview enables academics to view practice, education, and research from a variety of theoretical stances.

In our own work on investigating how medical students learn to prescribe, complexity insights enhance our interpretations of students' learning by allowing us to accept the realities of uncertainty in educational encounters. Our data show that learning processes are complex because of the social milieu in which interactions between participants, patients, teachers, and other clinicians take place. Actors within these environments employ their cognitive and metacognitive faculties, but are also tacitly influenced by their shared culture and social system. A complexity perspective draws our attention to interactions between phenomena, events, and actors and encourages consideration of the reciprocity between personal and social dimensions. We position students' experiences of learning to prescribe at the intersection of medical education and medical practice, and regard what would conventionally be conceptualized as "learning outcomes" as emergent properties of complex, socially situated interactions.

CONCLUSION

The paradigms we have discussed in this chapter are not mutually exclusive. The social and cognitive paradigms can be viewed as different levels of complexity within an overall system. While cognitive approaches explore the complexity of the human brain and individuals, social learning theories illuminate complexities within the social world. Our overall complexity view allows us to step back and observe tensions and resolutions between these paradigms in relation to a particular topic of enquiry. We suggest that education researchers embrace complexity as an overarching view and, within this perspective, select appropriate theories according to their chosen unit of analysis: at the level of individuals, communities, or whole systems. There will always be tension between individualistic and social paradigms. We suggest that these dissonances are not problematic but, instead, serve to expand our vision of the complex nature of the world.

10 A Sociotechnical Perspective of Electronic Prescribing

Jos Aarts

CONTENTS

KEY POINTS

- The introduction of electronic prescribing and electronic medication administration has been used to illustrate how sociotechnical approaches help to understand the interlinking between these technologies and their social and organizational contexts.
- Prescribing, whether done on paper or via a computer, is a social act.
- The use of any kind of technology will impact on that social act and nudge the health care professional towards different behaviors.
- Such sociotechnical impacts will influence the success or failure of the technological implementation.

INTRODUCTION

Because health information technology (IT) requires interacting with people and inevitably affects them and their workflow, understanding health IT requires a focus on the inter-relation between technology and its social context. A sociotechnical perspective aims to do just this and recognizes the complex interactions between people and technology and seeks to unravel these in order to improve the design and implementation of technological applications (Berg et al. 2003).

In health care, sociotechnical approaches have been almost exclusively used to investigate electronic prescribing (e-prescribing) and electronic medication administration records (eMARs). This chapter will therefore focus on these areas as an example of how a sociotechnical approach can help us to understand the interlinking between the technologies and their social and organizational contexts. It will address three main areas. First, it will consider how the adoption of e-prescribing can be explained from a sociotechnical perspective. Second, it will consider decision support systems, which are an integral component of many modern e-prescribing systems. Decision support systems are primarily aimed at improving prescribing by presenting reminders or alerts, yet their performance so far has been disappointing. Sociotechnical approaches have been used to investigate which norms have been built into such systems and how such systems can be improved in the context of daily use. Finally, we need to consider how the roles of users (including physicians, pharmacists, and nurses) have been changed by the implementation of e-prescribing and eMARs, how these changes affect patient safety, and what they mean for the design of the medication system.

THE SOCIAL CONTEXT OF PRESCRIBING

> The healing hand of the doctor reaches the patient through the prescription and the medicine. (Whyte et al. 2002)

Prescribing medication, whether on paper or via e-prescribing, is embedded in social norms and cultures. By definition, medications are substances that have the capacity to change the condition of a human for better or for worse, depending on whether a medication is a match or mismatch for the ailment (van der Geest et al. 1996). Their "concreteness" provides both the physician and patient with a means to deal with the problem at hand.

In Western medicine, physicians (and increasingly other health care professionals) are vested with the authority to prescribe medications. Through prescriptions, such health care professionals show the patient that they take their complaint seriously and are seeking to help them. Where medication is seen as the essence of medical practice, prescribing is the main activity expected from a doctor (Pellegrino 1976). Often, a non-prescribing physician is seen as a contradiction. A prescription seals a contract between the patient and physician and is mediated by a piece of paper. Taking this prescription is like taking along the doctor, with their knowledge and advice, to where the patient will have their prescription as dispensed by the pharmacist. Prescribing is, therefore, a social act mediated by a prescription. The paper with the prescription then becomes an active element that interacts with people. Understanding such relationships is at the core of sociotechnical analysis.

Evidence-based medicine (EBM) lies at the heart of the Western health care system. Rational scientific knowledge is at the core of medicine and is made visible through guidelines and protocols. Because of quality assurance, patient safety and the rising costs of health care, only treatments with proven patient benefits are increasingly reimbursed, to the exclusion of others. The discussion currently focuses on what constitutes evidence and how it is being applied (Greenhalgh et al. 2014). The accepted hierarchy puts evidence from randomized clinical trials at the top and gives little credibility to clinical case studies. However, it is estimated that perhaps less than 5% of medical treatments are based on sound scientific evidence, and many are still based on clinical traditions and habits that are passed on from generation to generation (Marciano et al. 2014). In the same vein, the quality of prescribing is often questioned. In their seminal study of physician motivations, Schwartz and her colleagues cited patient demand, the placebo effect, and clinical experience as the most important reasons behind non-scientific medication prescribing (Schwartz et al. 1989). They found that non-clinical motivations were dominant, such as maintaining a good patient–doctor relationship and preserving a professional self-image of being seen as a doctor who is acting within this relationship. Most studies of non-scientific prescribing have been conducted in primary care, where the doctor may have to deal with patient needs that are social rather than medical, may be practicing alone rather than with peers, and may have less access to clinical information. However, a later study also showed that, in hospitals, up to 50% of antibiotic prescribing was inappropriate (Hulscher et al. 2010), with professional habits, such as a dislike of "cookbook medicine" and ingrained clinical habits, playing an important role.

Various strategies have been deployed to influence medical practice. These include education, dissemination of scientific knowledge, and organizational and institutional measures. Organizational and institutional measures are designed to act as incentives, be they negative or positive. As already mentioned, decisions not to reimburse unproven medical procedures may discourage physicians from using them. On the other hand, physicians may receive financial rewards when they introduce quality-improvement programs (Rosenthal et al. 2004). Tools are also part of organizational and institutional measures. An example is the proliferation of guidelines that target the health care professionals caring for people with specific conditions and diseases. Professional medical societies often develop guidelines that represent the best available knowledge on how to diagnose, treat, and evaluate patients. They are seen as professional standards to which physicians are expected to adhere and can become part of the standards that governments set for appropriate care. The National Institute for Health and Care Excellence in England and Wales, the Quality Institute in The Netherlands, the Center for Medicare/Medicaid Services in the United States, and comparable government bodies in other countries publish guidelines for the use of health technologies (such as medicines, treatments, and procedures) within clinical practice. These guidelines may become visible in the form of flowcharts or the structure of a medical record that guides a physician on what to do. Even a "simple" prescription pad can have a structure that a physician must use to write down the legally required information, so that the appropriate medication can be dispensed. Phrased in sociotechnical terms, such tools discipline health care professional behavior.

E-PRESCRIBING

The rising costs of medicines prompted a hospital in California in 1972 to implement an information system that could support physician ordering, including prescribing (Hodge 1990). The doctors could electronically prescribe drugs from a drop-down list that contained the hospital's medication formulary. The drop-down list also showed the available dosages, pack sizes, and routes of administration. Thus, the doctors were enticed to use only drugs that the hospital kept in stock. If they wanted to prescribe a drug that was not in stock, they could still do this, but had to give a reason. Constraining prescriber behavior by means of this form of IT system brought the hospital substantial savings (Watson 1977). This type of IT system became widely known as computerized physician order entry systems or, alternatively, computerized provider or prescriber order entry (CPOE), indicating that other health care professionals with prescribing authority, such as nurse specialists, may also use the system.

In the 1990s, a number of studies appeared about medical errors, culminating in the Institute of Medicine's report *To Err is Human* (2000). The numbers looked apocalyptic: there were between 44,000 and 98,000 preventable deaths in the United States per year, meaning that medical errors (including medication errors) had become the eighth leading cause of mortality (Institute of Medicine 2000; Leape 2000). This figure has also been extrapolated to other countries, including The Netherlands (Leendertse et al. 2008). Improving patient safety and quality of care became central to national health care policies, and CPOE was considered as the IT system that could reduce errors and improve patient safety. In combination with electronic decision support, CPOE systems could alert the prescriber to drug allergies and drug–drug interactions and create order sets for particular patient groups. A decision support system typically contains rules that trigger a reminder or alert when a new drug is prescribed in addition to a drug that a patient is already on and may cause a serious interaction that would harm patient safety. These rules are derived from published scientific studies of drug interactions and toxicity and have to be regularly updated to accommodate new knowledge. Therefore, the contents of decision support systems are closely linked to the ideas of EBM. The impacts of these systems on patient safety are presented in Chapter 19; the sociotechnical implications are considered below, in order to illustrate the application of this approach.

ADOPTION OF E-PRESCRIBING SYSTEMS

The adoption of e-prescribing systems varies widely in Western countries. A recent study suggests that 13% of English hospitals use e-prescribing for all inpatients (Ahmed et al. 2013). In 2009, Aarts and Koppel conducted a study that showed a high adoption rate in the United States and the Netherlands, but virtually no such adoption in Germany (Aarts and Koppel 2009), with many countries somewhere in between. Though the technology is quite complex, the main reasons for this variation in adoption rates can be attributed to political, social, and cultural factors. In the United States, concerns about patient safety and quality of care led to the introduction of an IT stimulus package, which included financial incentives and "meaningful use" regulations (Blumenthal 2010). It is not sufficient for a provider to implement an electronic health record (EHR) system; for example, they must also meet requirements as to how it will be used to improve practice (Jha 2010). In the Netherlands, similar concerns about patient safety prompted the Dutch Health Care Inspectorate to mandate the use of e-prescribing. It has already shown an effect in primary care, where the adoption is still low. In Dutch hospitals, the landscape of EHR systems is changing rapidly. Hospitals are spending sizable funds on replacing their older systems, and e-prescribing systems are now more widely used than in 2009.

In Germany, however, physicians hardly use these systems. Again, a cultural issue plays an important role. German physicians enjoy a large degree of autonomy in their practice and feel that e-prescribing may compromise their professional authority to prescribe medications for their patients. Pharmacists are only allowed to recommend changes to prescriptions in exceptional circumstances; otherwise, they are obliged to dispense the prescription as it is written.

Although e-prescribing systems are often seen as "neutral" tools to help get a job done, in fact they represent social norms embedded in technology. However, the norms that developers choose to embed can clash with reality. For example, the accepted norm in many countries is that only physicians have the legal authority to prescribe medications. Therefore, only physicians are authorized to enter medication orders into the e-prescribing system. In some hospitals, however, it was (and perhaps still is) accepted practice that nurses on night and weekend shifts could order pain medications to relieve their patients' suffering, because of the limited availability of physicians (Aarts et al. 2004). For that purpose, there would be a pile of pre-signed paper prescriptions in nursing stations that nurses could complete when they were needed. Physicians trusted this work practice because they knew that the nurses were knowledgeable and experienced, and so were happy to facilitate it. The advent of e-prescribing systems made this way of working impossible and put the burden of entering orders completely onto physicians. In one instance, it caused physicians to protest formally at the increased workload (Massaro 1993). In some cases, users have created workarounds to make their work doable. In a study of the implementation of CPOE in a Dutch hospital, Goorman and Berg (2000) found that the problem of only physicians being permitted to enter medication orders was circumvented by using an "agent for" device, which was available for use in emergencies. A nurse could enter medication orders on behalf of a physician, which would then be signed off later by the physician. Realizing how practices are adapted in this way is important to understanding how a system is used and the potential patient safety implications.

DECISION SUPPORT SYSTEMS

Guidelines and protocols in the form of decision support are often embedded within CPOE, representing the "proper" practice of medicine. Entering a medication order, therefore, may prompt the physician to check specific patient data, and not allow completion of the order unless the physician has acknowledged that the data has, at least, been seen. In a number of specialties, this might be seen as an annoyance, taking precious time away from patient care, because the use of those specific pieces of patient information might be routine practice. This is also the case in situations where physicians receive reminders about drug interactions when they fully intend to prescribe those interacting drugs together, with appropriate monitoring. Reminders can prevent medication errors, but if appropriate reminders are drowning in a sea of less useful reminders, then their positive effects can be mitigated. The number of ignored reminders is over 90% in some studies (van der Sijs et al. 2006).

However, it has proven very difficult to reduce the number of inappropriate reminders. First, doctors might not agree among themselves as to which alerts can be turned off safely (van der Sijs et al. 2008). They report that, although they themselves might know what to do, doctors from other specialties or less experienced doctors would surely need them. Second, the technology is still far from perfect. Using medications as test items, van der Sijs and her colleagues found that six different CPOE systems responded in a variety of ways, with missing functionality requiring additional pharmacy review for prescribing to be done on some systems (van der Sijs et al. 2010). The issue of alerts is discussed further in Chapter 19.

Implementing CPOE can have unintended consequences as well. During ward rounds, medication orders can be verbally communicated by physicians to their colleagues, and these are sometimes corrected by nurses, who would know about the current health status of a patient. This is potentially both an effective way for nurses to know when to start administering new drugs and an effective safety net for preventing prescribing errors. A computerized system, instead, forces a doctor to find a computer to access, which is often installed in a separate office. The doctor then lacks this interaction with colleagues when prescribing the medicine. Koppel and colleagues identified a number of situations in which a CPOE system could introduce new errors, such as an inability to see immediately which patient's medications had been ordered for, because of a poor CPOE display, and delayed ordering because of the system being down (Koppel et al. 2005).

It is clear that e-prescribing systems are being designed and implemented with embedded intent and purpose. Often, they will *shift* workarounds in practice, instead of removing them completely. For example, because doctors prefer not to lose their interactions with colleagues when entering an order, new workarounds may arise to compensate for this, such as writing down the order on a piece of paper during the ward round and delaying entering the orders until afterwards. This intertwining between organizational context and culture shows how e-prescribing systems are, in essence, sociotechnical systems (Berg et al. 2003).

E-prescribing may reach its full potential for improving patient safety if a number of conditions in the short and long term can be met. In the short term, efforts should be directed towards mandatory interoperability using open standards. Interoperability ensures that data can be communicated across different systems without losing their meaning. The health IT market is fragmented and there are hundreds of vendors, thus encouraging many differing proprietary systems, few of which communicate with each other. Interoperability will enhance effective communication between health care professionals about the medicines taken by patients, as discussed in Chapter 17. Use of national standards for medication selection, with dosing ranges and routes of administration, in addition to decision support reminders and alerts, is expected to speed up the adoption of decision support systems. In The Netherlands, there is one national standard for drug–drug interaction alerts, for example, that is adopted by all IT vendors (van Roon et al. 2005).

Such standardization measures do not require large investments in developing new decision support technologies. Instead, their emphasis is on creating the necessary organizational conditions, which may also prove challenging. Customization that takes into account the context of a specific patient and medical expertise is therefore crucial for a successful system. At the crude level of medical specialties, distinctions may be made between surgery and general internal medicine, for example. Some groups of surgeons may have a limited selection of medications that they routinely use; it could be argued that it does not make sense to present alerts for medications that they hardly ever use, and even then only in very specific circumstances and with great care. It becomes more difficult, however, to take into account differences between subspecialties in internal medicine.

Another complicating factor is the level of expertise within a specialty. Junior doctors rely much more on factual knowledge, which they gradually incorporate to become experts. Prescribing systems that are able to incorporate medical expertise themselves require complex knowledge acquisition and machine learning technology that, for the foreseeable future, will not be available in practice (Uzuner et al. 2010). The other approach is linking e-prescribing with the data in EHRs and clinical guidelines. Alerting can be customized by extending the rules of decision support technologies to incorporate the individual patient's EHR data (Coleman et al. 2013b). Patient data often reside in diverse, heterogeneous databases, but modern data mining techniques seem to be promising in this regard (Frankovich et al. 2011). The next step is therefore likely to involve the extension of e-prescribing to the continuum of care, in the form of linking it with personal health records, which will make use of even more diverse and heterogeneous databases across different organizations.

THE MEDICATION LOOP

E-prescribing addresses only a part of the trajectory from prescribing to medication administration. In hospitals in the Netherlands, for example, after the implementation of e-prescribing, a prescription is usually received by the pharmacy, where the prescription will be transcribed, dispensed, and sent to the patient floor. Simultaneously, at the nurse's station on the patient floor, the nurse will print the order and affix it to the patient's paper chart. The nurses can then administer the medications to the patient. The steps in this process are increasingly being automated as well. Ward-based automated dispensing systems (ADS) have been introduced to ensure the safety of medication administration to individual patients (see Chapter 21 for more information). They are intended to replace medicine carts and can only be opened by nurses with the appropriate access rights. A field study of ADS use brought to light the practice of splitting medication doses (Balka et al. 2007)—a norm within the non-automated system that was not embedded in the new system, and was also contrary to relevant guidelines. It was a common nursing practice, nonetheless, when the required dose did not have an available formulation and tablets had to be split, for the nurse to save the remainder for a later time. Avoiding wasting of expensive medication is deeply engrained in nursing practice. Only careful registration of unused medications on the ADS would have alerted the pharmacy to the problem; such registration was not possible in this case.

To ensure that the right patient receives the right medication using ward-based ADS, patients receive wristbands with barcodes upon admission. This allows nurses to use barcodes to match medications with the correct patients before administering the medications. The technology has been reported to reduce medication administration errors substantially (Poon et al. 2010). Again, an apparently error-proof technology may still have unexpected consequences and cause new errors. Koppel and colleagues (2008b) listed a number of significant examples of the causes of such errors. A nurse may forget to visually inspect the medication being administered, instead relying on the system to provide the right one. The nurse may also not be able to scan the wristband, because it may become smudged or the patient may be sleeping and the scanning would wake them.

In a hospital, the medication loop can be monitored and potential sources of errors identified. Medication safety in nursing homes, on the other hand, is particularly problematic, mainly because of organizational fragmentation (Vogelsmeier et al. 2008). A nursing home will commonly not have an in-house pharmacy, but be dependent on a community pharmacy for delivering their medications. Again, ADS are seen as a solution by providing pre-packed medications labeled with barcodes to be given to patients. These pre-packed medications are picked up by nurse aides or delivered to the nursing home by pharmacy assistants. A Dutch study found that despite efforts to standardize the administration process, nurse aides often crushed medications to ease swallowing by patients, regardless of whether this was appropriate for the medication (van den Bemt et al. 2009a). Here, the technology and the stipulations of the administration instructions clashed with the perceived need to make a patient feel comfortable.

Managing medications for patients places a high responsibility on health care professionals, especially physicians. Although administering medications has serious safety gaps, the majority of medication errors still originate at prescribing (Dean et al. 2002b). The solution may not be technological, but could perhaps be found in the social sphere, by actively involving the patient in the medication loop. This would suggest that patients should not only be asked about the clinical outcomes they expect, but also about how they might manage their lives to include safe medication practices.

CONCLUSION

The sociotechnical perspective has shown us how the concept of rational prescribing can be embedded in e-prescribing and CPOE systems. These systems have been introduced with the aim of reducing costs and improving the quality of care and patient safety. They aim to nudge the provider into what is considered good behavior. In this sense, an e-prescribing system is a sociotechnical system. The technology, like a printed checklist, is not passive, but rather an active part of the system. The technology is still far from perfect and there are still many unknowns about its consequences. Users try to undo the shortcomings of the technology by workarounds, particularly where there is dissonance between the norms that are embedded in such systems and those of the users. In extreme cases, the potential users will not even employ the technology, as seen in Germany with e-prescribing. The introduction of technologies at the heart of health care requires a thorough understanding of their nuts and bolts and the social environment, an understanding that we can get from sociotechnical approaches. E-prescribing is a social act that tells as much about the users, their contexts, and the technologies.

11 Systems Perspective and Design

Tosha B. Wetterneck and Pascale Carayon

CONTENTS

KEY POINTS

- The systems perspective considers the medication use process as a complex interplay between people, tasks, technologies, organizations, and environments.
- The process is non-linear, with changes to any element potentially impacting on the entire system.
- Understanding the system design and how the elements interact can inform improvements in quality and safety.
- This perspective has been shown to support a non-punitive culture and benefit the implementation of new systems.

INTRODUCTION

Systems perspective and systems design are important for medication safety across the entire medication use process. This chapter will define the systems perspective from a human factors engineering viewpoint and consider the six major systems features in relation to medication safety: systems focus, interactions between systems elements, contextual issues, the whole-person perspective ("holism"), safety as an emergent property, and embedding of human factors/ergonomics-trained professionals into health care organizations as a safety strategy (Wilson 2014). We will highlight how system design flaws can lead to unanticipated medication errors and how human factors approaches can proactively identify and reduce or eliminate the hazards leading to these errors. Lastly, we will discuss two case examples of health care information technology (IT) implementations and how they benefitted from the application of the systems perspective.

SYSTEMS AND SYSTEMS APPROACHES

The medication use process is commonly represented as a linear process that starts with assessing the patient and prescribing or ordering a medication, proceeds on to transcription of the order (in some countries), preparation, dispensing and administration of the medication by a clinician or patient, and ends with monitoring of the medication's effect. However, from a systems perspective (Carayon et al. 2006; Karsh et al. 2006), the process is a complex interplay between clinicians, patients, tasks performed, technologies and medications themselves, and the health care or home setting environment in which it occurs and which is influenced by environmental (for example, noise) and organizational (for example, scheduling or patient safety culture) factors. The outcomes of the medication use process then feed back to influence this complex interplay. Errors that occur in one part of the process can also propagate down through the process. They may multiply as they propagate; for example, a single dosing error in ordering or prescribing may lead to multiple dosing errors at medication administration. Errors may also be detected and corrected in different parts of the system. For example, a wrong medication schedule prescribed by a physician may be identified and corrected during admission by the hospital ward nurse or after discharge by the community pharmacist. The understanding of systems design and how systems elements interact to lead to specific processes and outcomes can shed light on quality and safety for patients. Importantly, all safety—including medication safety—can be considered to be a system property (Institute of Medicine 2000).

A system can be defined as "a set of interacting, interrelated, or interdependent elements that work together in a particular environment to perform the functions that are required to achieve the system's aim" (Johnson et al. 2008). In order to understand system performance, one must examine the individual elements of the system more closely and the interactions between them. An important system characteristic is that any individual element of the system can affect the behavior of the entire system. This means that a change to one part of the system, such as the introduction of a new technology, invariably changes all other systems elements and overall system functioning. It also means that optimizing one element of the system in isolation, without consideration of the other elements and interactions, may not optimize the overall system functioning. System characteristics will be further discussed later in this chapter, along with the systems features of system focus and interactions.

The U.S. Institute of Medicine and National Academy of Engineering have collaborated to define and promote a systems approach in health care. They define a systems approach to health as "...one that applies scientific insights to understand the elements that influence health outcomes; models the relationship between those

TABLE 11.1
Systems Features in Medication Safety

Systems Features	Implications for Medication Safety
Systems focus	Need to understand all system elements involved in medication safety (see Tables 11.2 and 11.3)
System interactions	Increased focus on technology, but need to consider interactions between technology and other system elements
Context	Consideration of local context and variety in contexts that influence medication safety and contributing factors to safety
Holism	Consideration of health care professional work outcomes in addition to patient outcomes when implementing a safety intervention
Emergence	Importance of proactive risk assessment that needs to be complemented with organizational learning and continuous improvement
Embedding	Human factors professionals are employed by a health care organization and lead the design and implementation of new processes for the use of a new infusion pump for medication administration

elements; and alters design, processes, or policies based on the resultant knowledge in order to produce better health…" (Kaplan et al. 2013). It is widely recommended that the systems approach be used in patient safety programs for systems (re)design and incident/adverse event analysis. The World Health Organization (WHO, 2011) also promotes the teaching of a systems approach in its patient safety curriculum for health care professionals.

The systems approach to safety is important for both patients and health care professionals. A systems approach supports the understanding of how systems should be designed to support health care professionals to do the work they need to do to take care of patients in a high-quality and safe manner, while maintaining health care professionals' quality of working life (Karsh et al. 2006). The systems approach supports a non-punitive culture for errors that occur (Reason 1997). It focuses on building systems to maximize human performance and prevent errors, while building defenses to recover from errors that do occur, i.e., prevent them from reaching patients and causing harm (Wetterneck 2012; Wetterneck and Karsh 2011).

The systems perspective has six features (Table 11.1): a systems focus, interactions, context, holism, emergence, and embedding (Wilson 2014). Each feature will next be discussed in detail with relevant medication safety examples.

Systems Focus

It is intuitive that with the systems perspective comes a focus on systems and people's interactions with the systems in which they provide or receive care. There are several systems models used in health care to better understand quality and safety outcomes for both patients and clinicians that have been specifically applied to medication safety. The "Swiss cheese model" of complex systems (also called the accident causation model and described in Chapter 8) by Reason (1995, 2001) is probably the most well-known model. It describes how safety risks can be prevented by systems barriers; when an adverse event occurs, it is the result of multiple latent and active failures that penetrate these system barriers. Reason's model has been further extended by Vincent and colleagues (1998), again described in Chapter 8.

The Systems Engineering Initiative for Patient Safety (SEIPS) model of work systems and patient safety (Carayon et al. 2006; Carayon et al. 2014b) is another widely used model adapted from Donabedian's structure–process–outcome model for quality (1988). In this model, the work system has five elements: the person or people, tasks, tools and technology, organization, and environment (Figure 11.1). The way the system is designed (i.e., the characteristics of these system elements) leads to (care) processes that are performed and, therefore, the patient, clinician, and organizational outcomes. Table 11.2 displays the SEIPS model components, together with medication safety-related examples. In the SEIPS model, "people" include the clinicians and staff that care for patients and assist with patient care, as well as the patients and their families or caregivers.

People are the center of the systems model, acknowledging the importance of people within the health care system and the dependence of safety outcomes on their activities. Tasks are the activities done to achieve the system's goals. In medication use, activities include a physician entering a medication order into a computer system, a pharmacist double-checking a dose calculation, or a nurse scanning a barcode on a medication.

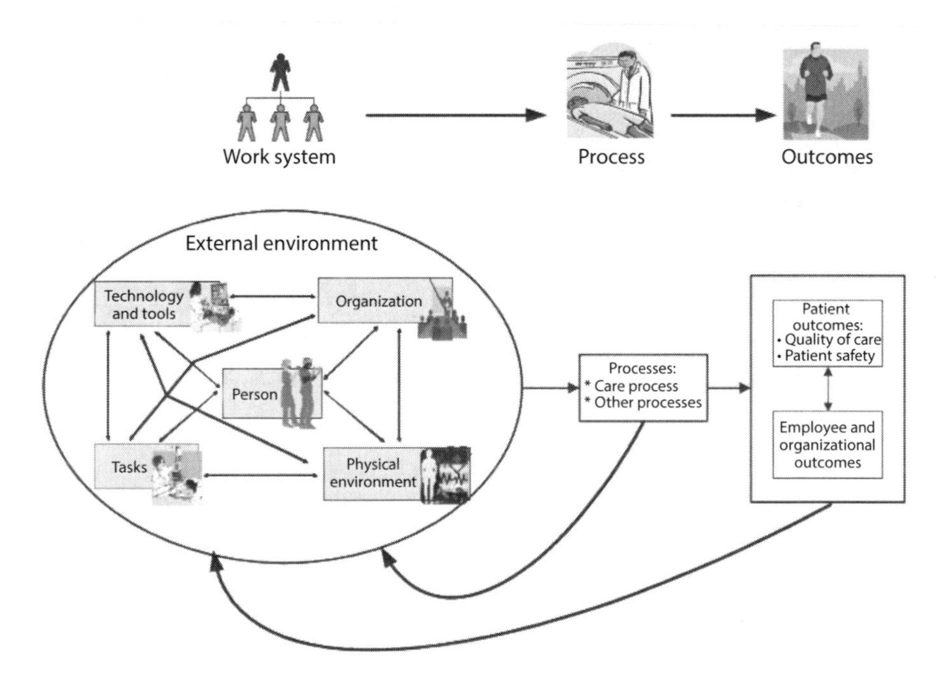

FIGURE 11.1 The Systems Engineering Initiative for Patient Safety (SEIPS) model of work systems and patient safety. (Reprinted from *Appl Ergon*, 45, Carayon, P. et al., Human factors systems approach to healthcare quality and patient safety, 14–25, Copyright 2014, with permission from Elsevier.)

TABLE 11.2
System Elements and Medication Safety Examples

Systems Element	Examples Relating to Medication Safety
Person	Pharmacists, nurses, physicians, patients, caregivers
Tasks	Entering medication orders into a computer system, double-checking dose calculation for chemotherapy, administering or taking medication
Tools and technology	Computerized provider order entry, robotics, intravenous pumps, barcode medication administration technology, syringes, decision support, dose calculators, patient pill boxes
Organization	Community pharmacies, hospitals, clinics, skilled nursing facilities, technology vendor corporations
Environment	Organization of patient medication drawers in hospital, organization of medications in pharmacy dispensing area, overhead lighting in medication use areas
Processes	Medication prescribing, transcribing, preparation, dispensing, administration, monitoring
Outcomes	Patient: adverse drug events, medication errors Healthcare professional: pharmacist burnout, nurse needle-stick injuries Organization: nurse or pharmacist turnover

Task performance is influenced by: job demands, such as time pressure, quantitative workload, or cognitive load; the variety of tasks to be done; and a worker's sense of job control and utilization of their skills. When a worker does not follow the prescribed procedure to complete a task, a violation or workaround has occurred. Workarounds have been shown to increase the likelihood of medication errors and workarounds themselves may have systems issues as their cause (Koppel et al. 2008b).

Tools and technologies are important assistive devices for workers to complete tasks. Examples in medication use include electronic prescribing (e-prescribing) or computerized prescriber order entry (CPOE), barcode systems for medication dispensing, and administration and dose calculators to aid in weight-based dosing and the determination of intravenous (IV) fluid rates. Importantly, the design and usability of the technologies and how they are implemented and fit into the workflow will affect their use, which ultimately determines their potential to improve medication safety or cause medication error. Technology design flaws will be discussed further under the "Interactions" subsection of this chapter.

Organizational characteristics of importance include teamwork, communication, supervision and management, organizational and patient safety culture, and schedules and staffing levels. Finally, the environment relates to the physical setting in which the work is done and includes the layout of the work area, noise, and lighting. Henriksen and colleagues (2008) have expanded the systems model to also include the external environment, which includes health care policy, public awareness, government initiatives, and economic pressures influencing safety. Table 11.3 lists common causes of medication errors categorized according to their system elements (Brady et al. 2009; Franklin et al. 2011; Keers et al. 2013a; Koppel et al. 2008b).

TABLE 11.3
Examples of Work System Factors Underlying Medication Errors

System Elements	Work System Factors
Person	Fatigue, stress, patient acuity, lack of medication knowledge, poor calculation skills
Tasks	High workload, high time pressure, lack of task standardization, emergency situations
Tools and technology	Poorly designed technology, confusing/damaged/lookalike medication packaging and labeling
Organization	Poor communication, interruptions, poor safety culture, lack of integrated information technology across care settings, inadequate patient handoffs, incomplete supervision of staff, overreliance on others to identify and correct errors
Environment	Unorganized medication drawers, loud ambient noise/distractions

All the system elements work together to produce the processes of care for patients and these elements subsequently lead to outcomes. The SEIPS model considers not only patient outcomes as relevant outcomes of health care, but also organizational and health care professional outcomes. Organizational outcomes include the financial status of the organization and employee turnover. Health care professional outcomes include quality of working life (for example, job satisfaction, stress, and burnout) and health and safety in the workplace. Notably, there are feedback loops between processes and outcomes and the work system. Feedback loops are active in "learning organizations" that evaluate their processes and outcomes and make changes to their work system to improve quality of care, resulting in cycles of improvement. Changes to work systems may also occur more organically and without intent. For example, prescribers may ignore some medication alerts from CPOE technologies because of time pressure, workload, and/or a high rate of false-positive alerts. They may not see any downstream effects from this action, such as the occurrence of medication errors or negative data in audits and feedback from the organization. Therefore, they continue to ignore alerts and teach new prescribers at the organization that they, too, should ignore these alerts to make their work go more smoothly.

In summary, the focus on the entire system and consideration of all of the system elements gives health care organizations the opportunity to gain a broader and deeper understanding of the factors contributing to care. This also affords organizations a more complete view of what needs to be changed to optimize safety outcomes (Catchpole and Wiegmann 2012).

Interactions

The interactions between the systems elements are an important consideration in systems design. In fact, much of the focus of systems design should be on designing these interactions (Wilson 2000). With regards to medication safety, as in most industries, an interaction that receives a lot of attention is the human–technology interaction; for example, a nurse programming an IV pump (Lin et al. 2001) or using barcode medication administration (BCMA) technology (Carayon et al. 2007; Koppel et al. 2008b; Smetzer et al. 2010). In this interaction, the usability of the technology is focused upon (i.e., does the technology design meet the needs of the user to perform specific activities?). However, the complexities of health care also lead us to examine other systems interactions in this relationship. For example, the nurse may be working with a nursing student as a team to care for a patient, thus technologies are used by teams of users, as well as individuals. Loud ambient noise from patients or other health care professionals talking may lead to nurses not hearing BCMA technology alarms caused by a mismatch between the medication and the medication order in the system (an interaction between the user, technology, and environment). The adequacy and comprehensiveness of the organization's training for nurses to use BCMA technology may influence their ability to use the device in novel situations or to problem-solve when the device is not working as expected (interaction between user, technology, and organization). A lack of fit between the user, technology, and other system elements may lead to workarounds in the use of the technology, such as the nurse not using the technology as intended by the organization, and a risk of medication administration error occurrence. Thus, when using a systems approach to design a system or investigate an adverse event, one needs to look at the individual systems elements, the interactions between them, and the potential influence of other elements on these interactions.

The aim of tools and technologies in health care is for them to function as assistive devices for health care professionals' performance (i.e., to assist them in doing their job in a more efficient manner and with fewer errors, Karsh et al. 2006). Designing for users means designing for both physical performance (for example, pushing a keypad on an IV pump interface) and cognitive performance (for example, perceiving and correctly interpreting a duplicate medication alert from the CPOE system and then problem-solving in terms of how to proceed correctly, Karsh et al. 2006). Specific to cognitive performance is designing to match a user's mental model of how the device should operate in their setting (Sawyer 1996). However, sometimes tools and technologies are designed in a manner that makes it more likely for errors to occur. This is known as design-induced error, in which a "user's behavior is directly influenced by operating characteristics of the equipment" and leads to an error or adverse outcome (Sawyer 1996). Design flaws occur when the interaction between the user, the tasks they perform, and their environment are not considered. We next discuss several examples of design flaws and design-induced medication errors.

Electronic health record and CPOE technologies have been shown to improve medication safety in multiple studies, primarily through decreasing medication prescribing errors (Chaudhry et al. 2006; Reckmann et al. 2009). However, many studies have also reported new errors or a worsening of known errors. Wetterneck and colleagues (2011) reported an increase in duplicate medication prescribing errors after CPOE implementation and performed a systems analysis to determine the contributing factors. Among these were design flaws that allowed two people to enter medication orders for the same patient at the same time, which led to problems with reviewing existing orders while prescribing a new medication and a subsequent high rate of false-positive duplicate medication prescribing alerts, resulting in users ignoring these alerts. These design flaws also had a contextual basis. During team-based patient rounds in the intensive care unit (ICU), medication orders were discussed and multiple people had access to computers and could enter orders at the same time, compared to paper processes when only one person could write orders in the chart (i.e., the medical record) at once. Cheung and colleagues (2014) evaluated medication incidents submitted to a Dutch reporting system and found that 16% were related to the use of a range of information technologies used in hospitals and community pharmacies, some of which had design flaws (for example, a medication name could not be found in the CPOE system, which was needed to order a medication). In the United States Bates and colleagues (1999) reported on an error in the design of the computer orders for prescribing potassium chloride, which led to overdose prescribing errors.

Many design flaws have also been reported about IV pumps, related both to the hardware and the technology interface. Schroeder and colleagues (2006) reported a serious medication administration error related to the free-flow of a medication in the operating room. The error was related to a tubing misload in the pump, which changed the pump's perception of medication flow, allowing free-flow to occur. The issue was eventually designed out by the vendor. "Key bounce" (i.e., pressing a number on the pump interface and having two of the same number appear on the programming screen, leading to overdose administration errors) has been reported with multiple IV pumps, leading to patient harm (Institute for Safe Medication Practices 2006). On some pump interfaces, this error was related to the intensity and duration of pushing the keypad number, but was not always reproducible. Rather than training users on how to touch the keypad numbers in a way that may not produce key bounce, the systems approach would call for a redesign of the interface that would not allow key bounce to occur, given the likely variation in patterns of keypad use or touching by users.

Other common design flaws concern the packaging and labeling of medications. Often pharmaceutical companies use the same type of font, coloring, and text formatting on all of their medication products, which can lead to lookalike errors in their preparation, dispensing, or administration. For example, within the United States, repeated errors of heparin administration have occurred, related to health care professionals mistaking high-concentration heparin vials (for example, 5000 units/mL) for lower concentrations used for flushing IV lines (for example, 100 units/mL), with some errors leading to bleeding and death. This led to the U.S. Food and Drug Administration requiring that vendors make a labeling change of both color and text formatting for heparin to help prevent these wrong drug concentration errors.

Perhaps one of the most disturbing design flaws in health care and medication delivery are tubing misconnections that result from the interconnectivity of syringes, tubing, lines, and catheters, which can be used to deliver oral or enteral tube feeding, IV medications, and epidural or intrathecal medications (Franklin et al. 2014a). This universal interconnectivity is a convenience for both health care professionals and vendors. Unfortunately for patients, it can lead to wrong route medication errors, in which medications intended for the intravascular, oral, epidural or intrathecal route are given via a different route. Often, these wrong route administrations (such as giving fentanyl and bupivacaine IV, rather than epidurally) are deadly or cause serious, permanent harm (Smetzer et al. 2010). Root cause analyses have identified many potential solutions to prevent these wrong route errors, such as the color coding of tubing, labeling of tubing and catheters with brightly colored stickers, staff education, and tracing of tubing back to its point of origin on the patient body. Despite these interventions, wrong route errors continue. The design flaw here relates to not considering the design of these devices in the context of their use for the patient who has many different tubes and catheters on their body and the deadliness of the tubing misconnection errors when they occur. An international working group is now developing standards to design out the potential for this error, such that connectors on devices are specific to their application (i.e., intravascular or enteral delivery, etc.) and cannot be interconnected (Association for the Advancement of Medical Instrumentation 2014).

Context

Health care professionals perform work within a care setting or context that can greatly influence system design requirements and medication safety. Contexts in health care vary greatly in terms of patient-related variables, such as acuity (for example, an ICU with critically ill patients versus a general surgical unit with patients recovering from routine surgery) or patient populations (for example, a geriatric clinic with elderly adults versus a pediatric clinic with children and their families) or, in some countries, insurance status. Context also varies based on organizational features (for example, large integrated health systems versus a small-sized independent primary care clinics), geography (for example, states and countries with different rules governing health care), and different health care coverage structures (or no government health care coverage). From a systems approach, a change in health care context usually implies differences in the work systems and processes in place. Therefore, when implementing an intervention to improve medication safety (for example, CPOE technology or pharmacist-led medication histories on hospital admission), the design of the intervention and how it is implemented will need to be adapted to fit the local context. To do this requires knowledge of the elements of the work system and the local processes or workflows related to the intervention.

From a medication safety perspective, contextual differences involve variation in the types of medication used, the technologies used to assist health care professionals and patients in medication use, the tasks performed, etc. Context changes the risks in the system for medication errors to occur and may change their likelihood of occurrence. Patients in ICUs experience more preventable adverse drug events (ADEs) than patients in general medical and surgical units in the same hospital (Bates et al. 1995b). This is likely to be due in part to the acuity of patient illness, requiring a higher number of medications to be used per patient (Cullen et al. 1997b) and the use of higher-risk medications (for example, vasopressor medications with narrow therapeutic windows where doses are frequently manipulated according to patient condition). Other health care settings in which high-risk medications are commonly used include the operating room (for example, paralytics, anesthetics, and vasoactive medications) and hematology, oncology, or bone marrow transplant units (for example, with the use of chemotherapeutic medications).

Dosing is different across patient populations and therefore may require different clinical decision support tools to assist with proper prescribing. For example, in pediatrics, most medications are dosed based on the weight or surface area of the child, so dose calculators are important to prevent calculation errors in dosing. In older people, medication doses may need to be decreased from the common adult doses, due to differences in weight, body fat, and kidney function. Also, as a person ages, they accumulate more acute and chronic diseases requiring medication treatment; thus, older adults are commonly taking more than five medications a day and the likelihood of drug interactions increases. Decision support for prescribing needs to consider all of these factors to assist with medication dosing and use in children and the elderly. Whatever the intervention to improve medication safety, context will determine in part the design and implementation of the intervention, how it is used by health care professionals and patients, and its impact on medication safety.

The external environment is also a major contextual factor in medication safety. Local, regional, or country-wide regulations, laws, and safety goals may be in place. These encourage or mandate health care organizations to take action on various medication safety issues. For example, in the United States, the Joint Commission (2014) that accredits health care organizations has three national patient safety goals that pertain to medication safety: labeling of medication delivery containers; medication reconciliation; and anticoagulation medication use safety. The goals of the Australian Commission on Safety and Quality in Health Care (2012) include medication safety as a priority area, focusing on five areas: medication reconciliation, safety in warfarin use, safety in venous thromboembolism prophylaxis, correct medication dose calculation for children, and prevention of hospital admissions related to ADEs in the elderly. Medication safety priorities may also vary based on the setting's socioeconomic status. WHO convened a panel of experts to discuss priorities to improve patient safety in the primary care sector (Cresswell et al. 2013) and identified counterfeit medications as a major safety issue in low socioeconomic settings only.

When considering contextual issues in the (re)design of a health care system, it is useful to gather input of health care stakeholders in the process. Human factor implementation principles highlight the importance of front-line health care professional, manager, organization top-level management, and patient participation and commitment in the design and implementation of interventions to improve safety (Holden et al. 2013a; Karsh 2004). Their input helps to identify important contextual issues at multiple levels (i.e., health care professional–patient interaction, clinic/unit, organization and the external environment, Karsh et al. 2006). A participatory ergonomics approach—used widely in industry—is a formal approach to include stakeholders in the design and implementation process (Carayon et al. 2014c; Xie 2013).

Holism

In health care, it is common to focus on one part of a patient's health, such as a chronic disease (for example, diabetes) or an acute complaint (for example, back pain). Yet the patient's acute and chronic health issues are influenced by their total health—and not only their physical health, but also their mental health and psychosocial well-being. The same idea also applies to systems. Multiple physical, cognitive, and organizational or social aspects influence health care professional performance and interact with other systems' components. The systems approach encourages a holistic approach, looking at the whole system and all of these influences. At times, it may be necessary to focus on just one aspect of the system interaction, such as focusing on the physical ergonomic needs of nurses when designing a hand-held BCMA scanning device. However, to focus solely on the physical needs would miss important issues that may influence design, such as social influences. For example, nurses may work in teams and need to use the device together, in which case the screen should be designed for two people to view at the same time. It is also important to consider organizational influences; for instance, the device may need to be used on patients who are in protective isolation and may therefore be put into—and need to function in—plastic bags or be washed repeatedly (Carayon et al. 2007).

A holistic approach not only refers to looking broadly across influencers of systems elements, but also to considering broadly the outcomes of interest and maximizing the outcomes of patients, health care professionals, and organizations together as a goal. Too often, organizations implement safety interventions to improve patient safety that have an untoward effect on health care professionals' work, as described in Chapter 10. Novak and colleagues (2013) describe the implementation of BCMA technology at two hospitals and how the rigidity of the organizational procedures around BCMA use negatively affected nursing work, resulting in nurse workarounds and duplication of work; others have noted similar workarounds in the use of BCMA technology (Koppel et al. 2008b). Workarounds and inefficient work have immediate and downstream implications for patient safety, health care professional productivity, and organizational outcomes. Therefore, it is important to consider the impact of any safety intervention on the system as a whole and on clinician outcomes, with a goal of achieving safer care while maintaining rewarding, productive work for health care professionals.

Emergence

Emergence is the fifth systems design feature and a feature of complex systems. It deals with "unexpected behaviors that stem from interaction between the components of an application and the environment" (Johnson 2006). A caveat to this notion is that emergent behaviors are those that cannot be predicted, even though one has a good understanding of the complex system. In health care, emergence manifests when health care professionals use tools and technologies in a manner that may not have been expected by the maker or vendor or organization, or an unexpected technology response occurs in a situation. This unexpected (emergent) use may be beneficial; it may be the result of a health care professional adapting a tool or technology for use within their work context. It may also be harmful; unintended negative safety consequences may result from a tool or technology not working as expected or from a health care professional working around the tool or technology in a fashion that degrades patient safety.

As an example, Wears and colleagues (2006) describe a case of a critically ill woman with delays in care due to the unexpected failure of automated dispensing systems (ADS) in the emergency department. The failure was related to a software upgrade process in an integrated hospital information system, which caused a storm of medication information messages to be sent to the ADS, overwhelming them and preventing proper function. The organization did not anticipate that the emergency department ADS would be affected by the upgrade because the source of the messages—an order-checking software feature—was not used by the units in the emergency department. Most of the unintended consequences related to technology use in health care, including those previously discussed in this chapter, are not so dramatic, and yet they can have profound safety impacts.

Human factors professionals are now focusing more on system resilience, or "the ability of systems to anticipate and adapt to the potential for surprise and failure" (Woods and Hollnagel 2006). By definition, the term error implies preventability. However, given the human factors in our health care systems, it is near impossible to design and implement systems that will prevent the occurrence of all medication errors. Therefore, it is important to build systems and processes that can detect errors when they occur and correct them either before they reach the patient or afterwards to mitigate patient harm (Wetterneck 2012). This is called error recovery. Built-in error recovery is common in medication safety. For example, physicians' medication orders are commonly checked by pharmacists. In some cases, computer clinical decision support systems check the orders for correctness and appropriateness, while also checking the order with other medications the patient is taking to identify duplications in therapy and potential drug interactions. Errors that occur in medication use systems can therefore be analyzed to identify opportunities to include error recovery processes or to evaluate the effectiveness of existing processes (Habraken and van der Schaaf 2010). Odukoya and colleagues (2014) evaluated e-prescribing processes in community pharmacies to understand how pharmacists and pharmacy technicians detect, explain, and correct prescribing errors. To correct the errors, pharmacists described faxing the electronic medication order back to the prescriber and highlighting the error and the needed clarification to show the prescriber the error, in the hope of preventing future errors of the same type by that prescriber (Odukoya et al. 2014).

Proactive risk assessment is a tool that can be used to predict systems failures before they occur and to redesign systems, tools, technologies, tasks, and so on to prevent the failure from occurring or to mitigate the effects of the failure if it does occur (Carayon et al. 2011; Lyons 2009). Very often, so-called unintended consequences could have been prevented if sufficient time and resources had been invested in a proactive risk assessment. However, it is also important to recognize that not all situations can be anticipated; emergence is a key feature of complex systems.

Embedding

Embedding is the last of the six system features and involves having a professional with expertise in using the systems approach, such as a human factors or ergonomics professional, as part of the organization. The ability of an organization to hire someone with this expertise as opposed to using a consultant service will depend upon the size and financial resources of the organization, but the return on this investment from a quality and safety perspective can be immense (Hendrick 1996; Tompa et al. 2013). The embedding of human factors professionals within health care organizations can foster the concept of "bilingualism," whereby human factors professionals understand the health care context and needs and the health care professionals understand the systems approach, human factors concepts and the methods needed for the design and implementation of safety interventions (Carayon 2010; Xiao and Fairbanks 2011). Many examples exist of this embedding in health care (Kamath et al. 2011).

MEDICATION SAFETY EXAMPLES

Implementation of Smart IV Infusion Pumps

A multidisciplinary team including physicians, nurses, pharmacists, and human factors and biomedical engineers used proactive risk assessment, specifically failure mode and effects analysis (FMEA), to evaluate the IV medication administration process and the new technology before implementation (Wetterneck et al. 2006). To better understand the current system and its vulnerabilities, human factors engineers and pharmacists performed observations of the current process and the hospital event reporting system was queried for IV medication administration errors. In a process that took 46 hours over 4.5 months, the team identified over 200 failure modes. As part of the FMEA process, usability testing was performed on the new pump to inform process redesign and health care professional training. Recommendations were developed for moderate- to high-priority failure modes and categorized according to the SEIPS model work system elements as changes to organization policy and procedure, changes to training or education for health care professionals, short- and long-term technology software or hardware changes, physical environmental changes, and people changes. The new pump implementation was delayed to allow for the recommendations to be implemented. After implementation, the team evaluated the success of the recommendations for preventing failure mode occurrence. Three weeks after pump implementation, a free-flow medication event occurred with the use of the pump (Schroeder et al. 2006). Given the ongoing monitoring activities by the FMEA team, the pump was immediately sequestered and an event analysis occurred. A failure in tubing loading into the pump was discovered and in the laboratory the free-flow of medication could be reliably reproduced. The identification and resolution of this event were facilitated by the supportive environment and culture created by the participatory, multidisciplinary implementation process.

Overall, the implementation process was felt to be highly successful by the organization. The presence of human factors engineers and health care professionals with human factors experience greatly contributed to this. In addition, the activities leading to the implementation of the smart IV pump technology (for example, FMEA with a multidisciplinary team, usability evaluation, or training) created a culture of collaboration and openness that contributed to the identification and correction of the free-flow medication event (Carayon et al. 2008). This is in line with the concept of continuous technology implementation and the process of continuous improvement, which allow for "emergent" safety problems to be detected and corrected.

Implementation of Computerized Prescriber Order Entry Technology

Beuscart-Zephir and colleagues (2007) used a systems approach to redesign a hospital system and its medication processes for the implementation of CPOE technology. They developed and used several human factors methods in four stages (Beuscart-Zephir et al. 2010):

1. Analysis of the medication use work system, with a particular focus on physician–nurse and nurse–nurse communications about medications.
2. Use of the results for cooperatively designing the CPOE technology and its associated work system. This cooperative design process involved both the health care organization and CPOE designers.
3. Iterative evaluation and redesign. Five human factors professionals evaluated the CPOE user interface using a set of design criteria (Bastien and Scapin 1993). This enabled the identification of several human factors issues (for example, workload), which were then prioritized and dealt with in several simulations.
4. Assessment of the new work system and its impact on patient safety and the overall performance of the sociotechnical system.

The human factors analysis of the medication administration work system performed in the context of CPOE implementation identified issues at various system levels, including the organizational, collective, and individual. Therefore, implementation of a health IT system such as CPOE technology represents "a key opportunity to redesign the work system" (Beuscart-Zephir et al. 2010). Changes in the work system require cooperation with the leadership of the health care organization, as well as the designer and vendor of the technology. This collaborative design process with significant input from users can bring benefits to all actors involved.

CONCLUSION

In summary, a systems approach to medication safety is important for both patients and health care professionals. Six features of the systems approach have been described: a systems focus, interactions, context, holism, emergence, and embedding. Use of the systems approach for system (re)design and event analysis supports both a non-punitive culture and the work that health care professionals do to care safely for patients.

12 Improvement Science

Anne Lesko, Peter Lachman, Stephen E. Muething, David Vaughan, Pamela J. Schoettker, and Uma R. Kotagal

CONTENTS

KEY POINTS

- Improvement science and reliability science have been introduced to health care from other high-risk industries and have been used to ensure safe medication use.
- Improvement science provides a flexible framework for the implementation of interventions to improve patient safety.
- The framework has four key elements that are used cyclically until the intervention has achieved—or has been judged not to have achieved—the desired change.
- Reliability science has also been used to improve safety in medication use by calculating system reliability and developing tools and methods to compensate for human fallibility.

INTRODUCTION

A variety of strategies and techniques have been suggested to improve medication safety and reduce adverse drug events (ADEs) and medication errors, as described in Section III of this book. Increasingly, quality improvement science and reliability science methods are being used to improve patient safety, reduce unwarranted variations in care and outcomes, and produce sustainable changes in systems of care delivery (Bigham et al. 2009; Britto et al. 2014; Crandall et al. 2011; Iyer et al. 2011; Kugler et al. 2009; Luria et al. 2006; Muething et al. 2012; Ryckman et al. 2009a, b; Tofani et al. 2012; Vermaire et al. 2011; Wheeler et al. 2011; White et al. 2011, 2012, 2014). This chapter introduces the concepts of improvement science and reliability science and considers how these can be applied to the prevention of medication errors and ADEs, together with specific examples.

IMPROVEMENT SCIENCE

The science of quality improvement was developed by statistician and management consultant W. Edwards Deming (1993). This work laid the foundation for many of the philosophies, approaches, and methods used to make improvements in health care today. Deming's System of Profound Knowledge focuses on four basic components: understanding how the diverse components of a system interact; understanding variation (random or common-cause variation versus non-random or special-cause variation); the psychology of change (understanding what motivates people to change); and the theory of knowledge, which emphasizes the use of a formal, structured application of the scientific method to learn about complex systems.

Building on Deming's approach, the Model for Improvement provides a flexible framework for developing, testing, revising, and extending interventions for improving systems (Langley et al. 1996). The model has four key elements: aim, measurement, ideas for change, and tests of change. The first three elements are characterized by three key questions:

1. What are we trying to accomplish?
2. How will we know that a change is an improvement?
3. What changes can we make that will result in an improvement?

The final element is a disciplined approach to rapid testing and learning from change called the Plan–Do–Study–Act (PDSA) cycle.

The Plan step of a PDSA cycle starts with stating the specific objective for the cycle. Questions are raised and answers predicted based on current knowledge. A plan is developed to test the theory (who, what, where, and when), including methods for data collection and analysis. During the Do step, the plan is then carried out. Observations made in carrying out the plan are documented, observations that were not part of the plan are identified, data are evaluated for changes over time, and anything that went wrong is noted. In the Study step, data are analyzed using the qualitative and/or quantitative methods as specified during the Plan step, the results are compared to the predictions, and current knowledge is modified if the data contradict certain beliefs about the process. If the data confirm existing knowledge about the process, there will be an increased degree of confidence that the current knowledge provides sufficient basis for action. In the Act step, action is based on the new knowledge gained from the test. Responsibilities for implementing and evaluating changes to processes are assigned, forces in the organization that will help or hinder the changes are listed, the organizations and people affected by the changes are identified, changes are communicated and implemented, and the objective of the next cycle is established. Thus, large improvement projects can be broken down into manageable pieces using the Model for Improvement and addressed through a series of small-scale PDSA cycles.

Many improvement teams also create and use what is called a key driver diagram to guide and prioritize their interventions and PDSA cycles. An example is presented

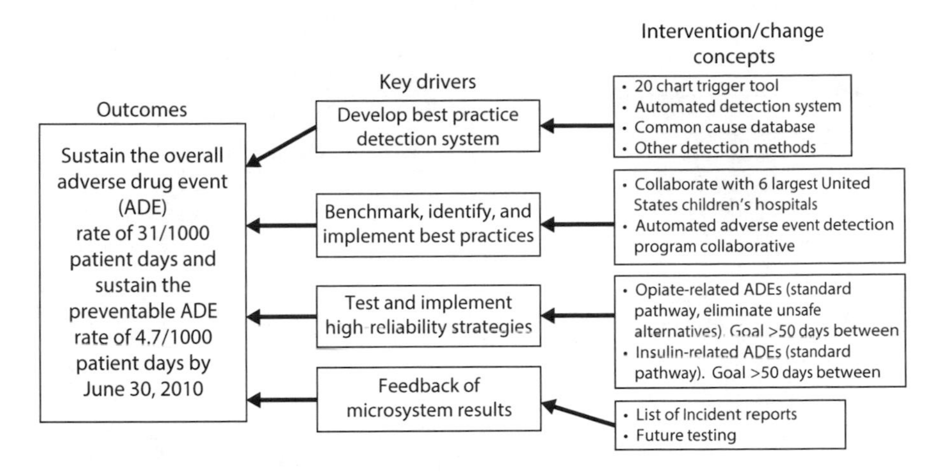

FIGURE 12.1 Key driver diagram from Cincinnati Children's Hospital Medical Center.

in Figure 12.1. This includes a project aim that is specific, measurable, achievable, realistic, and timely (or "SMART"; Doran 1981), addresses the theory or hypothesis driving the aim, and lists what needs to be in place and working in order to achieve the aim.

Measurement is an important part of testing and implementing changes (Langley et al. 1996; The Health Foundation 2013; Vincent et al. 2013). Donabedian (1988) suggested grouping measures into structure, process, and outcome. Structural measures evaluate the features of the environment in which care takes place. Process measures evaluate the particulars of care that a patient receives. Outcome measures evaluate the end results of care. A balanced set of measures should be useful, easy to obtain, linked to customer (or patient) promises, and indicate whether a change made actually leads to improvement.

Pareto charts are frequently used to quickly examine and compare the effects of problems or causes (Tague 2005). A Pareto chart (Figure 12.2) is based on the theory that a relatively small number of issues will contribute significantly to a desired outcome (the "80/20" rule: 80% of problems come from 20% of the causes). The goal of a Pareto chart is to discover which few issues should be concentrated on to achieve maximum impact. This type of chart includes both bars and a line graph. Individual values are represented in descending order by bars and the cumulative total is represented by the line.

Statistical process control charts can be used to examine the impact of interventions on outcomes over time (Amin 2001; Benneyan et al. 2003). These are commonly used in industry to analyze process improvement efforts and can enable the identification of common-cause variation (the usual, historic variation seen in the system) or special-cause variation (unusual variation possibly introduced into the system by an intervention or a special circumstance or situation). The charts are annotated with a median or mean centerline (depending on the type of chart), upper and lower control limit lines (usually representing three standard deviations

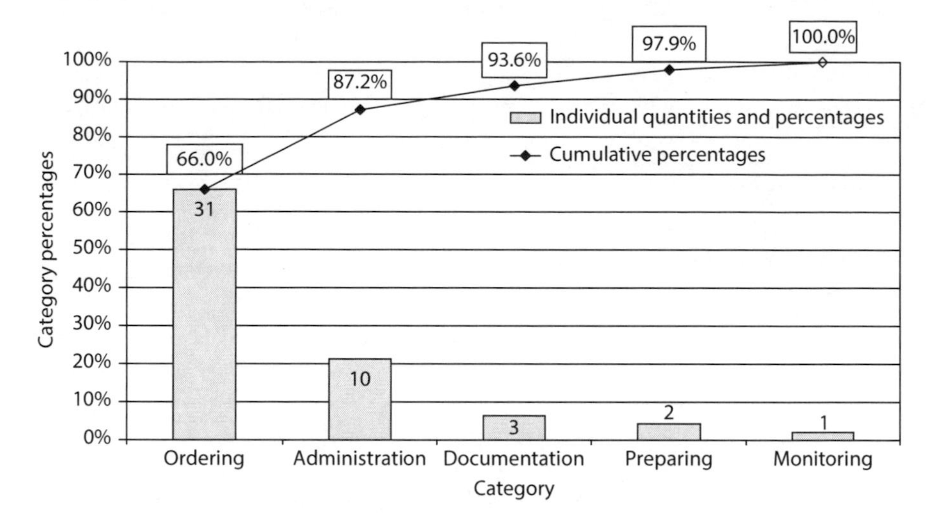

FIGURE 12.2 Pareto analysis of causes of medication errors.

either side of the centerline), and a goal line, and they indicate the desired direction of change. For independent observations, an observed change in the data is considered a special cause when any of the following are true:

- There is a "shift" (i.e., eight or more consecutive points are either above or below the centerline).
- There is a "trend" (i.e., there are six consecutively increasing or decreasing data points).
- There are "clustered points" (i.e., there are at least fifteen consecutive points close to the centerline).
- There is a single point outside the control limits (three standard deviations) or two of three consecutive points near a control limit (in the outer third).

RELIABILITY SCIENCE

The principles of reliability science, developed by industrial engineers (Ebeling 1997), can be used to examine a complex system and its processes, calculate its overall reliability, and develop tools or methods to compensate for the limits of human ability. Reliability science has been effectively used in complex industries such as manufacturing, aviation, and nuclear power to improve safety and the rate at which a system consistently produces appropriate outcomes (Bierly and Spender 1995; Bourrier 1996; Roberts 1990a; Schulman 1993; Weick 1987). Within health care, the definition of reliability has been modified to be the measurable capability of a process, procedure, or health service to perform its intended function in the required time under commonly occurring conditions (Berwick and Nolan 2003).

Reliability science has the potential to help health care providers reduce defects in care or care processes, increase the consistency with which appropriate care is delivered, and improve patient outcomes (Nolan et al. 2004). Reliability is often quantified as the number of actions that achieve the intended results divided by the total number of actions taken (Amalberti et al. 2005; Berwick and Nolan 2003; Nolan et al. 2004). For example, if five of every 1000 patients undergoing a surgical procedure die or suffer significant morbidity, the reliability of the procedure would be $(1000 - 5)/1000 = 0.995$, or a rate of 99.5%. Unreliability, or the defect or failure rate, is 1 minus the reliability. For our example, the failure rate would be $1 - 0.995 = 0.005$, or 0.5%.

The failure rate is often expressed as an index based on an order of magnitude. For example, level 1 reliability, or 10^{-1}, is achieved when there is an 80%–90% success rate or one or two failures per every ten opportunities. Similarly, level 2 reliability, or 10^{-2}, is achieved when there are five or fewer failures per 100 opportunities. Level 3 reliability, or 10^{-3}, is when there are five or fewer failures per 1000 opportunities, and so on. A "chaotic" process is defined as one in which the desired result occurs in fewer than 80% of the opportunities.

Designing health care systems for reliability requires a three-tiered strategy: (a) preventing failure; (b) identifying and mitigating the failures that inevitably occur; and (c) redesigning the system to decrease the failures that occur (Nolan et al. 2004). Preventing failure usually begins with standardizing care (Resar 2005; Rozich et al. 2004) and improving communication and teamwork. Level 1 or 10^{-1} reliability strategies might include the use of education/awareness, standardized equipment, order sets, memory aids, and feedback regarding compliance (Luria et al. 2006). However, it is considered impossible to design a reliable process using only 10^{-1} strategies. Moving a process to the next level of reliability requires a more sophisticated approach to process design. The successful application of level 2 for 10^{-2} reliability strategies can decrease the opportunities for humans to make mistakes by identifying and mitigating the failures that are inherent to all processes. Examples include making the desired action the default, redundancy, taking advantage of habits and patterns, scheduling, differentiation, constraints, affordances and building decision aids, and reminders into the system (Luria et al. 2006). Further improvement to 10^{-3} performance or greater requires an understanding of which failures are still occurring, how often they occur, and why they occur. For process design to be successful, measurement of critical failures needs to be part of the initial design strategy. Once failures are quantified, they can be prioritized and the system redesigned to avoid them in the future.

Failure mode and effects analysis is another potentially useful tool for designing a new process or evaluating a current process (Institute for Healthcare Improvement 2004; Woodehouse et al. 2004). It is conducted by carefully reviewing the different failure modes (which can be done prospectively), how frequently they are likely to occur, the consequences of each failure, and how easily each failure can be detected. Once these factors are known, there are various standardized approaches to prioritizing the failure modes. Once prioritized, mitigation strategies can be appropriately applied. The failure modes can be identified by both review of the evidence and the expertise of frontline caregivers.

Researchers in the fields of organizational behavior and management have studied the capabilities of certain potentially hazardous systems to perform with extremely high rates of reliability (Rijpma 1997; Roberts 1990b; Weick 1987; Weick et al. 1999). Systems such as nuclear power plants and aircraft carriers are considered high-reliability organizations (with 10^{-6} performance) as they are exceptionally consistent in reducing the probability of potentially catastrophic errors and managing unexpected events when they do occur (McKeon et al. 2006). They do this by establishing a culture of mindfulness that focuses on five key concepts:

- Preoccupation with failure (small, seemingly inconsequential errors are regarded as a symptom that something is wrong)
- Reluctance to simplify (diversity in experience, perspective, and opinion is encouraged)
- Sensitivity to operations (attention is paid to what is happening on the front line)
- Commitment to resilience (capabilities are developed to detect, contain, and recover from errors that do occur)
- Deference to expertise (decision-making is in the hands of the people with the most expertise and knowledge of the situation, regardless of seniority, rank, or status) (Weick et al. 1999)

EXAMPLES OF USING QUALITY IMPROVEMENT METHODS TO REDUCE ADVERSE DRUG EVENTS AND MEDICATION ERRORS

Different methods of identifying ADEs and medication errors are highlighted in Chapter 7. However, detection of ADEs and medication errors is not enough to prevent them. Successful use of improvement methods to reduce ADEs requires an understanding of their causes and patterns. Rapid identification and analysis may reveal patterns that are likely to cause future ADEs (Rozich et al. 2003). This information can help prioritize and direct efforts to redesign systems to prevent future occurrences (Bagian et al. 2001, 2002). For example, after implementing a change package of interventions that included standardization of medication ordering, reliable medication-dispensing processes, reliable medication-administration processes, improvement of patient safety culture, and clinical decision support, a collaboration of 13 children's hospitals reported a 42% decrease in total ADEs (Tham et al. 2011). We next present some examples in more detail.

EXAMPLE 12.1 REDUCING ADEs AT A CHILDREN'S HOSPITAL

Cincinnati Children's Hospital Medical Center implemented a series of system-level and drug-targeted interventions over a number of years and phases, designed to dramatically and reliably (Luria et al. 2006) reduce their ADEs. The key driver diagram is shown in Figure 12.1.

In phase 1, the medical center began tracking and analyzing ADE rates using a pediatric-focused ADE trigger tool (Takata et al. 2008). Starting in January 2002, each month, two trained nurses used trigger tool methodology to examine 20 randomly selected medical records of patients hospitalized for at least 48 hours to identify ADEs. ADEs—defined as any injury, large or small, caused by the use (or non-use) of a drug—were confirmed by two physicians. Harm for each ADE was classified according to levels F through I, established by the U.S. National Coordinating Council for Medication Error Reporting and Prevention (2011). Level F designates temporary harm requiring initial or prolonged hospitalization, G designates permanent harm, H designates intervention necessary to sustain life, and I designates patient death. Using a standardized algorithm, each ADE was further classified as either preventable (for example, an overdose of a sedative) or non-preventable (for example, an allergic reaction in a patient not previously known to be allergic). A preventable ADE was defined as an event where a definite breach of standard professional behavior or technique was identified, necessary precautions were not taken, or the event was preventable by modification of behavior, technique, or care (Child Health Corporation of America 2009).

Initial improvement efforts in phase 1 were focused on system-level changes identified by hospital leadership and the medication safety committee to reduce prescribing- or ordering-related ADEs. These included: additional safety checks for high-risk medications (Institute for Healthcare Improvement 2012); next-generation automated dispensing cabinets in patient care areas; implementation of a clinical information integration system that included computerized prescriber order entry (CPOE), clinical documentation and an electronic medication administration record; standardized medication concentrations; dose range checking and absolute blocks added to the CPOE system; smart infusion devices (Snodgrass 2005); and the addition of clinical pharmacists to the liver transplant team, pediatric intensive care unit (ICU), and emergency department.

In phase 2, additional improvement efforts focused on reducing preventable ADEs concerning specific types of drugs. For example, insulin was moved to locked storage at the bedside, sterile product preparation for some intravenous (IV) antibiotics was automated, guidelines were developed for concentrated 10 mg/mL IV morphine in the pediatric ICU, and the CPOE system provided prescribers with optional stool-softener orders that could be activated when ordering opiates. In April 2005, Cincinnati Children's Hospital Medical Center joined 13 other children's hospitals in a 12-month improvement collaborative project sponsored by the Child Health Corporation of America to reduce narcotic-related ADEs (Sharek et al. 2008). In 2006, the medical center also benchmarked against five other similar pediatric organizations to identify additional targets for intervention and improvement.

In phase 3, the focus was expanded to ADE detection. In July 2006, the trigger tool approach (Rozich et al. 2003) was combined with electronic clinical information system capabilities to establish an automated ADE detection program (Muething et al. 2010). A real-time, in-depth analysis of each event identified by one of seven pediatric-specific triggers enabled the determination of apparent cause (Takata et al. 2008). In 2007, a multidisciplinary ADE steering team was formed to identify ADE causal trends and to prioritize and direct efforts to redesign processes to prevent future occurrences. ADE steering team members also served on various hospital committees, such as the clinical practices committee and the pharmacy and therapeutics committee. The team routinely analyzed the hospital's ADE data to identify improvement initiatives and set improvement goals. At least two medication improvement teams were chartered per year, based on ADE trends identified through analysis of all ADE data sources, including incident reports, random manual chart reviews, and automated trigger tool reviews. The improvement emphasis was on preventing ADEs that occurred frequently or had a high potential for harm.

Improvement teams were co-led by a physician and another health care provider, such as a pharmacist or nurse. The hospital chief of staff provided oversight, monitored progress toward goals, and helped to overcome barriers. Quarterly reviews of results by senior leadership and the Board of Directors kept the team focused and moving forward. Graphical displays of outcome measures and monthly reports of improvement initiatives were made available via a hospital intranet site to engage all employees.

During this phase, insulin use guidelines and the post-operative pain management order set were revised, Ramsay sedation scoring (Ramsay 1974) was adopted as an objective sedation assessment tool, situation–background–assessment–recommendation (or "SBAR") communication techniques (Kaiser Permanente of Colorado and Evergreen Colorado 2011) were adopted, a night-time nurse was added to the pain service, and barcode scanning of medications at the patient bedside began.

Phase 4 began in February 2010 and improvement activities were focused on reducing all patient harm in National Coordinating Council for Medication Error Reporting and Prevention levels F to I. Review of these data showed that opiate-related over-sedation was the most frequent potentially harmful ADE (Muething et al. 2010). Variations in practice, multiple clinical teams writing orders, multi-drug regimens, and variations in the interpretation of patient assessments were noted as potential contributing causes. Interventions included hard-stop upper dosing limits for prescribing all central nervous system depressants, a general constipation treatment/laxative order set, a general pain management order set to encourage use of acetaminophen (paracetamol) and non-steroidal anti-inflammatory drugs before morphine when clinically appropriate, a change in pain team practice to use morphine as the first choice and hydromorphone as the second choice, a weaning protocol in the pediatric ICU to decrease methadone use, and requiring long-acting opiates to be approved by the pain team.

Following implementation of the interventions in all phases, the ADE rate declined from an average of 87 ADEs per 1000 patient days to 32 ADEs per 1000 patient days and then remained steady (Figure 12.3). Additionally, preventable

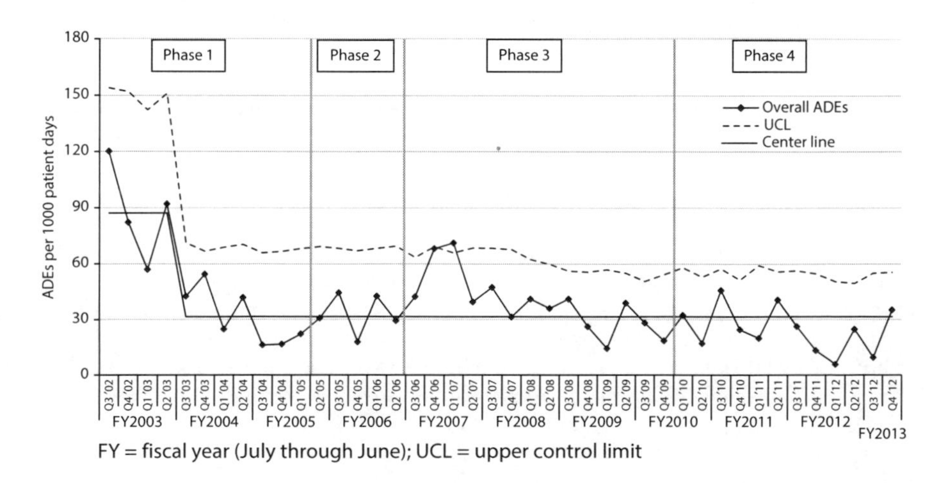

FIGURE 12.3 Control chart showing the change in adverse drug events (ADEs) over time.

ADEs decreased from 22 per 1000 patient days in phase 1 to 4.7 per 1000 patient days in phase 3 and to 1.4 per 1000 patient days in phase 4.

EXAMPLE 12.2 IMPROVING MEDICATION RECONCILIATION

Process improvement at The Adelaide and Meath Hospital, Tallaght, Dublin, Ireland, has improved medication reconciliation performance, with clinical pharmacists carrying out a high-quality medication reconciliation review process (Fitzsimons et al. 2011) post-admission for 30%–60% of adult medical/surgical inpatients, depending on staffing levels. The pre-admission medication list and pharmacist's recommendations were documented in the patient's clinical notes and also verbally communicated to a doctor or nurse. However, the pharmacist's interventions were effective (i.e., they reached the patient at any time during the admission) in only 54% of cases (Galvin et al. 2012). This project aimed to improve effectiveness and decrease time to effect, without increasing resource requirement.

Process mapping (with verification by process experts, including junior doctors, clinical pharmacists, and nurses) was supplemented by determining timelines for a convenience sample of ten patients. This involved assessment of the value that each step added to the process (i.e., whether there was a benefit to the patient or not). This analysis identified a number of potentially significant delays and failures that were detrimental to the desired outcome. The failure to adjust the patient's inpatient prescriptions in line with the pharmacist's recommendations was chosen as the focus of the project due to the potential for significant improvements to patient care and staff satisfaction.

The aim for this project was, by August 2012, for at least 75% of patients receiving a clinical pharmacist medication reconciliation review to have the recommended changes made to their prescription chart (or a documented reason for not making the change) within 12 hours of the recommendation. Ideas for change were gathered by examining previous in-house and external research and interviewing stakeholders, including patients, junior doctors, clinical pharmacists, nurses, and senior pharmacy managers. The approach was based on a previous project in the hospital (Egan et al. 2012). The most popular ideas for change were ranked by impact, work, and readiness for testing. A series of PDSA cycles was then implemented:

1. Existing data sets were examined. It was found that the communication method and the pharmacist involved influenced effectiveness, but the medical/surgical team caring for the patient and the potential severity of the discrepancy identified did not.
2. One pharmacist's documentation of the pre-admission list and clear documentation of recommendations was highlighted by doctors as being optimal. A PDSA cycle found that this pharmacist's recommendations were no more effective than other pharmacists' recommendations, thus optimizing documentation was unlikely to increase effectiveness.
3. Adding verbal communication to the existing process showed promising results, with improvement of the effectiveness of the recommendations made by the pharmacist increasing from 46% to 75% and the time to implementing the recommendations decreasing from a mean of 22 to 5 hours.
4. A larger test of adding verbal communication found that effectiveness of the pharmacists' recommendations increased from 55% (notes only) to

79% (notes and verbal communication) and decreased the mean time to effect from 34 to 9 hours. It added less than two minutes per patient to the process as a whole.

5. Documentation in the notes initially, with verbal communication if the changes had not been made 24 hours later, was effective in 73% of cases, but the mean time to effect was 50 hours, and this was found to be more time-consuming for the pharmacists.

The addition of verbal communication to the pharmacist's process was agreed as an effective and resource-neutral change and was formally added to the pharmacists' process in July 2012. Measurement to track implementation was included as part of the pharmacists' workload statistics. Figure 12.4 shows the decreasing proportion of patients receiving a pharmacist medication reconciliation review during 2012 and into 2013 because of reduced pharmacist staffing. Initially, pharmacists added verbal communication to their medication reconciliation process (point "a"); however, this was not sustained. At a stakeholders' meeting of clinical pharmacists, the pharmacists' reservations were addressed. This led to consensus on the need to communicate verbally only if there were any changes recommended following the medication reconciliation review (point "b"). Following this meeting, the change was adopted by all clinical pharmacists, resulting in a consistent proportion of medication reconciliation

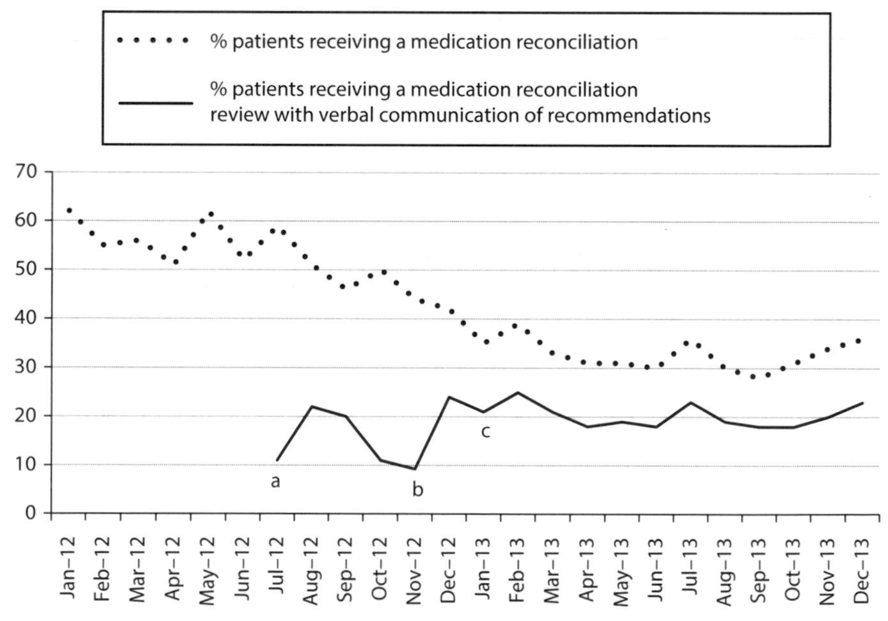

FIGURE 12.4 Percentage of adult inpatients getting two types of pharmacist medication reconciliation review.

reviews involving verbal communication with a doctor (point "c"), subsequently sustained for more than a year.

This project demonstrated that improved efficiency and effectiveness of the process can result from a simple process change and without additional resources. Small tests of change can quickly validate whether ideas for change are likely to be effective and identify those worthy of further testing. Willing volunteers, rather than those nominated by their line manager, were found to be effective, with these volunteers supporting the change with their peers. The involvement of process experts, including more junior health care professionals and patients, provided a richer understanding of the process and how it could be improved. Resistance to change was overcome by bringing the patient's or doctor's perspective to the pharmacists and by proving the effectiveness of the change. The most significant lesson was of the need to track the implementation of an agreed change, which identified the need for further intervention and ensured a full spread and sustainability.

EXAMPLE 12.3 REDUCING PRESCRIBING ERRORS

The theories of high reliability present a challenge to those working in health care, where most care would fall into the highly unreliable category. Staff at Great Ormond Street Hospital NHS Foundation Trust in London have attempted to address this in relation to prescribing. Patients on the hematology, oncology, bone marrow transplant, immunology, and infectious diseases wards are prescribed a large number of high-risk medications for which the potential consequences of prescribing errors are significant. Principles of high reliability were implemented to reduce the incidence of prescribing errors.

Initial assessment of the problem was based on incident reporting and anecdotal evidence. Two ward-based pharmacists were recruited to record the number of prescribing errors on every ward, every day, for every patient. These data were used to determine an accurate baseline rate of errors. The denominator was the total number of drugs prescribed on the wards. The prescribing error rate per 100 items prescribed on inpatient wards was 7.6 errors per 100 drugs for the hematology/oncology wards. The improvement work was then spread to the transplant, immunology and infectious diseases wards, whose baseline error rate was 12.6 errors per 100 drugs prescribed.

The five principles of high-reliability theory (Weick et al. 1999) were used to provide a framework for interventions and ideas. PDSA cycle methodology was used to implement changes, and the pharmacist team implemented and monitored the impact of these changes. The first two interventions focused on the first key concept of preoccupation with failure:

Intervention 1: Pharmacists monitor prescribing and look for errors for every child on the ward, every day, using an electronic data-collection tool.

Intervention 2: Medicine reconciliation by the pharmacists by checking the medication history within 24 hours of admission for every child to ensure that the right drugs are prescribed by the doctors from the start of a child's inpatient stay.

The next two interventions then focused on reluctance to simplify interpretations:

Intervention 3: When the pharmacist finds an error, they discuss it with the doctor to explore the full set of reasons why the error occurred.

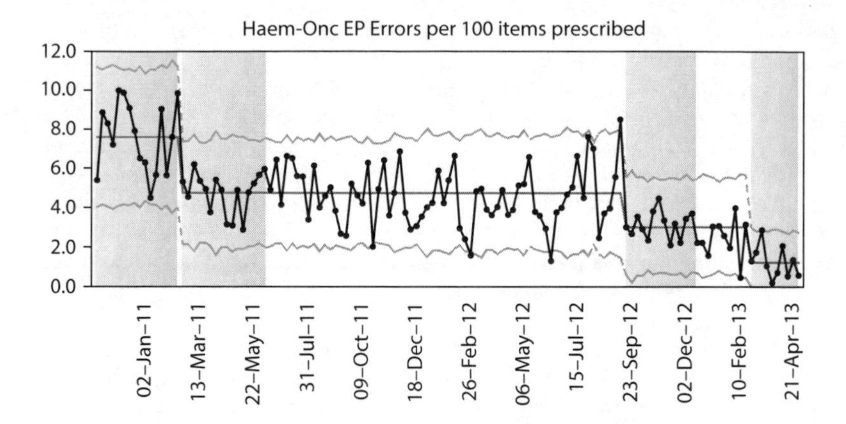

FIGURE 12.5 Hematology/oncology electronic prescribing (EP) errors.

Intervention 4: Themes around prescribing errors are highlighted with the larger group in ward rounds and a number of different local interventions are implemented to address any root causes identified.

This was followed by an intervention demonstrating sensitivity to operations:

Intervention 5: The pharmacists act as active observers of the process and have a constant awareness of potential weaknesses due to staffing, patient complexity, pressures on the ward, and other factors that may affect prescribing on the ward. This often results in interventions by the pharmacists or extra support for doctors who are known to make prescribing errors.

Commitment to resilience was the focus of the next two interventions:

Intervention 6: The pharmacists contain prescribing errors by ensuring that the doctor who has made the error rectifies the error immediately, before it can reach the patient and potentially cause harm.

Intervention 7: The pharmacists make proactive prescribing interventions, which are suggestions made before the drug is prescribed.

The final key concept covered was deference to expertise when the themes highlighted by the pharmacists are fed back during ward rounds and the prescribers decide the solutions required. As shown in Figure 12.5, the interventions led to a sustained decrease in prescribing errors.

CONCLUSION

Improving patient safety is an ongoing challenge. Quality improvement methods emphasize testing theories about organizational processes and systems through experimentation and replication to produce a detailed understanding of the factors affecting system performance (Margolis et al. 2009). Reliability science can be used to examine a complex system and its processes, calculate its overall reliability, and develop tools or methods to compensate for the limits of human ability (Luria et al. 2006). Together, quality improvement science and reliability science methods show great promise as tools to help improve patient safety, reduce unwarranted variations in care and outcomes, and produce sustainable changes in systems of care delivery.

ACKNOWLEDGMENTS

The authors of this chapter would like to thank Ciara Kirke, Drug Safety Coordinator, The Adelaide and Meath Hospital, Dublin, and Toral Pandya, Anupama Rao, and Judith Delaney, Great Ormond Street Hospital NHS Foundation Trust, and their clinical teams for the case material.

13 Safety Culture

Denham Phipps, Todd Boyle, and Darren Ashcroft

CONTENTS

KEY POINTS

- Safety culture in health care is generally defined with reference to the values, beliefs, and practices that an organization holds with respect to patient safety.
- We suggest four types of models for understanding safety culture, which underpin numerous tools used to assess it.
- Assessment can be used for diagnosis, raising awareness, benchmarking, fulfilling regulatory requirements, or evaluating change.
- Attempts to improve safety culture should involve staff at all levels of the organization and should ideally start with the most senior members of staff.

INTRODUCTION

In health care, as in other safety-critical endeavors, the notion of "safety culture" has become a familiar reference. What, though, does this mean? And why does it matter? In this chapter, we will explain what safety culture is and review some of the different ways in which it has been formulated. We will then discuss how safety culture relates to the policies and practices of health care organizations, such as their engagement with harm disclosure and incident reporting. Finally, we will suggest methods for assessing and improving safety culture. In order to illustrate how such methods can be used in practice, we will consider a case study from community pharmacies.

DEFINING SAFETY CULTURE

A cursory glance at the books and articles written about safety culture will reveal that various definitions have been proposed. However, a key theme across these definitions is a reference to the values, beliefs, and practices that an organization holds with respect to safety matters. Guldenmund (2010) describes safety culture in terms of three levels (the three "A"s). From the most intrinsic to the most extrinsic, these are

- The collective *assumptions* within the organization about its identity and what it is doing. These are largely implicit and reflect general values such as how power should be distributed across the social hierarchy.
- The *attitudes* that the organization's members hold towards people, processes, equipment, and risks. These tend to be relatively explicit and expressed through policies and procedures.
- The manifest *artifacts* that demonstrate how the organization deals with safety matters in practice. These include the routines and communications that are encountered on a day-to-day basis.

Gelfand and colleagues (2011) provide a more detailed overview of the cultural values that could be related to organizational safety and the ways in which these influence attitudes and behaviors. As Guldenmund (2010) implies, safety culture permeates much of what an organization believes and does regarding safety; this is illustrated by the frequency with which references to it appear in accounts of organizational accidents across various industries (for example, Reason 1990). In health care in the United Kingdom, "*An Organization With a Memory*" drew attention to the importance of organizational culture in patient safety: "... [it] is perhaps the only aspect of an organization that is as widespread as its various defenses; as such, it can exert a consistent influence on these barriers and safeguards—for good or ill" (Department of Health 2000). This observation is borne out in studies such as that by Hofmann and Mark (2006), who found a relationship between nurses' perceptions of the local safety climate (defined below) and the occurrence of medication errors in their respective hospitals.

UNDERSTANDING SAFETY CULTURE

Various models have been proposed to explain how a safety culture is formed and maintained. We suggest four general types of models for understanding safety culture: component, developmental, work climate, and system models.

Component Models

In component models, safety culture is seen as the sum of its parts. One well-known example, which is referred to in the U.K. Department of Health's report (2000), is

Reason's informed culture model (Reason 1997). An "informed" culture is one that proactively collects and acts on safety data (for example, incident reports). In order to achieve an informed culture, an organization needs to have four elements in place:

- *Reporting culture*—staff who are encouraged to be open about safety-related issues and to engage in incident reporting
- *Just culture*—clarity about what is acceptable and unacceptable conduct, with fair and consistent application of professional standards
- *Flexible culture*—a capacity to adapt one's structures and/or practices in the face of hazards
- *Learning culture*—a willingness to learn lessons from safety reports and to implement these lessons in order to improve safety

If reporting culture, just culture, flexible culture, and learning culture are the reactants that create an informed culture, then *blame culture* is its inhibitor. Box 13.1 outlines the problems that can ensue when a blame culture is allowed to take hold within an organization.

Another example of a component model has been proposed by Cooper (2001), who borrows from Bandura's notion of "reciprocal determinism" (Bandura 1986). According to Cooper, safety culture arises from the interaction between cognitive factors (staff attitudes and beliefs), behavioral factors (staff behavior in practice), and situational factors (organizational policies and practices). In this view of safety culture, a key issue is the setting of social norms among staff that encourage safe behaviors and discourage unsafe behaviors (an approach known as "behavioral safety").

BOX 13.1 BLAME CULTURE

"Blame culture" refers to the situation in which incidents that occur within an organization are automatically attributed to the personal failings of individual members of staff. The problem is not the assignment of responsibility to an individual *per se*, but that it is assigned unfairly, when it would be more appropriate to focus on the circumstances in which the individual is working (for example, excessive workload or poorly functioning equipment). The negative effect of a blame culture is the discouragement of staff from being open about incidents, thus depriving the organization of opportunities to correct safety problems (for example, see McKee and colleagues' [2010] case study of organizational change in National Health Service acute trusts).

One antidote to a blame culture is Reason's culpability decision tree, which the English National Patient Safety Agency (NPSA) adapted for use in health care. At the time of writing, a copy of the NPSA's "incident decision tree" is freely available from their website (National Patient Safety Agency 2014) or, alternatively, Meadows and colleagues (2005). Working through this tree will help to determine whether or not it is appropriate to hold an individual responsible for the occurrence of a given medication safety incident.

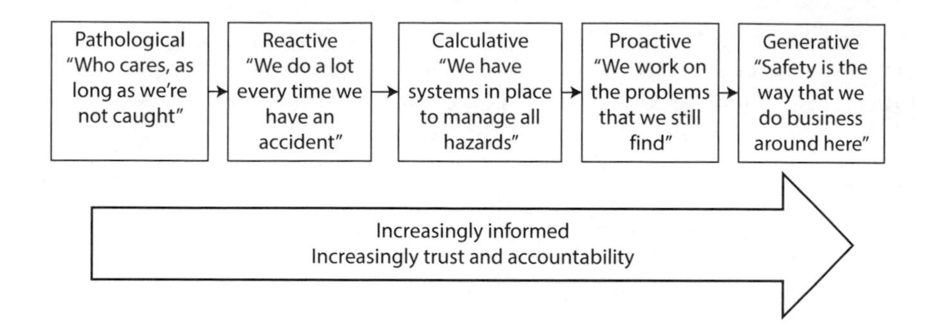

FIGURE 13.1 Developmental stages of safety culture. (From Hudson, P. 2003. *Qual Saf Health Care* 12(Suppl 1): i7-12. With permission.)

Developmental Models

Developmental models depict safety culture as a developmental process, in which the organization moves between different "levels" or "strengths" of safety culture. In one such model, originally proposed by Westrum (2004) and later extended by Parker and colleagues (2006), the organization's approach to dealing with safety matters develops through a series of stages, as shown in Figure 13.1. The least developed safety cultures are considered to be "pathological," where the organization ignores or suppresses information about safety matters. The most developed are "generative," where the organization proactively seeks and acts upon safety-related information. Consider the two examples of health care organizations described in Box 13.2. Which do you think is most likely to be pathological and which generative? Which is most similar to the way that your organization deals with safety concerns?

Work Climate Models

In work climate models, safety culture is a product or interpretation of the ways in which the work is organized, managed, and carried out. Clarke (2000) describes three aspects of the working environment that give rise to safety culture:

- *Management commitment to safety:* staff perceptions of managers' and supervisors' safety-related attitudes and actions
- *Safety management system:* staff experiences of safety management activities (such as training, policies and procedures, reporting systems, and safety equipment provision)
- *Risk perceptions:* staff perceptions of workplace hazards, risk to personal safety and feelings of safety

According to Zohar (2010), a key issue in staff perceptions of safety is the extent to which the organization's espoused safety policies match the way that safety issues are dealt with in practice. For example, an organization might declare itself to be committed to safety, but does it then invest sufficient resources in staffing, training,

BOX 13.2 EXAMPLES OF LEARNING IN HEALTH CARE ORGANIZATIONS

EXAMPLE 13.1 STURDY MEMORIAL HOSPITAL

"A urologist ... discovered [that] one of his patients was diagnosed with prostate cancer yet had a negative biopsy. After a second similar example ... the hospital then carried out an internal audit of 279 biopsies done over two years, which showed 20 of them to be in error. Rather than attempting to cover up the situation, or downplay it, Sturdy announced it would hire consultants to audit some 6000 additional biopsies. The hospital reported the biopsy problems to state authorities, sent regular updates to its staff, and wrote 88,000 letters to hospital patients explaining the situation."

EXAMPLE 13.2 THE CANADIAN BLOOD AUTHORITIES

"[There were] suspicions in the early 1980s that the national blood supply might be contaminated with HIV and hepatitis ... [but] doubts were muted, critics were muzzled, response was slow, and the public was misled. For instance ... while epidemiologists were discussing the potentially compromised blood supply, hemophiliacs were being told that blood fractions they used were safe. Supplies of potentially contaminated blood were still being distributed even after supplies that had been successfully heat treated were available."

Westrum, R. 2004. Qual Saf Health Care *13(Suppl 2): ii22–ii27.*

and maintenance needed to achieve and sustain such a commitment? An organization may have an incident reporting system, but are incident reports recognized and acted upon, or are staff simply reporting incidents for the sake of reporting? Discrepancies in areas such as these are likely to have a detrimental effect on safety culture.

At an individual level, the conditions that influence the development of safety culture can be described in terms of *job characteristics*—that is, features of the working environment that affect psychological and/or physical well-being. Such features include the challenges and rewards of one's job, the demands and conflicts encountered, the relationship between staff and leaders, and opportunities to communicate and innovate.

Clarke's (2010) meta-analysis of safety climate studies led her to propose a hypothetical causal chain from job characteristics, through safety climate, to safety behavior and performance. While few studies as yet have sought empirical evidence for this set of relationships in health care settings, Virtanen and colleagues (2009) found that a measure of job characteristics in a hospital ward (such as working hours and the perceived ratio of effort invested in work to the rewards gained from it) predicted the risk of hospital-acquired infection among patients on that ward.

System Models

Some approaches to safety culture are based primarily on open systems theory—the idea that an organization is a collection of technical and social elements that interact to create work processes and products in a given environment (Waterson 2009). Here, safety culture is seen as an ongoing interplay between goals, relationships, and activities within an organization. Phipps and Ashcroft (2014) describe safety culture as a dynamic relationship between maintaining safety and achieving other organizational goals. This relationship is affected by the level of emphasis placed on the respective goals, the degree of conflict between them, and the extent to which the environment constrains or facilitates the achievement of each goal. For example, pharmacy staff can be faced with a trade-off between safety and productivity in their work, with the level of risk incurred depending on how they negotiate this trade-off (Phipps et al. 2009).

ASSESSING SAFETY CULTURE

Achieving a safety culture, regardless of the model, requires ongoing assessment and reflection. There are several reasons for assessing the safety culture within an organization (Nieva and Sorra 2003), including:

- Raising the general awareness of safety among staff
- Fulfilling regulatory requirements (for example, to conduct a quality or safety audit)
- Benchmarking between different departments or organizations
- Identifying specific problems
- Evaluating the effect of changes or interventions

Given the variety of ways in which safety culture has been defined and understood, there are several approaches that could be taken to assess it. In general terms, a safety culture assessment can be based on qualitative data (which asks "what" and "why"), quantitative data (which asks "how much"), or a combination of the two. The data could come from a range of sources, including interviews, observations, discussion groups, audits, or surveys. Many assessments use *safety climate* as a proxy for safety culture, as explained in Box 13.3.

To give the reader a flavor of the tools available for safety culture assessment, we will next describe some examples of those used in health care. This is by no means an exhaustive list of what is available; however, it will serve to illustrate the typical format and content of assessment tools.

Manchester Patient Safety Assessment Framework

Parker and colleagues' (2006) developmental model of safety culture, described earlier, provides the basis for the Manchester Patient Safety Assessment Framework (MaPSAF, Kirk et al. 2007). This qualitative instrument contains eight dimensions of safety culture in health care settings, which are

BOX 13.3 SAFETY CLIMATE

In terms of Guldenmund's (2010) "three As," it is usually easier to assess artifacts and attitudes, and link these to safety performance, than it is to do so for assumptions. Therefore, it is common for safety culture assessments to focus on what is called *safety climate*, which is the manifest form of an organization's safety culture (that is, the collection of artifacts and attitudes that are presumed to emanate from it) and is typically measured through a quantitative survey. Sometimes the terms "safety culture" and "safety climate" are used interchangeably, but strictly speaking, the latter is a subset of the former.

1. Commitment to patient safety
2. Perceptions of the causes of incidents and their reporting
3. Investigating incidents
4. Learning following an incident
5. Communication
6. Staff management
7. Staff education and training
8. Team-working

For each dimension, the organization is classified as pathological, reactive, calculative, proactive, or generative. Table 13.1 shows how an organization at each stage would be described in terms of commitment to patient safety.

The framework is intended to stimulate discussion between members of an organization about the prevailing safety culture and what could be done to enhance it, using the descriptors provided as a guide (Parker 2009). The original version was

TABLE 13.1
Manchester Patient Safety Assessment Framework (MaPSAF) Descriptors for "Commitment to Patient Safety"

Level	Excerpt from Descriptor
Pathological	Risks are worth taking; if an adverse event occurs, insurance schemes are there to bail us out
Reactive	Patient safety only becomes a priority once an incident occurs and the rest of the time only lip service is paid to the issue apart from meeting legal requirements
Calculative	Patient safety has a high priority; however, safety systems are not widely circulated to staff or reviewed
Proactive	Patient safety is promoted throughout the organization and staff are actively involved in all safety issues and processes
Generative	Patient safety is integral to the work of the organization and its staff and is embedded in all activities

devised for British primary care organizations (Kirk et al. 2007), but it has since been adapted for hospitals, ambulance services, mental health services, and community pharmacies in Britain (for example, Ashcroft et al. 2005a). A German language version has also been produced (Hoffmann et al. 2014). A recent survey of National Health Service (NHS) primary and secondary care organizations in England found that, among those that conducted any safety culture assessment, MaPSAF was the most popular (Mannion et al. 2009). As a qualitative rather than quantitative assessment, it lends itself well to diagnosis and awareness-raising. However, it is not clear how well suited MaPSAF is for measuring changes in safety culture over time, or for benchmarking.

Medication Safety Self-Assessment

The Medication Safety Self-Assessment (MSSA, Institute for Safe Medication Practices 2001) does not explicitly refer to safety culture or climate. However, as an assessment of organizational practices with regard to medication safety, it captures the artifacts associated with safety culture. It is intended primarily for use in hospital and community pharmacies and asks either a staff representative or all staff members as a group to rate the pharmacy's practice in ten areas:

1. Patient information
2. Drug information
3. Communication of drug orders and other drug information
4. Drug labeling, packaging, and other nomenclature
5. Drug standardization, storage, and distribution
6. Medication device acquisition, use, and monitoring
7. Environmental factors, workflow, and staffing patterns
8. Staff competency and education
9. Patient education
10. Quality processes and risk management

In each area, there are a number of characteristics that relate to specific safety-related activities (for example, "when taking orders over the telephone, the prescriber is specifically queried about comorbid conditions, allergies and the patient's weight"). Each characteristic is rated on a five-point scale from "A" ("there has been no activity to implement this") to "E" ("this has been fully implemented for all staff and patients"). The outcome is a detailed understanding of which aspect of practice the pharmacy should address in order to improve safety. To help the users of MSSA in their efforts to improve safety, the U.S. Institute for Safe Medication Practices (ISMP) has produced a set of guidance notes covering the areas assessed in the MSSA (Institute for Safe Medication Practices 2001). The MSSA has been used most extensively in the United States, where it was devised (for example, Smetzer et al. 2003). However, alternative versions have been developed for use in other countries, such as Canada (for example, Greenall et al. 2005). The MSSA generates a profile of ratings rather than a composite score; therefore, like the MaPSAF, it appears better suited to diagnosis than to comparing organizations or time-points.

Safety Attitudes Questionnaire

The Safety Attitudes Questionnaire (SAQ) is a quantitative survey instrument that is based on its authors' studies of organizational safety in aviation and health care (Sexton et al. 2006). It consists of 60 questions, which together measure the following factors:

- Teamwork climate (perceived quality of collaboration between personnel)
- Job satisfaction (positivity about the work experience)
- Perceptions of management (approval of managerial action)
- Safety climate (perceptions of a strong and proactive managerial commitment to safety)
- Working conditions (perceived quality of the work environment and logistical support)
- Stress recognition (acknowledgement of how performance is influenced by stress)

The SAQ assesses both the safety climate and related job characteristics. Therefore, it is consistent with the work climate model of safety culture outlined earlier in this chapter. It was developed in the United States and subsequently validated in various hospital departments in the United States, United Kingdom, and New Zealand (Sexton et al. 2006). It has since seen relatively widespread use among English NHS organizations (Mannion et al. 2009) and, more recently, has been adapted for use in Swedish community pharmacies (Norden-Hagg et al. 2010). Given that it generates a set of overall scores, it is a better choice of instrument for benchmarking organizations or evaluating the effects of organizational changes.

Pharmacy Safety Climate Questionnaire

The Pharmacy Safety Climate Questionnaire (PSCQ) was originally developed as a quantitative survey instrument to complement the pharmacy version of the MaPSAF (Ashcroft and Parker 2009). Unlike the latter, but like the SAQ, the PSCQ is intended for benchmarking organizations and measuring changes over time. The first version comprised 34 questions, each of which related to one of the eight dimensions in the MaPSAF. For the purposes of the questionnaire, two of the MaPSAF dimensions were combined, so only seven dimensions (six original and one composite) are measured by the PSCQ. A subsequent validation study, using data from five European countries (Phipps et al. 2012), led to the PSCQ being revised. This second version retains 24 of the original questions and measures the following factors:

- Organizational learning (the pharmacy's engagement in incident reporting and learning from incidents)
- Blame culture (the tendency to unfairly blame individuals for safety incidents)
- Working conditions (the extent to which the working environment supports safe working)
- Safety focus (the extent to which safety is treated as a priority)

BOX 13.4 TIPS FOR ASSESSING SAFETY CULTURE

- Ensure that the assessment instrument is valid and reliable—that is, it assesses what it intends to assess and does so in a consistent manner. Details about reliability and validity can usually be found in the instrument's user guide or in research articles.
- Consider as broad a range of issues as possible (for example, job characteristics in addition to safety culture).
- Assess as many parts of the organization as possible, starting with individual work units. Note that safety culture or climate can vary across work units within an organization.
- Aim for a high participation rate among staff.
- Present the findings to both senior managers and staff.
- Repeat the assessment periodically.
- Link the assessment to intervention.

Pronovost, P. and Sexton, B. 2005. Qual Saf Health Care *14: 231–233.*

The focus of the PSCQ is narrower than that of the SAQ. However, data from British community pharmacies suggest that it similarly links safety climate to job characteristics (Phipps and Ashcroft 2011; Phipps et al. 2012). While the PSCQ has been used mainly in community pharmacies, some studies have applied it to hospital pharmacies (for example, Lalor et al. 2015; Phipps et al. 2011).

While all the instruments presented can provide an assessment of an organization's safety culture, their usefulness and accuracy depends greatly upon their appropriate use within the specific health care setting. In Box 13.4, we provide some words of advice for those who wish to use these or any other safety culture assessments in their workplace.

IMPROVING SAFETY CULTURE

While a safety culture assessment could be an end in itself, it is often a precursor to efforts at improvement. How is this initiated? Some basic guidance is provided in Box 13.5. Weaver and colleagues (2013) reviewed 33 studies of safety culture interventions in acute care and found that they used three main types of activity:

- Team training or the provision of a tool to improve communication within teams.
- Executive or interdisciplinary walk-rounds.
- Comprehensive unit-based safety programs. This is an iterative process of safety education, staff identification of safety concerns, adoption of the work unit by a senior manager, implementation of improvement actions, documentation and sharing of results, and reassessment of culture (for example, Pronovost et al. 2005).

BOX 13.5 GETTING STARTED WITH SAFETY CULTURE IMPROVEMENT

- *Start at the top:* Get commitment from upper-level management and provide proper feedback to staff.
- *Make it clear to everyone that safety matters:* Get all employees involved and beware of habitual deviations from protocols and guidelines.
- *Encourage learning from errors:* Establish lines of communication and develop an error-reporting and learning system.
- *Search for solutions:* Make use of prospective and retrospective hazard analyses.
- *Prepare people to work safely:* Identify training needs, develop programs to address them, and evaluate the effectiveness of these programs.

Wilson-Donnelly, K.A. et al. 2005. Hum Factor Ergon Man *15: 135–176.*

Weaver and colleagues (2013) recommend combining more than one activity for best effect. In addition, they recommend that the choice of specific activities is guided by an understanding of safety culture theories, such as those outlined earlier in this chapter. For example, the work climate model suggests a need to ensure that the job characteristics within an organization are conducive to safe working (Phipps and Ashcroft 2011), while Westrum's (2004) developmental model (and hence MaPSAF) highlights the importance of encouraging incident reporting and learning activities.

It is worth emphasizing that efforts to improve safety culture need to take into account not just the beliefs and attitudes of staff, but also the organizational environment within which they work. As a case in point, Phipps and Ashcroft (2012) found that, when community pharmacists were grouped according to the pattern of ratings on the PSCQ and a job characteristics questionnaire, their respective pharmacies appeared to differ in terms of how they balanced the workload imposed on pharmacy staff with the provision of resources to support the staff. Furthermore, one group of respondents reported having little influence over the safe operation of their pharmacies, while a pervasive concern among all of the respondents was being unfairly blamed for incidents. Meanwhile, Ashcroft and colleagues (2006) and Williams and colleagues (2013) have suggested that, when deciding whether or not to report a medication incident, pharmacists take into account (among other things) the potential impact on interprofessional relationships versus the likelihood that a positive change will result from reporting an incident. In short, safety culture improvement should not be seen solely as the responsibility of individual members of front-line staff. It also requires the moral and practical support of senior managers and mutual trust between the different parts or levels of the organization.

In the next section, we will examine a safety culture improvement program that has recently been conducted in Canadian community pharmacies. This case study shows how the models and methods discussed so far might be applied to an individual health care setting, and provides examples of the issues encountered in practice.

SafetyNET-Rx: Towards a Generative Safety Culture

Health care organizations face significant challenges in enhancing their safety culture. Establishing a generative safety culture, for example, often requires input and support from a variety of stakeholders, ranging from regulatory authorities to technology providers. One initiative, known as SafetyNET-Rx, provides an interesting case study of how the development of support mechanisms by a variety of health care stakeholders has enabled community pharmacies to work towards a generative safety culture (Boyle et al. 2012, 2014a,b).

SafetyNET-Rx focuses on enhancing the safety culture of Canadian community pharmacies. It does so by removing the fear of discussing medication incidents and providing pharmacy staff with the training, tools, and support needed to learn from such errors and plan and implement meaningful changes. The Nova Scotia College of Pharmacists developed SafetyNET-Rx through a partnership with university researchers, community pharmacies, and ISMP Canada. In 2010, the Nova Scotia College of Pharmacists revised its standards of practice related to quality improvement in community pharmacies to include, among other things, anonymous reporting of medication incidents to a national and independent third party, annual community pharmacy safety self-assessments, evidence of a documented continuous quality improvement (CQI) program, and quarterly meetings to discuss key incidents (Boyle et al. 2014b). To assist community pharmacies with the meeting of standards, the SafetyNET-Rx program was developed and tested with 78 of Nova Scotia's 300 community pharmacies.

As presented in Figure 13.2, the SafetyNET-Rx program (Boyle et al. 2012, 2014b) begins with a day-long training session for program facilitators. These facilitators—who are the quality improvement and safety champions in the pharmacy—help to train staff on SafetyNET-Rx and are the key points of contact for staff with questions regarding the program. To ensure full staff involvement in the program, each pharmacy has two facilitators; a staff pharmacist and a pharmacy technician. The training session focuses on removing the taboo of talking about medication incidents, introducing concepts related to safety culture, quality management and CQI in community pharmacy practice, and providing hands-on training on the SafetyNET-Rx tools and related technologies. In addition to training, prior to starting the SafetyNET-Rx program, community pharmacies are expected to complete a safety self-assessment, such as ISMP Canada's version of the MSSA (described above). Results of the assessment serve as a benchmark so that improvements can be tracked over time. As part of the SafetyNET-Rx program, community pharmacies are expected to complete the MSSA once every year in order to assess how the safety culture changes over time.

Following the SafetyNET-Rx program, pharmacy staff return to their usual working practices until a medication incident or safety issue occurs. When this happens, anonymous details of the incident are reported to a national database using an online reporting tool, such as ISMP Canada's Community Pharmacy Incident

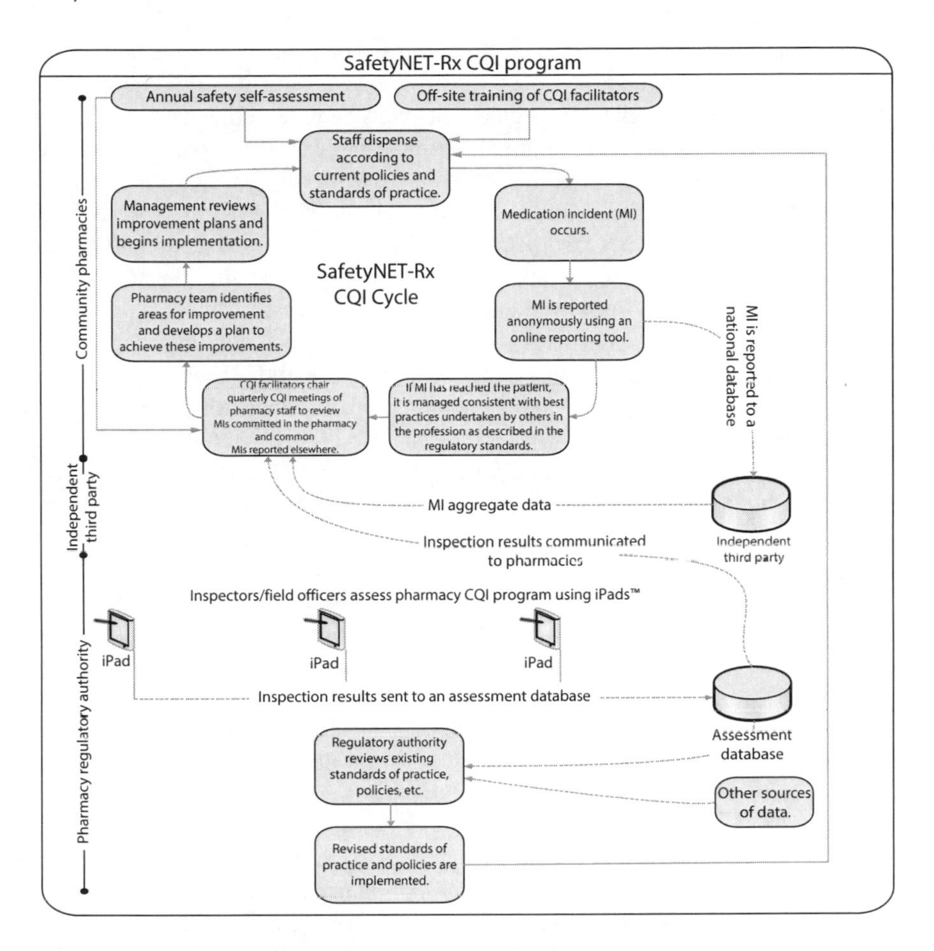

FIGURE 13.2 SafetyNET-Rx program. (Reprinted from *Res Social Adm Pharm*, 10, Boyle, T.A. et al., Keeping the "continuous" in continuous quality improvement: Exploring perceived outcomes of CQI program use in community pharmacy, 45–57, Copyright 2014, with permission from Elsevier.)

Reporting system. Once every three months, a staff meeting is held to discuss medication incidents and patient safety. There are a number of important goals to this meeting. First, community pharmacy staff examine key medication incidents that have occurred since the last meeting. By applying root cause analysis, pharmacy staff identify the key factors that led to the incidents and plan subsequent changes. Second, pharmacy staff, using the incident reporting tool, examine key incidents (for example, those with significant impact or that occur frequently) that have occurred at other community pharmacies in Canada and assess the likelihood of similar incidents happening to them. Third, with input from the Nova Scotia College of Pharmacists, staff explore how to improve their SafetyNET-Rx process and its success in enhancing the safety culture within the pharmacy. In cases of a severe incident, such as an error actually leading to patient harm, a staff meeting to discuss the incident and identify the root causes is held right away. Suggested changes to key processes from the quarterly meetings are then reviewed by pharmacy management and, if approved, are implemented. Pharmacy staff dispense according to the new policies and procedures until a medication incident happens again and the cycle starts over.

The SafetyNET-Rx program requires that pharmacy staff not only examine and discuss medication incidents, but that they also discuss the *process* used to report and learn from such incidents and enhance safety. Therefore, the Nova Scotia College of Pharmacists' pharmacy inspectors or field officers will, as part of their routine inspections, assess community pharmacy compliance with the standards of practice related to quality improvement and suggest areas that need improvement. Using an iPad application, pharmacy inspectors provide each community pharmacy with a grade for the completeness of their CQI process and a report on key areas that need to be improved in order to help achieve and sustain a culture of safety. The pharmacy manager receives a copy of the report, which is expected to be discussed at quarterly meetings. The tool also allows the Nova Scotia College of Pharmacists to better understand the challenges that community pharmacies may face with sustaining a safety culture and, therefore, enables the College to develop more relevant and targeted strategies.

Migrating to a culture where safety is openly discussed and supported by all members of the organization can be a daunting task for many health care organizations. Therefore, it is not surprising that community pharmacies that adopted SafetyNET-Rx to enhance their safety culture faced a number of initial challenges. A key challenge was finding the time to integrate key components of SafetyNET-Rx into the normal work practice. For example, community pharmacies had difficulty finding the time to report incidents to the national database as soon as they occurred and in all staff involved in the quarterly safety meetings. Some pharmacy staff were unwilling to accept the technological components of SafetyNET-Rx, such as using an online system to report incidents. Community pharmacies also highlighted some initial apprehensiveness with openly discussing incidents, having all staff fully participate in the program and, in a few cases, worsening staff relationships from discussing medication incidents (Boyle et al. 2012, 2014a).

A number of these issues were addressed by community pharmacies customizing SafetyNET-Rx to better fit their needs and resource limitations. For example, to address issues surrounding time and technology acceptance constraints, community pharmacies developed a hybrid online–manual reporting process. When an incident occurred, key details were written down on a manual form, with details being entered online in batches or at the end of the day. Other pharmacies scheduled quarterly meetings after hours, with staff time paid for by the employer. Issues regarding staff apprehensiveness were addressed by pharmacy managers openly discussing specific medication incidents or safety issues in which they were personally involved. The pharmacy staff then addressed these incidents as a group to better understand the key elements of the SafetyNET-Rx program, in particular its focus on the process, rather than individuals, as the source of the error (Boyle et al. 2012).

Over time, community pharmacies were able to customize SafetyNET-Rx to fit their specific needs and resource limitations, and in doing so, displayed key elements of a generative safety culture. For example, staff felt that they had a better understanding of the key processes and workflows within the pharmacy. In addition, staff highlighted an increased openness in discussing incidents with others in the pharmacy. The program was eventually accepted as part of the normal work routine, with staff becoming more confident and aware of their individual responsibilities with regards to patient safety, and believing that the dispensing processes in the pharmacy were safer and more reliable as a result of openly discussing incidents. Overall, staff reported a reduction in the number of medication incidents that were occurring in the pharmacy (Boyle et al. 2014a).

CONCLUSION

In this chapter, we have introduced the reader to the notion of a safety culture, suggested some ways of understanding what it is, and provided some advice on the assessment and improvement of safety culture. Our case study further illustrates how the theories and methods introduced earlier in this chapter can be applied to medication safety improvement in practice.

However, we have only scratched the surface of this topic. There remain questions and debates regarding the nature of safety culture, how to best assess it, and how useful it is in understanding and improving patient safety (for example, Cox and Flin 1998; Vincent 2010; Waterson 2014). Nevertheless, safety culture can be seen as an issue of relevance to the investigation of patient safety issues.

Section III

Putting Solutions into Practice

14 Introduction

Bryony Dean Franklin and Mary P. Tully

This final section presents various approaches to introducing interventions and solutions to enhance medication safety. These range from more general approaches, such as changes in educational programs (Chapter 16) or communication (Chapter 17), to interventions focusing on specific stages of the medication use process, both high tech (Chapters 19 through 21) and low tech (Chapters 18 and 21). We also consider the role of incident reporting systems in supporting medication safety (Chapter 15) and, very importantly, the roles of patients and their carers (Chapter 22).

These various approaches are most certainly not mutually exclusive and a mixture of some or all is likely to be required, depending on local context. We are also aware that the idea of interventions being "innovative" is a relative concept—what may be highly innovative in one setting may be already standard practice in another. Nevertheless, we hope this final section will highlight ideas for consideration for individuals, teams, and organizations on every part of the innovation spectrum.

There are clearly many other broad strategies that are relevant, and this section is not intended to be exhaustive in this respect. For example, while we consider various "non-technical" skills, such as communication in Chapter 17 or teamwork and error awareness in Chapter 18, there are many other categories of fairly general non-technical skills that we do not have space to address in depth. These include issues such as handovers, situation awareness, decision-making, and leadership (Gordon 2013; Patey et al. 2005; Ross et al. 2014).

Whichever solutions are adopted, to get the best out of these, other skills and techniques will also be required. These are likely to include skills in change management, project management, and other aspects of implementation science in order to integrate research findings and evidence into policy and practice.

Each chapter in this section includes one or more expert commentary. Given the importance of local context in selecting appropriate medication safety solutions, these are intended to provide a broader perspective than those of the chapter authors. We have, therefore, invited commentaries from people from other countries and/or other disciplines. We have also taken a broad view as to who is an "expert" in the area of safe medication use. Overall, more doses are consumed when the medication is in the control of the patient alone than when it is in the control of health care professionals, and so three of our commentaries have been written by expert patients, drawing on their personal experiences of living with the need to ensure safe medication use for themselves.

15 Incident Reporting and Feedback Approaches

Sukhmeet S. Panesar, Bruce Warner, and Aziz Sheikh

CONTENTS

KEY POINTS

- Incident reporting and learning systems are used internationally as ways of ensuring that health care organizations can learn from the mistakes of the past.
- Health care professionals and patients can report errors they are aware of into such systems either individually or via their organization.
- Collated data can then be used to alert health care professionals locally, regionally, or nationally both about the risks and about what to do to mitigate them.

INTRODUCTION

There have been significant advances in health care over the last decade, and patients now receive treatments that were unlikely in the past. However, these significant strides in therapeutic developments have not been matched by a corresponding reduction in iatrogenic harm. Some positive steps have been made: for example, patient safety has gained global importance and is a key program of the World Health

Organization (WHO) (World Health Organization 2002). This chapter described the rationale and approaches to using incident reporting systems as learning tools to provide feedback to organizations to mitigate the risk of error. Feedback to individuals is covered elsewhere in Chapter 18. At the end of this chapter, there is an expert commentary about the potential value of applying this approach internationally.

In seeking to reduce the unacceptable levels of iatrogenic harm, active efforts must be made to report and learn from errors in health care. One of the most frustrating things for patients and health care professionals is the apparent failure to learn from the same mistakes. Very rarely do mistakes occur in isolation. The case of British teenager Wayne Jowett, who received intravenous vincristine via the wrong route, is a classic example of a systemic failure that has occurred since the early 1960s and continues today (Franklin et al. 2014a; Toft 2001).

A useful approach to learning from errors is the creation of a patient safety reporting system, which can help us learn from errors and should lead to the development of interventions aimed at mitigating such errors. Learning can occur locally, nationally, or internationally, depending on the approach chosen to disseminate the lessons learnt from such incident reports. Furthermore, as one reports an incident, there is an opportunity to reflect on the incident, discover if such incidents have occurred elsewhere, and identify points of action to prevent the error from occurring again. Reporting must be coupled with measures of action, and there are several methods by which patient safety reporting systems can help to achieve success: generation of alerts on complications associated with new drugs; dissemination of lessons learned by aggregating and analyzing patient safety incidents; and the revelation of previously unrecognized problems, such as bone cement implantation syndrome (Leape 2002; Panesar et al. 2009a).

Methods of identification of incidents vary significantly and include the use of reporting systems, surveillance using routine data, pharmacovigilance, and significant event audits, to mention but a few. Any or all of physicians, pharmacists, nurses, patient safety officers, patients, and their families may be involved. Some systems center on certain types of events (such as medication errors), others focus on events where serious harm has occurred, and some are more all-encompassing. Methods of analysis and risk prioritization also vary widely between different systems. A common approach is an in-depth root cause analysis of one event, with the production of a report that is then widely disseminated. Other approaches include the online publication of the virtually unaltered description of anonymous incidents, as shown in Box 15.1.

BOX 15.1 EXAMPLE OF AN UNALTERED DESCRIPTION OF AN ANONYMOUS INCIDENT

Patient was prescribed 15,000 units of Fragmin [dalteparin], but when weighed on admission [date] was only 46 kg. Treatment dose for this weight is only 10,000 units, so a 50% overdose was prescribed and administered.

National Reporting and Learning System. 2014. Reducing treatment dose errors with low molecular weight heparins. http://www.nrls.npsa.nhs.uk/alerts/?entryid45=75208.

Rates of harm are also calculated in different ways, making the summation and comparison of incident report data problematic. Differences arising from the use of checklist reporting forms versus free-text reports, web-based versus paper recording, and inconsistent approaches to assessing and/or prioritizing the significance of events all add to these problems. Methods of incident mitigation range from individual feedback through local educational meetings to the production of national alerts and media or web-based releases. Evaluation of the effectiveness of reporting systems is also under-developed, and in many cases, no evaluation has been conducted. Anecdotally, it is assumed that safety is improving through the use of reporting systems, but data supporting this are lacking. In fact, due to the increased number of reports collected, data can often be misinterpreted as suggesting the opposite of what they actually do indicate (Noble et al. 2011).

APPROACHES TO CONCEPTUALIZING, POPULATING, AND INTERROGATING DATA

WHO has a key role in patient safety reporting systems. Guidelines were developed in 2005 by WHO's Patient Safety Program (World Alliance for Patient Safety 2005); these outline areas such as the purposes and methods of reporting, components of a reporting system, and a guide and checklist for setting up a national system (Donaldson and Fletcher 2006). Potentially controversial issues such as voluntary versus mandatory reporting, anonymous or confidential reporting, and resource allocation are all covered. However, realizing an "orange-wire" system that translates critical information internationally, as advocated in the guidelines, is still a distant goal. The concept of an "orange-wire" refers to a hypothetical situation in which all aircrafts of a particular type would be grounded if one of them were found to have a faulty orange wire that could cause harm; in health care, many such faulty orange wires exist, but the attention paid to them is limited (Donaldson 2004). A global patient safety reporting system could help with the better identification of such incidents and provide the impetus required to bring about the necessary learning and sharing of solutions to promote safer care.

One future approach may be to imitate the aviation industry's approach to such a problem. Following a number of airline crashes some years ago, the aviation industry brought together all relevant partners (including regulatory bodies and technical experts) under one banner: the Commercial Aviation Safety Team. This body has demonstrably reduced fatalities in aviation, and the creation of a similar body in health care has been advocated (Pronovost et al. 2009).

WHO holds the following hope for reporting systems:

> The currency of patient safety can only be measured in terms of harm prevented and lives saved. It is the vision of [WHO] that effective patient safety reporting systems will help to make this a reality for future patients worldwide. (World Alliance for Patient Safety 2005)

The true value of a reporting system is dependent on its input, and if clinicians fail to supply useful information, the outputs may be meaningless. National systems might be better developed than regional and local systems, but evaluation of their effectiveness and translatability into action is significantly under-developed.

Reporting systems can operate at several levels: local, regional, and national (see Table 15.1 for examples). National systems have been more ambitious in their aims than local systems. One of the most well-known national systems is the National Reporting and Learning System (NRLS) for England and Wales; this includes all types of patient safety incidents, and at the time of writing, it has received 11 million reports since 2003. However, the oldest national system is the Australian Incident Monitoring System, which has been used in Australia for over 20 years, having originated from an adverse event reporting system first used in anesthesia in the 1980s. The system was engineered by the Australian Patient Safety Foundation, who were given the responsibility of developing a system for reporting adverse events and near-misses in Australian public sector hospitals in 1996. It uses specific software for collecting confidential data and for classifying and producing reports, which also allows individual units to compare the frequency of incidents with other organizations (Australian Patient Safety Foundation 2010). The system was recently renamed the Advanced Incident Monitoring System and the software is now marketed internationally. Data are anonymous and can be used to identify trends that can be compared nationally, as well as communicated through reports produced by the Australian Patient Safety Foundation.

Medication errors are often cited as a particularly common type of adverse event reported using patient safety reporting systems. Specific medication error reporting systems, such as the anonymous MedMARx system in the United States, have also been very successful in quantifying and responding to incidents through organizational feedback. Almost 900 hospitals have submitted information, and over one million reports have been received to date. The system provides a way of sharing data between health care organizations and has also contributed information in specific areas, such as operating theater medication errors (Beyea et al. 2003; MedMARx 2002; Santell et al. 2003).

Some clinical specialties have opted for specialty-specific reporting systems. An example of this is the Confidential Reporting System for Surgery (CORESS) for Britain and Ireland, which relies on surgical trainees and consultants reporting untoward incidents on a standard proforma. These "stories" then undergo editorial review and are published in the form of quarterly reports with advice on preventing the error from occurring again. Anonymity of the patient and the clinician is maintained throughout. The person who submitted the case receives a certificate. However, up until 2010, fewer than 100 incidents had been reported (CORESS 2010). Another example of a specialty-specific system is a voluntary and anonymous online system for neonatal units set up in 54 hospitals and used by 739 health care workers from the Vermont Oxford Network in the United States. Analysis revealed that the most frequent type of reported incident was medication errors (Suresh et al. 2004).

We next discuss the NRLS for England and Wales in more detail, as an example of a national system used for reporting and learning from errors, highlighting key points of international relevance.

TABLE 15.1
International Examples of Incident Reporting Systems

Reporting System	Notes
Wimmera Base Hospital, Australia	A longitudinal survey of adverse inpatient events over 8 years of progressive implementation of a risk-management program. Screened 49,834 inpatient admissions by retrospective record review from 1991 to 1999 at a rural base hospital in the Wimmera region of Victoria, Australia.
Australian Incident Monitoring System	This system provides a mechanism for any incident or accident in health care to be reported, using a single standard form. Via a web-based system, incidents are classified using software that uses two unique classification systems developed by the Australian Patient Safety Foundation. This system is now called the Advanced Incident Monitoring System.
Australian General Practice Sentinel Event Survey	A study testing incident reporting in Australian general practice, which was published in 1997 and 1998. It involved 324 general practitioners. Reports were anonymous and 805 reports were received over 20 months.
National Register of Medical Incidents, Sweden	Every health care organization in Sweden is obliged to report all adverse events or risks of them to the National Board of Health and Welfare. Reports are registered and classified in a national database, and anonymized information bulletins are regularly circulated to all organizations.
Critical Incident Reporting System, Switzerland	This is a reporting system for anesthetists set up in 1996. Contents of the detailed online form are automatically inserted into a central database, which enables easy compilation of the data and anonymous discussion of each case.
Medical errors and adverse events, United States	The Medication Error Reporting program is a voluntary system for front-line practitioners. Information is shared with the U. S. Food and Drug Administration and any relevant manufacturers. MedMARx is a medication error program that began in 1998. It is voluntary, internet-based, and anonymous. MedWatch is a reporting system for serious adverse events associated with medical products.
Emergency Care Research Institute, United States	This institute collects information on medical device incidents and hazards. It uniformly codes the information using its proprietary Universal Medical Device Nomenclature System.
State-wide incident reporting systems, United States	At the time of the Institute of Medicine report (2000), 19–23 of 50 states had mandatory reporting, but only 15 complied with Institute of Medicine criteria, and these related only to acute and general hospitals. Under-reporting was common. Several comparisons and analyses have identified wide variations in definitions, processes, results, and costs between states.
Osaka University Hospital, Japan	A web-based incident reporting system that incorporates daily analysis and trend and pattern analysis. A Clinical Risk Management Committee directs solutions development, involving staff education and risk managers acting on incidents.
National Reporting and Learning System, England and Wales	A web-based reporting system with structured and narrative input. An alert system results in the publication of reports and aims to change practice due to the act of reporting.
Jeder Fehler Zachlt web-based incident reporting for general practice, Germany	A web-based reporting system with structured and narrative input. Analysis is carried out by patient safety experts. Selected reports are available online and in the medical press. Users can comment on and discuss incidents online.

THE NATIONAL REPORTING AND LEARNING SYSTEM FOR ENGLAND AND WALES

Origins

Just over a decade ago, there was a call for the urgent development of a national database of medical errors in England and Wales; the vision was that this would help the medical community to better understand the epidemiology of errors that caused harm and those that did not, define research priorities in this area, and develop error-reduction strategies (Sheikh and Hurwitz 1999). This call arose out of the recommendations of two key reports from the United States and Australia, which highlighted the need for patient safety to be an integral part of health policy considerations (Institute of Medicine 2000; Wilson et al. 1995).

Structure

When first envisioned, the underlying model for the NRLS was simple: it would be a fully mandatory reporting scheme for medical errors experienced by patients under the care of the National Health Service (NHS) (Department of Health 2000). The main argument for mandatory systems is that these should enable a truly comprehensive picture of the patient safety landscape to emerge and, furthermore, improve health care professionals' sense of accountability. However, it has subsequently been noted that mandatory systems deter practitioners and hospitals from reporting incidents, as they fear that disclosure could lead to possible repercussions for the reporting physician or their organization (Leape 2002). At first, the National Patient Safety Agency (NPSA), which hosted the NRLS, did consider a mandatory reporting model, but in the end opted for a voluntary, anonymized reporting structure in the hope of enabling fuller disclosure of incidents without fear of reprisal on the part of the individual making the report (Department of Health 2000). However, in April 2010, almost a decade later, reporting of serious untoward incidents (those that constituted severe harm and death) became mandatory in England and Wales (National Patient Safety Agency 2010). Perhaps surprisingly, the reporting of all incidents (including serious ones) has since increased.

The NRLS allows patient safety incidents to be reported directly via a web-based open-access system or via a more popular option in which reports are submitted in an anonymized fashion via the individual organization's local risk management system. Data are arranged categorically in 75 separate data fields. These include incident categories at two levels, specialty and location of the incident and a free-text description (Catchpole et al. 2008). The largest proportion of incidents originates from medical specialties (34%), followed by surgical specialties (16%), mental health (13%), and obstetrics and gynecology (10%). Of note is that the proportion of reports from primary care has been particularly low (5%), for reasons that as yet remain poorly understood (Panesar et al. 2009b).

Approaches to Data Generation and Analysis

Each NRLS report refers to an unintended or unexpected incident that could have or did lead to harm for one or more patients receiving NHS-funded care in England or

Wales. The NRLS also includes incidents that reached the patient but did not lead to harm and those that did not lead to harm because an incident was prevented from reaching the patient. Incidents are further stratified into different levels of patient harm as perceived by the reporter. When a patient safety incident report is made via an NHS organization, a record of it is stored digitally in a safety management system in the organization concerned. The information is then de-identified before being stored in the NRLS.

Each incident reported as leading to death or serious harm is reviewed individually by trained clinical staff and a range of outputs are produced to provide solutions to patient safety problems. Data from the NRLS are published in a number of formats, including summative quarterly reports for reporting organizations in England and Wales. Individual organizational reports are also produced, showing local reporting rates benchmarked against other similar organizations (Panesar et al. 2009b). This allows for a middle-ground approach, allowing for a certain degree of disclosure and yet also maintaining the anonymity of individual patients, reporters, and the specific identity of comparator organizations.

There is ongoing consultation with subject-matter experts, including professional organizations such as the Royal Colleges. NHS organizations also have deadlines imposed on them by which time they must implement any recommendations made.

Analysis of the reported incidents by the NPSA has helped lead to the identification of possible solutions to these problems (see Table 15.2 for examples). While these have proved useful, there remain several challenges associated with the analysis and interpretation of data, which largely reflect issues with the architecture of the NRLS. The approaches used for analyses include stratified sampling of frequently occurring incident types and free-text data mining for specific topics. The very large number of case reports being received renders it difficult to undertake detailed analysis of all incidents (Panesar et al. 2009b). Such analysis is also compromised by the lack of detail in many of the reports received and, because the reports are anonymized, there is no opportunity to go back to those making the reports or to case notes to glean further information.

Initially, the NPSA produced detailed patient safety solutions that, when evaluated, proved difficult for organizations to implement. Simpler solutions were then developed using the one-page "rapid response report" format that outlined the problems and described actions that could be taken to help prevent other patients being similarly harmed. NHS organizations were also provided with supporting information describing in detail the relevant contextual data from the reporting system, together with advice on implementation.

Developing Alerts

Within an NHS organization, patient safety incidents reported relating to a specific clinical area are forwarded to a risk manager. These are then reviewed locally to identify any areas where action could be taken to make services safer—for instance, a ward reporting repeated errors with certain high-risk injectable medicines. Patient safety incidents reported locally are also uploaded to the NRLS. This is designed to

TABLE 15.2
Examples of Medication-Related Rapid Response Reports Produced from the National Reporting and Learning System for England and Wales

Date	Topic	Title
November 28, 2011	Intravenous equipment	Minimizing risks of mismatching spinal, epidural, and regional devices with incompatible connectors
March 20, 2011	Insulin	The adult patient's passport to safer use of insulin
January 31, 2011	Intrathecal, epidural, and regional medicines	Safer spinal (intrathecal), epidural, and regional devices
December 16, 2010	Ambulatory syringe drivers	Safer ambulatory syringe drivers
November 25, 2010	Loading doses	Preventing fatalities from medication loading doses
August 26, 2010	Intravenous fluids and medicines	Prevention of over-infusion of intravenous fluids and medicines in neonates
July 30, 2010	Low-molecular-weight heparins	Reducing treatment dose errors with low-molecular-weight heparins
June 16, 2010	Insulin	Safer administration of insulin
February 24, 2010	Omitted or delayed medicines	Reducing harm from omitted and delayed medicines in hospital
February 9, 2010	Intravenous gentamicin	Safer use of intravenous gentamicin for neonates
January 21, 2010	Immunization	Vaccine cold storage
December 1, 2009	Lithium	Safer lithium therapy
September 29, 2009	Oxygen	Oxygen safety in hospitals
February 19, 2009	Bowel-cleansing solutions	Reducing risk of harm from oral bowel-cleansing solutions
December 9, 2008	Midazolam	Reducing risk of overdose with midazolam injection in adults
October 21, 2008	*Haemophilus influenzae* type B (Hib) vaccine	Risks of omitting Hib when administering Infanrix–IPV + Hib (a combined preschool booster vaccine)
August 11, 2008	Vinca alkaloid minibags (small-volume infusion bags)	Using vinca alkaloid minibags (adult/adolescent units)
April 24, 2008	Intravenous heparin flush	Risks with intravenous heparin flush solutions
January 22, 2008	Oral anti-cancer medicines	Risks of incorrect dosing of oral anti-cancer medicines
November 26, 2007	Paraffin skin products	Fire hazard with paraffin-based skin products on dressings and clothing
September 3, 2007	Injectable amphotericin	Risk of confusion between non-lipid and lipid formulations of injectable amphotericin
June 18, 2007	Cytarabine	Risk of confusion between cytarabine and liposomal cytarabine (Depocyte®)

occur at the push of a button by local risk managers to minimize the burden on NHS organizations, so that at a national level, all incidents are received, however trivial. These are mapped against data fields, which are updated to align with international patient safety classification terms. Most incidents come from locally uploaded data, as described above. A minority (less than 1%) are received directly from web-based reports from individual clinicians. Some also come through particular specialty-specific initiatives, such as the reporting scheme set up in partnership with the Royal College of Anaesthetists. This scheme includes more detailed taxonomy around particular areas of clinical interest, such as difficult airways.

Pre-2011

Prior to 2011, the challenge for the NPSA was to identify the most pressing risks and issues from a vast database of over 10 million incidents. In addition to wider trends and patterns, each incident reported as resulting in severe harm or death was scrutinized by clinical staff, in particular the free-text description of the incident. These were screened to focus on incidents that suggested wider system problems that could affect a number of organizations. Over 300 serious incidents were carefully reviewed in this way each month, and a few were selected for further work. More evidence from the wider database was sought at this stage, together with data from other sources, such as litigation, as well as international sources, such as the Pennsylvania Patient Safety Authority (2010). NRLS staff also sometimes went back to organizations reporting serious harm events at this stage for more information about what action was taken locally. A wide range of clinical advice was sought for further understanding of the problem and possible actions to reduce harm. Criteria for urgent action were

- Evidence of substantive harm from incident data or other sources
- Risks not well recognized by staff
- Clear actions available to prevent harm

Issues that met these criteria were developed as rapid response reports (Table 15.2). These were usually produced within 2–4 months of the original incident report, although some were produced in a matter of weeks when swift action was needed (for instance, to prevent risks to hemodialysis patients from additives to the hospital water supply).

2012 Onwards

From early 2012, the NPSA was shut down and its functions transferred to NHS England. A new system was launched in January 2014 for alerting the NHS to emerging patient safety risks. The system allows for the timely dissemination of relevant safety information to organizations, as well as acting as an educational and implementation resource. It builds on the best elements of the NPSA system and is known as the National Patient Safety Alerting System. This is used to disseminate patient safety information at different stages of development so that newly identified risks can be quickly highlighted to health care organizations. It allows rapid dissemination of urgent information, as well as encouraging information sharing between organizations and providing useful education and implementation resources for use by health care organizations.

The National Patient Safety Alerting System is a three-stage system based on that used in other high-risk industries. Stage 1 is "Alert: Warning," which warns organizations of an emerging risk. It can be issued very quickly once a new risk has been identified to allow rapid dissemination of information. Stage 2 is "Alert: Resource," involving the provision of resources, tools, and learning materials to help mitigate the risk identified in Stage 1. Stage 3 is "Alert: Directive," when organizations are required to confirm they have implemented specific actions or solutions to mitigate the risk. Alerts are issued on the basis of a set of agreed principles and may cover issues including:

- New or under-recognized patient safety issues
- Widespread, common, and challenging patient safety issues, not solved by alerts in isolation
- Improving systems for clinical governance, reporting, and learning

The new system gives organizations the opportunity to tackle emerging risks in their own way and to establish a sense of ownership. Through Stage 2 alerts, organizations are provided with potential solutions and resources to mitigate the risks. The system encourages voluntary compliance for the early adopters, which allows organizations to find solutions that best suit their individual organizations and minimizes the requirement for directives. Health care organizations are issued with required actions to be signed off in a set timeframe in accordance with a central sign-off process. The actions are tailored for each patient safety issue. All three stages of alert are used for issues representing a major risk. However, on occasion, it may only be necessary to use part of the alert process. For example, issues of a widespread and well-known nature may not require Stage 1, while those where a clear and specific solution exists may be addressed only with a Stage 3 directive.

By April 2014, monthly data were being published on any NHS organizations that had failed to declare compliance with any National Patient Safety Alerting System alerts by their due date. Failure to comply can be used by regulators and by commissioners with responsibilities for improving quality. Failure to comply with a Stage 3 alert within the deadline can be a cause for significant concern on the part of regulators, commissioners and, most importantly, patients. Alerts are targeted as narrowly as possible in order to keep them relevant to those receiving them. The target audience for the alerts is identified in consultation with the sponsoring Patient Safety Expert Group and relevant experts. In some cases, it can be difficult to identify a definitive audience, and therefore it may be necessary to issue an alert to a wider audience. Alerts are disseminated via the Central Alerting System, which is used to share alerts with area teams, clinical commissioning groups (who register for the Central Alerting System), secondary and tertiary care organizations, and primary care organizations (via area teams). The relevant organizations are advised of the "expected" audience, but have the freedom to share the alerts more widely as they see fit, recognizing local variation in the provision of services. Responsibility for the development of alerts rests with the NHS England Patient Safety Domain, in collaboration with subject experts and relevant organizations.

STRENGTHS AND LIMITATIONS OF INCIDENT REPORTING SYSTEMS

Whereas the NRLS has provided what, in some cases, are the first insights into key areas and offered possible solutions to preventing their recurrence, several limitations of this database (and similar databases) exist. These include those relating to comprehensiveness and those relating to quality.

COMPREHENSIVENESS

The comprehensiveness of the data in the NRLS varies from organization to organization. The NRLS comprises data collected by over 400 different NHS trusts, each with varying systems used to collect and store data. Counts of incidents are simply incidents reported to the NPSA—they are likely to represent only a small subset of the number of incidents that actually occur. Incident reports are often made soon after the incident occurs, but before the incident has been investigated locally. Therefore, reports to the NRLS may not contain complete information about the incident, especially the findings of more detailed investigations, such as root cause analyses (Lamont and Scarpello 2009; Wu et al. 2008).

Some incidents recorded in local risk management systems and subsequently forwarded to the NRLS may not technically be patient safety incidents. For example, deaths from natural causes that occurred in hospital and deaths where patients died unexpectedly without any associated patient safety incident are sometimes reported to local risk management systems for local audit purposes, and hence forwarded to the NRLS. The data are also likely to include incidents where the impact on the patient is not clear, or where it is not clear if the incident could have been avoided. For example, suicides are often reported to local risk management systems, including cases where the event could not have been prevented by health services (Lamont and Scarpello 2009).

It is also important to note that a higher number of reported incidents from an organization, specialty, or location does not generally mean that the organization, specialty, or location has a higher number of incidents; it may instead reflect greater levels of reporting. An increase in the number of incidents reported should not therefore be taken as an indication of a worsening of patient safety, but rather an indication of increasing levels of awareness of safety issues among health care professionals (Mayo and Duncan 2004; Stratton et al. 2004; Wolf and Hughes 2008). While it might, theoretically, indicate a worsening safety profile, experience in other industries has shown that as an organization's reporting culture matures, staff are more likely to report incidents. Nonetheless, even in high-reporting organizations, many incidents do not get reported (Sari et al. 2007); this is a particular issue for medication-related incidents, as discussed in Chapter 7.

Reporting levels and rates from trusts vary greatly. Some trusts report daily, while others do not report at all. In many cases, incidents are batched and then sent in large numbers. It should never be assumed that the number of reported patient safety incidents is representative of actual numbers across the NHS. In addition, reporting in secondary care is far greater than that in primary care. Ambulance and mental health organizations have the most varied reporting patterns.

The gross under-reporting to the database has been a cause for concern and, as such, use of the data is often limited to the warning, communication, and detection of rare patient safety incidents (Vincent et al. 2008). While this may be a valid criticism, it is clear that reporting is increasing as clinicians become more aware of its presence and furthermore develop confidence that there will not be any personal repercussions to making reports. Also of relevance in this context is the varying degree of engagement by different professional groups. Convincing clinicians of the usefulness of the data they contribute should, in due course, further increase the frequency and quality of reporting. Clinical problems tend to be under-reported, while other potentially less serious, non-clinical problems, such as slips, trips, and falls, are perhaps over-represented. It is still proving difficult to engage senior doctors in a generic reporting system. In order to try and overcome these problems, the NPSA engaged frontline and senior clinicians, and undertook two pilot projects aimed at improving reporting from general practice and anesthesia. Working with the Royal College of Anaesthetists and the Royal College of General Practitioners, two bespoke reporting systems have been developed that are incorporated into the architecture of the NRLS. Encouragingly, the reporting system involving the Royal College of Anaesthetists has been a success, and there has been a significant improvement in the level of reporting from anesthetists (National Patient Safety Agency 2008). The impact of the work with the Royal College of General Practitioners is currently being assessed.

Quality

Within the NRLS, steps have been taken to maximize the quality of the data held by checking for duplicate reports and feeding back to individual trusts if there are problems with their reports. The reporting of the degree of harm in the NRLS is intended to be the actual harm suffered by the patient (National Patient Safety Agency 2004). However, there is often confusion between the *potential* degree of harm associated with an incident and the *actual* degree of harm that occurred, such that near-misses (where no actual harm resulted) are sometimes erroneously coded as severe harm or death. NRLS data cannot be checked with the reporter, as information on the identities of individual staff or patients is not held centrally, so there is no way to verify or clarify incident details.

When performing analyses on patient demographics, it is important to take into account missing data. Gender is completed for approximately 70% of patients, age in 66% of patients, and ethnicity for only 20% of patients (Personal communication with Information Analysis Team, National Patient Safety Agency). It should not be assumed that the missing data are evenly distributed, and levels of missing data should be stated in any output. The lack of a denominator also limits epidemiological work (Lamont and Scarpello 2009).

CONCLUSION

The challenges for improving patient safety in health care remain significant. Patient safety reporting systems represent an important step and are an important resource in ensuring that information about adverse events is both learned from and shared widely. All health care staff, regardless of specialty, can contribute to these efforts by reporting patient safety incidents to a local or national patient safety reporting system. While important challenges remain in relation to encouraging fuller, franker, and more comprehensive reporting, and then meaningfully analyzing these data, it is fair to conclude that substantial progress can be made.

COMMENTARY

Itziar Larizgoitia
Team Leader Reporting Systems and Learning, Service Delivery and Safety, World Health Organization

As the authors of this chapter clearly describe, reporting and learning systems are considered fundamental tools to improving the safety of health care. WHO, through its Patient Safety Program, adopted a priority objective to stimulate global learning through enhanced reporting of patient safety incidents. Most of WHO's subsequent actions in this area have focused on strengthening the foundations and infrastructure of reporting systems to facilitate the global exchange of information. The absence of a common language to name and classify patient safety incidents was the main challenge. In 2009, WHO therefore published a Conceptual Framework depicting the knowledge domain or conceptual universe related to the occurrence of patient safety incidents. This has subsequently helped many health care organizations and countries in shaping their reporting systems around common elements (World Health Organization 2009b). This was complemented with key terms and definitions in what was considered the first stages of an International Classification for Patient Safety. Developments in the upcoming International Classification of Diseases 11th Revision, which incorporates the essence of this conceptual framework within the coding of complications from health care, together with further understanding of its advantages, have led to the framework being considered as a key information model. Ongoing WHO-led work is also exploring the notion of a "minimal information model," or minimal dataset, which can be used as a common entity to facilitate global exchange of information between reporting systems, and thus shared learning.

Attention to patient safety reporting systems tells us only part of the picture of what goes wrong in health care. For true global learning, it is important to triangulate various sources of evidence, including routine reporting systems, case note reviews and other methods, given the different and partial perspectives that each provides. In fact, opportunities for interesting synergies lie in the various vigilance systems intended to prevent adverse events within specific health care areas, such as pharmacovigilance, hemovigilance, medical devices vigilance, radiation safety, and others. WHO and partners are exploring such opportunities around the concept of the minimal information model to develop communication across a range of specialty and generic reporting and vigilance systems, as well as for patient safety reporting systems.

The concept of a global patient safety reporting system is still a distant prospect. For reporting to be an effective source of learning, it also needs to be integrated in an enabling environment that includes a framework of transparency, accountability, and the pursuit of excellence. A framework that facilitates and protects the disclosure of failures and unintended patient harm is needed, while protecting the rights of patients and relatives to know about the incident and attributing just responsibilities and liabilities to those involved, all within a culture of safety and improvement. These are among the most important directions of travel for safer health care systems and thus are some of the priorities for WHO and its partners in the near future.

16 Educational Interventions

Pauline Pearson, Lesley Scott, Jessica Hardisty, and Alan Green

CONTENTS

KEY POINTS

- There is widespread recognition that undergraduate training for health care professionals should include a strong emphasis on patient safety, including medication safety.
- A number of educational approaches can be used, including simulation, interprofessional learning, and the involvement of service users and carers.
- It is important to address human and organizational factors, particularly at an undergraduate level, as these are the predominant causative factors of medication use errors in clinical practice.

INTRODUCTION

This chapter will focus on educational interventions to improve the safety of medication use. First, we quantify the degree to which a lack of practical knowledge and skills around medication use contributes to error, using evidence drawn from the national and international literature. The chapter will then describe the role of educational interventions at both undergraduate and postgraduate levels in facilitating the development of the knowledge, skills, and confidence to ensure that health care professionals adopt safe and effective attitudes to medication use in clinical practice.

Descriptions of educational interventions that have been implemented in a range of settings will be provided, along with guidance as to how they can be delivered. There will be references made to traditional teaching methods, such as mass instruction through didactic lecture-based teaching, although these can be criticized for promoting a passive approach to learning. Greater attention will be given to interventions that encourage learners to be active and engaged in self-reflection, increasing their preparedness to enter the clinical workforce. Educational interventions that support the professional practice of those who are established in their careers will also be described. Examples will include case-based learning, interprofessional learning, simulation, continuing professional development initiatives, and the role of service users and carers in the training of health care professionals. The evidence base for the effectiveness of these interventions will be discussed, drawing on research that has explored both their desired and actual outcomes. Conclusions will be drawn regarding future implications for educational and health care policy. At the end of this chapter, there is an expert commentary about the potential value of applying this interprofessional approach.

DEFINING THE PROBLEM

Errors can occur during the prescribing, dispensing, and monitoring of medication, as well as during administration to the patient, as described in Chapters 2 through 4. A range of professionals, notably doctors, pharmacists, nurses, and allied health care professionals (including those with prescribing rights), together with technicians and assistants, can be involved in these processes.

There is evidence to suggest that individuals entering the clinical workforce from undergraduate and pre-registration training may be ill-equipped with the knowledge and skills pertinent to patient safety and, more specifically, in relation to the safe use of medication. A recent study investigated the formal and informal ways pre-registration students from four health care professions in the United Kingdom (medicine, nursing, pharmacy, and physiotherapy) learned about patient safety in order to become safe practitioners (Pearson et al. 2009). It found that the definitions of patient safety and key learning topics in curricula reflected the professional group involved. For example, pharmacy focused primarily on medication errors and medicine was mainly concerned with diagnostic errors, prescribing, and high-risk procedures. Nursing programs addressed more broad issues of hands-on care, including safe drug administration. It was not common, however, for any pre-registration course to include teaching about underpinning causal factors relating to patient safety. Alper and colleagues (2009) highlighted that only a quarter of United States and Canadian medical schools had a specific patient safety curriculum, with the majority of these using lectures and small-group work to deliver the sessions. Following suggestions that junior doctors contribute disproportionately to errors in prescribing in the United Kingdom, the EQUIP study was commissioned to look at the causative factors of this (Dornan et al. 2009). As described in Chapter 2, unfamiliar working environments and prescribing charts, communication problems, workload, tedium, interruptions, and lack of knowledge were all identified as factors contributing to errors. In addition, there was a lack of a tangible "safety culture" for junior prescribers in relation to medication management.

DEVELOPING SOLUTIONS

Milligan (2007) suggested that, in order to shift health care towards a stronger patient safety culture, there was a need to change the way health care professionals were educated and trained. It was suggested that health care curricula should be changed to incorporate clear teaching on the importance of looking at whole systems, and by including human factors theory from the beginning of educational programs. It was proposed that the Human Factors Analysis and Classification System (HFACS), a human error taxonomy developed as a tool for investigating and analyzing the human causes of aviation accidents (Wiegmann and Shappell 2003), could guide the development of educational programs and enable students to understand the origins of system errors.

The World Health Organization (WHO) has also responded to calls for a stronger focus on patient safety by publishing a curriculum guide (World Health Organization 2011). A specific section that focuses on improving medication safety outlines eight main knowledge competencies that are expected of all health care professionals. The student should know:

- The scale of medication errors
- That the use of medication is associated with risk
- Common sources of error
- Where in the process errors can occur
- The responsibilities associated with prescribing and administering medication
- How to recognize common hazardous situations
- Ways to make medication use safer
- The benefits of a multi-disciplinary approach to medication safety

There has also been recognition that education around patient safety issues needs to continue throughout the health care professional's career. The Berwick report (2013) on improving the safety of patients in England highlighted learning in the workplace as a key recommendation to delivering better patient safety. The report recommended that the learning process should provide all health care professionals, involving lifelong education, and that the U.K. National Health Service (NHS) should become a learning organization to allow this change to occur. In addition to postgraduate education, standardized and reliable responses to errors with subsequent appropriate and actionable feedback to staff is necessary. The HFACS has also been highlighted as an appropriate methodology to collect and analyze data around serious adverse events and display critical risk factors to staff (Diller et al. 2014).

MEDICATION SAFETY IN UNDERGRADUATE CURRICULA

The integration of medication safety learning into health care students' curricula is a directive from the regulatory bodies of health care professionals internationally. In response to WHO's curriculum guide (World Health Organization 2011), the adoption of specific learning outcomes and standards for education focusing on medication safety has been encouraged. For example, the U.K. Medical Schools Council's Safe Prescribing Working Group (2008) recommended the adoption of the following learning objectives into undergraduate medical students' curricula to ensure that every new doctor has the appropriate knowledge and skills on their first day in clinical practice:

- The ability to establish an accurate drug history
- The ability to plan appropriate therapies for common indications
- The ability to write a safe and legal prescription
- The ability to appraise critically the prescribing of others
- The ability to calculate appropriate doses
- The ability to provide patients with appropriate information about their medication
- The ability to access reliable information about medication
- The ability to detect and report adverse drug reactions

Similarly, the British Royal Pharmaceutical Society (2014), in its publication of the Foundation Pharmacy Framework, has outlined the core roles and activities expected of a newly qualified pharmacist after undergoing undergraduate and pre-registration training. This framework is built on four clusters, and the first cluster, "Patient and Pharmaceutical Care," focuses on the need, provision, and selection of medication, as well as medication-specific issues, information, and patient education. The U.K. Nursing and Midwifery Council (2007) has also provided extensive guidance around the safety of medication administration.

In addition to these medication-focused learning objectives and competencies, educational interventions regarding medication safety should act to assist students to develop the skills and attitudes required by all health care professionals to enhance patient safety. They must, therefore, give sufficient attention to how human factors contribute to error (Russ et al. 2013). Safety competencies that may be used to identify these broader themes have also been developed. The Canadian Patient Safety Institute (Frank and Brien 2009) proposes that all health care professionals should:

- Contribute to a culture of patient safety
- Work in teams for patient safety
- Communicate effectively for patient safety
- Manage safety risks
- Optimize human and environmental factors
- Recognize, respond to, and disclose adverse events

To ensure teaching and learning adequately prepares students to enter the clinical workforce, it is important that students recognize the relevance of their training to their future practice and are taught in a practical way using realistic examples. The creation of supportive and effective learning environments that provide students with the opportunity to develop and apply medication knowledge and skills is the challenge for educators.

INTERPROFESSIONAL EDUCATION

It is acknowledged that a significant proportion of health care professionals' undergraduate teaching and learning is delivered in a uniprofessional environment. This allows them to develop knowledge and skills specific to their profession, which will be valued by and complementary to those of other professions when they enter clinical practice. However, there is also a growing emphasis among policy-makers highlighting the need for improved education and training involving shared learning for all health care professions (Medical Schools Council 2008; Nursing and Midwifery Council 2007; Royal Pharmaceutical Society 2014). In the United States, the Institute of Medicine called for recognition of the fact that, in practice, professionals work in interprofessional teams, yet they are not educated together (Greiner and Knebel 2003). They called for accreditation bodies to revise program standards and co-ordinate efforts across the professions.

The nomenclature differs between commentators on this subject, with the term "interprofessional" being used most frequently in the literature. Harden (1998) proposed that there are several steps between "isolation" (whereby each profession organizes its own teaching and is unaware of what is taught in other professions), through "nesting" (in which aspects relating to the work of other professions are included in otherwise uniprofessional courses) and "multi-professional" (where the emphasis in the course is on multi-professional education and each profession looks at themes from the perspective of its own profession), to "interprofessional" (whereby each profession looks at the subject from the perspective of its own and other professions). The Centre for Advancement of Interprofessional Education (2014) defines interprofessional education (IPE) as occasions when two or more types of health care professionals learn with, from, and about each other, in order to improve the quality of care.

The aim of IPE is to drive, a move away from the professional "silos" that are encouraged and exacerbated by the current system (training health and social care professionals separately) to a model that facilitates and encourages teams of health care professionals who collaborate to provide safe, effective, and evidence-based care. IPE is expected to be embedded within the curricula (Health Care Professions Council 2012) to ensure newly qualified graduates are aware of and understand how they can contribute to the broader team, in terms of delivering a high-quality service for patients, as well as to "prepare a collaborative 'practice-ready' health workforce that is better equipped to respond to local health needs." This is also an important aspect of the English Department of Health's "joined-up care" (2013), which looks to promote patient-focused attitudes to medication use.

There is also the potential for undesired outcomes, which need to be given equal consideration in order to avoid the reinforcement of negative stereotypes. Providing opportunities for students to work collaboratively with students from other health care professions should enable them to

- Improve their understanding of professional roles and boundaries
- Understand and value the expertise and values of other team members
- Enhance their communication skills with other professionals
- Experience working in a multidisciplinary team to solve clinical problems

- Adopt a patient-centered approach within an interprofessional team
- Adopt a willingness to share goals
- Understand and take on leadership roles

To ensure IPE is successful in delivering an effective learning experience, the teachers from the two or more professions involved need to be aware of what is covered by the other professions' curricula. Opportunities for IPE need to be carefully coordinated to ensure the interprofessional sessions have meaning and context relating to their uniprofessional teaching. The facilitators of the sessions need to be well prepared, represent the professional mix of the students, and be supported through the process (Barr and Low 2012).

In the United States, an expert panel outlined principles that should be considered when constructing IPE (Interprofessional Education Collaborative Expert Panel 2011). These were purposefully general to allow greater applicability and contextualization to the professions involved, and included the following:

- Community/population oriented
- Relationship focused
- Process oriented
- Linked to learning activities, educational strategies, and behavioral assessments that are developmentally appropriate for the learner
- Able to be integrated across the learning continuum
- Sensitive to the system context/applicable across practice settings
- Applicable across professions
- Stated in language that is common and meaningful across the professions
- Outcome driven

Teaching material used for IPE should use examples that are realistic, relevant, and simulate the complexity of decision-making experienced in the clinical working environment. Innovative approaches are needed to make patient safety issues "real" for students. Examples of IPE initiatives around medication safety include projects that have aimed to enhance students' knowledge of high-risk medications (those identified by the U.K. National Patient Safety Agency as contributing disproportionately to patient harm, such as opioids, anticoagulants, and insulin), as well as prescribing decision-making in common conditions and for "high-volume" prescription items, such as statins, antihypertensives, and proton pump inhibitors (Hardisty et al. 2014). In order to bring realism to the classroom, clinical cases should reflect the future roles and working environments of those who are participating. Table 16.1 outlines how case-based learning with a high level of clinical complexity, involving patients with multiple co-morbidities, can be designed to follow the journey of patients from their admission to a hospital, through the management of the acute presentation to their eventual discharge, exploring all prescribing decisions that need to be taken into account at each stage and the contribution of and interactions between professional groups. Such approaches evoke the challenges faced by individuals when they join the ward-based health care team as newly qualified practitioners and, therefore, may be particularly relevant to pharmacy, medical and nursing students during the

TABLE 16.1
Encouraging a Team Approach for Interprofessional Education to Explore the Prescribing Issues during a Patient's Hospital Stay

	Admission to Hospital	Managing Acute and Chronic Conditions	Discharge from Hospital
Learning objectives	The ability to establish an accurate drug history	The ability to plan appropriate therapies for common indications Deduce appropriate treatments for acute illnesses based on symptoms, signs, and investigations Review patients' existing medication in light of the current condition, appraise previous prescribing The ability to write safe and legal prescriptions Access information regarding prescribing and calculate appropriate doses Detect and manage adverse drug reactions	The ability to critically appraise the prescribing of others Ensure prescriptions meet appropriate standards, are legible, unambiguous, and complete Ability to review and cancel prescriptions if appropriate
Contributions by professional group	Medical students: discuss a patient's past medical history and work with other students to link these to the medications prescribed Nursing students: interpret information received from the patient and determine the implications of this for the administration of medications Pharmacy students: clarify medications prescribed and give advice regarding the therapeutic appropriateness and dosing	Medical students: interpret the results of investigations and communicate the findings and therefore a diagnosis to other students Nursing students: advise the group on the correct prescribing of regular and as required medications to give ward nursing staff adequate information to ensure safe administration Pharmacy students: calculate and review patient parameters relevant to the medication regimen, such as renal function and blood pressure	Work together to review the discharge prescription and ensure this is accurate and contains adequate information to ensure a safe discharge to primary care or alternative services
Human factors explored	Time and workload pressures Competing priorities and distractions System factors leading to inadequate or incomplete information due to a lack of shared medical records between primary and secondary care Sound-alike and look-alike medications Differing prescribing practices, stationary (e.g., prescription forms) and guidance between health care organizations and between primary and secondary care		

latter stages of their undergraduate training. Teaching sessions can be arranged to enable factors other than therapeutic knowledge and medication management skills to be explored, including dealing with time pressures, distractions, and incomplete information. This provides an opportunity for students to experience these issues and helps them to develop confidence and coping mechanisms prior to entry into the clinical workforce, an approach that is encouraged by those with an interest in the contribution of human factors to medication risk (Russ et al. 2013).

Stages of Training

Consideration should be given to when, in students' undergraduate training, IPE and patient safety teaching should be introduced. There is less evidence for the introduction of patient safety material early in professional curricula (Stewart et al. 2010), although some studies have explored students' experience of patient safety through IPE and found that receiving this content early in their education "appears to provide students with a better idea of their place in the overall health care picture and frames their future experiences in the context of patient safety" (Fuji et al. 2010).

In addition to examples such as that given in Table 16.1, which may be used at later stages of curricula, the basic principles of safe prescribing may also be introduced in the early stages of medical training (see Table 16.2). In order to achieve this, clinical cases may be developed around a single therapeutic area, which can be linked and integrated to uniprofessional teaching and the learning content covered during that stage of training. This enables knowledge gained through lectures or directed study, for example, to be applied in an IPE context, providing students with an opportunity to add practical prescribing skills to their academic knowledge of a subject and also raising awareness early in their training of medication safety issues and the roles of other professional groups (Ker et al. 2003).

Organizational and Contextual Factors

Providing an IPE opportunity for students requires educators to give attention to several organizational and contextual factors to ensure the success of any initiatives, as outlined in Table 16.3. Overcoming logistical barriers, including timetabling to enable the participation of the relevant professional groups and integrating IPE into the schedules of busy staff and students, are crucial to the implementation of such initiatives (Hammick et al. 2007; Nisbet et al. 2008; Thistlethwaite and Nisbet 2007).

Outcomes and Evaluation

There are a small number of studies demonstrating the impact of IPE on improving medication safety, all of which outline students' self-reported changes in knowledge and skills after educational interventions. Evaluation of the impact of IPE focusing on medication safety has shown it to significantly increase therapeutic knowledge in common acute and chronic disease states for medical, nursing, and pharmacy students in the United Kingdom (Hardisty et al. 2014). Students recognized the importance of IPE and demonstrated behaviors associated with interprofessional learning,

TABLE 16.2

Exploring Prescribing Skills around a Specific Therapeutic Area

A Woman with a History of Recurrent Urinary Tract Infections **Tasks**	**Learning Objectives**
A woman attends her general practitioner having noticed symptoms of burning pain when passing urine. Urine dipstick is positive for nitrites, protein, leukocytes, and hemoglobin. Empirical treatment is to be started with trimethoprim. *Write up the antibiotic prescription (on a primary care prescription form) using appropriate resources to determine an appropriate dosage and duration of treatment.* The following day, the patient's condition deteriorates and a diagnosis of pyelonephritis was made. Culture and sensitivity results are available that show resistance to the antibiotic prescribed empirically. *Using the information provided, write up a new prescription for the patient and discuss factors for choosing one antibiotic over another.* The patient's condition does not improve with the new antibiotic and the patient is admitted to the hospital. Blood results are available and require interpretation for the safe prescribing of the intravenous antibiotics gentamicin and ciprofloxacin. *Review the inpatient prescription chart to determine whether the antibiotics have been prescribed correctly.*	Deduce appropriate treatment for acute illness based on signs and symptoms and the results of investigations. Access and appraise the appropriate information sources to inform prescribing and the calculation of dosages. Ensure prescriptions are legally valid, legible, unambiguous, and complete in a variety of settings (primary and secondary care) and for oral and intravenous administration of medication. Critically appraise the prescribing of others. Interpret patient parameters relevant to prescribing.

TABLE 16.3

Organizational and Contextual Factors Needed for Interprofessional Education

Engagement	Funding and support
	Collaboration between educators and higher-education institutions
	Student recruitment
	Timing in curriculum and timetabling
Development	Case development and seminar content
	Piloting
	Learning objectives
Structure	Seminar length
	Group dynamics
Facilitation	Style of facilitation
	Facilitator recruitment and training
	Role modeling

such as an increased confidence and willingness to question each other, and recognized the ability of these behaviors in practice to improve patient safety. In the United States, undergraduate sessions bringing together medical and nursing students for a workshop on medication safety relating to pediatric medicine demonstrated that the self-reported scoring by the students increased for their knowledge and awareness of the topic (Stewart et al. 2010). Another pediatric-based workshop, involving medical and pharmacy students, demonstrated an improvement in attitudes towards IPE and in confidence when prescribing in the pediatric setting (Taylor et al. 2012). Further research is required to investigate whether IPE opportunities during undergraduate training will translate into fewer medication safety issues in practice.

SIMULATION IN MEDICATION SAFETY TRAINING

Clinical case studies can demonstrate the complexity of prescribing for chronic and acute conditions by replicating the decision-making processes required in a clinical working environment and highlighting human factors that can have implications for patient safety. However, these methods do not allow students to *experience* the potential impact of, and strategies for, coping with time pressures, fatigue, distractions, and cognitive overload (Frank and Brien 2009). Simulation, defined as "an educational technique that allows interactive and at times immersive activity by recreating all or part of a clinical experience without exposing patients to associated risks" (Maran and Glavin 2003), can allow the impact of some of these factors to be experienced by students.

Simulation as an educational modality is being used increasingly within preregistration health care courses, particularly around patient safety topics. It encompasses the use of role-players or part-task trainers, through to state-of-the-art patient simulators (Gaba 2004). Increasingly, simulation is used to replicate near-misses or serious untoward incidents and, by using patient simulators, it is possible to demonstrate what can happen with regard to medication errors in a safe environment. U.K. government policy has called for the greater use of educational technologies to provide opportunities for health care students to acquire, develop, and maintain the essential knowledge, skills, values, and behaviors needed for safe and effective patient care (Department of Health 2011). There is evidence to show that simulation improves generic skills, such as interpersonal communication skills, critical thinking, and clinical reasoning in complex care situations (Lewis et al. 2012).

The term "high fidelity," in relation to simulation, is used when a clinical situation is simulated to be as closely representative of "real life" as possible. In some definitions of high-fidelity simulation, it is assumed that, to achieve this, it must incorporate computerized manikins with realistic anatomy and clinical functionality (Cant and Cooper 2010), although standardized patient actors may be used if the health care intervention being taught does not involve physical or pharmacological intervention. There is evidence that the use of standardized patients is particularly beneficial when teaching issues such as consent or addressing psychosocial matters (Bramstedt et al. 2012).

Whether employing manikins or actors in educational interventions, significant attention must be given to the context of their use, to ensure that their benefit is maximized. Clinical scenarios need to be developed that use the functionality of the technology, but also promote the development of wider knowledge and skills. Table 16.4 gives a case example that may be used with a patient simulator to highlight specific medication safety issues, as well as providing students with an experience of the simulated clinical environment, with the aim of enhancing their resilience to its stressors. Simulation enables re-runs of the scenario if the students fail to identify and process the relevant information in a reasonable time frame. It also enables the demonstration of the consequences of medication errors if they are not rectified promptly. Simulation provides the educator with a powerful tool to purvey the importance of the learning outcomes and strengthens the students' understanding of the profession they will soon enter.

Evaluations of the use of simulation in the training of health care professionals show that students value simulation-based learning for its ability to "bridge the gap"

TABLE 16.4
Using Patient Simulators in Medication Safety Training

Dosage Error with an Opiate

Medication safety issue: in 2008, one of the U.K. National Patient Safety Agency's Rapid Response Reports (see Chapter 15) documented 4223 incidents involving the wrong use of opioids. Of these events, approximately 21% led to harm to the individual patient. The consequences of medication errors relating to opioids can be potentially fatal.

Step 1

Students are provided with the patient's past medical history, current medication, and details of the presenting complaint.

Step 2

A patient simulator is used to exemplify the presenting signs and symptoms of an opiate overdose.

Early recognition of the symptoms of opioid toxicity is of extreme importance so that swift action can be taken to prevent respiratory depression. The students are required to assess the patient's signs and symptoms; for example, pupillary examination would note pin-point pupils.

Step 3

Students are required to deduce the acute diagnosis from the patient's presenting symptoms, medications prescribed, and past medical history.

Step 4

Students are required to deduce the appropriate treatment for the acute condition by accessing and appraising the appropriate information sources to inform the prescribing and the calculation of dosages.

The choice of drug, dose, frequency, route of administration, allergies, and patient parameters all need to be considered in a short period of time to prevent the patient suffering from fatal respiratory depression. Naloxone can be administered via injection; discussion of which route and how to administer the medication and at what rate is conducted.

Step 5

Once the acute situation is stabilized, students are required to reflect on the causative factors (dosage error due to lack of therapeutic knowledge by another member of staff) leading to this acute event and how these could have been avoided/rectified.

between theory and practice (Weller 2004). Simulation has also been recognized by students to provide them with an opportunity to carry out tasks that they would not be able to do prior to qualification, particularly if practical placements in undergraduate training only allow observation rather than participation in key tasks. Nursing students, for example, have been shown to have limited access to medication management training during their clinical rotations. Simulation education around medication administration provides an opportunity to rectify this and has been shown to increase students' knowledge and confidence (Mager and Campbell 2013; Sears et al. 2010). Simulation-based learning has also been shown to translate into a reduction in medication error rates in clinical practice compared to lecture-style education (Ford et al. 2010).

ENGAGING SERVICE USERS AND CARERS

Regulatory bodies in the United Kingdom now expect service user and carer engagement to be explicit and embedded in all undergraduate courses for health care professionals. The Health and Care Professions Council (2013) has added service user and carer engagement into their requirements for the accreditation of programs and requests that higher-education institutes demonstrate this from the 2015–2016 academic year onwards. The NHS vision is that of the patient being the center of the services that they use (Darzi 2008; Department of Health 2010; Goodrich and Cornwell 2008) and, in order to embed this philosophy in pre-registration students, it is essential to provide opportunities for them to engage with service users/carers.

The service user or carer may have many different ways of interacting with students. Developing the communication skills relevant to medication use with undergraduate health care students can be done by incorporating service users and carers into teaching activities, with the aim of highlighting the patient's perspective on their medication, long-term conditions and other specific issues, such as adverse effects and polypharmacy.

POSTGRADUATE EDUCATION AND WORKPLACE LEARNING

All of the strategies described previously in this chapter, including traditional didactic lectures through to IPE and simulation, can be implemented in postgraduate education to support the professional practice of those established in clinical careers. In determining how and when educational initiatives should be employed, careful consideration is required of the needs of the learner and the factors that lead to medication errors in practice. It should also be recognized that, while undergraduate training is aimed at providing students with the underpinning knowledge and skills and the opportunity to see how these are implemented in clinical practice, the context of postgraduate education is very different.

Traditional education and training techniques are recognized as valuable when the goal is for individuals to become familiar with new knowledge or technologies. The challenge for educators working with qualified and experienced health care staff, however, is to provide educational opportunities that enable individuals to develop sustainable abilities that are appropriate for continuously evolving working environments (Fraser and Greenhalgh 2001). This requires approaches that address the features of the clinical working environment leading to medication errors. Lack of knowledge and skill is not the sole cause of error, with complex human and systems features contributing and, in many cases, actually being the predominant factors (Unver et al. 2012). Russ and colleagues (2013) describe situations in which further training and development may not address a patient safety issue. They suggest that training is inappropriate if it is attempting to change innate human characteristics or address an error that is occurring across multiple people, or in circumstances where individuals have been previously trained about a safety issue and the problem persists. Fraser and Greenhalgh (2001) argue that the focus of education for qualified professionals should not just be competence (knowledge, skills, and attitudes), but also capability, which is defined as the ability to adapt to change, generate new knowledge, and continuously improve performance. Alternative learning methods are needed to develop capability, and these must focus on processes. They give examples of process-oriented learning methods that can be used, which include:

- Experiential learning—shadowing, apprenticeship, and rotational attachment
- Feedback—responses that provide the learner with information on the real or projected outcomes of their actions
- Peer-supported learning groups—the small-group process used for mutual support and problem-solving
- Simulations—opportunities to practice unfamiliar tasks in unfamiliar contexts by modeling complex situations
- Team-building exercises—activities focused on the group's emergent performance rather than that of the individuals

There are limited studies that robustly evaluate the effectiveness of postgraduate educational interventions at improving medication safety. WHO's Good Prescribing Guide has been evaluated in randomized controlled trials in a variety of international settings and has been shown to improve knowledge, although it does not address the impact of human factors (Kamarudin et al. 2013; Ross and Loke 2009). In particular, there has been little research on educational interventions targeted at non-medical prescribers (Kamarudin et al. 2013). The use of simulation-based learning, for example, has been shown to provide significant advantages to patient care through the reduction of medication errors when used with qualified nursing staff caring for critically ill patients (Ford et al. 2010). There is a need for further research in this area. However, many of the suggestions for educational interventions and responses to errors, such as the application of the HFACS, have not been followed up with thorough evaluations to investigate whether improvements in safety can be gained.

CONCLUSION

There is widespread recognition from policy-makers, regulators, and educators that undergraduate training for health care professionals should place a strong emphasis on patient safety. Therefore, as medication contributes significantly to patient harm, medication safety should be a priority area. There are a number of educational interventions that may be employed to deliver teaching and learning around medication safety, including simulation, IPE, and the engagement of service users and carers. Many of these interventions have only been evaluated in terms of their ability to improve students' self-reported knowledge and skills and their awareness of risks. Further robust evaluation is needed to determine whether these translate into reductions in patient safety issues in practice. Although it is acknowledged that it is important to address human and organizational factors at an undergraduate level, these should also be the focus of postgraduate teaching and learning, as these have been found to be the predominant causative factors of medication use errors in clinical practice.

COMMENTARY

Maria Cordina
Associate Professor, Clinical Pharmacology and Therapeutics, Faculty of Medicine and Surgery, University of Malta, Msida, Malta

The obligation to ensure patient safety is generally well acknowledged by health care professionals. However, the factors that lead to a threat to patient safety are not necessarily appreciated by all. Effective educational strategies to address medication safety need to be guided by solid evidence. In countries where health service research is not well established (which I have seen in many Mediterranean, Central European, and Asian Countries), where the appropriate infrastructure for reporting medication-related incidents is not in place and where the philosophy of a "no blame" culture is lacking, it is rather difficult both to identify the problem and to establish the magnitude of the problem caused by medication errors. This therefore leads to a lack of evidence by which to define the problem within a local context. The absence of convincing evidence to present to policy-makers, regulators, and educators is a primary stumbling block to addressing the issue effectively. The inability to adequately and convincingly illustrate the impact of medication errors on patient safety may lead to an element of denial of the existence of the problem. This makes it tremendously difficult to convince the relevant stakeholders of the need to implement educational strategies to address an issue that is not perceived as being a significant or even actual threat to patient safety.

Educators at both preclinical and clinical levels for the different health care professionals involved in medication need to have the capability to recognize the problem fully and be prepared to commit to integrating measures to improve the safety of medication use as part of their curricula, as is described in this chapter. A first step for these institutions could be to identify and convince key educators from different professions, who may be already sensitized to the issue, regarding the importance of placing increased educational emphasis on improving medication safety. These educators in turn could act as multipliers, gaining the support of colleagues within their own profession and within their educational institution. This could also contribute to generating interest in medication safety across the different professions.

Using an interprofessional approach in designing the required curricula is highly desirable. While general concepts should be addressed, the specific culture and practices within the health care system/environment where the students will work need to be taken into account. The more the curriculum is tailored to meeting the requirements of a particular health care system, the more relevant it will be perceived by educators and students, thereby enhancing the probability of them engaging with the topic. This could be done by setting up a working group consisting of both educators and service providers from all health care professions. In a small country like Malta, this is relatively easy to achieve, since most of the time, the service providers are the educators. Then, the educators should design a curriculum that incorporates and prioritizes educational interventions to address the contributing factors they have identified as threatening patient safety within their health care system.

It is essential that both the educators from the different professions as well as the various student groups recognize the importance of their own contribution to the safety of medication use. This would enable them to assume ownership of the processes involved in ensuring patient safety in medication use in their future careers, thereby contributing to improved outcomes in medicines management.

17 Communication between Health Care Professionals

Patricia M.L.A. van den Bemt, Erica M. van der Schrieck-de Loos, and Fatma Karapinar-Çarkıt

CONTENTS

KEY POINTS

- Improvements in the transfer of medication information at hospital admission can be achieved by using as many sources of information as possible, by employing pharmacists or pharmacy technicians, and by involving patients and/or carers.
- Improvements in hospital discharge communication need to focus on patient education and involvement, the timely and accurate communication of medication information, and accurate documentation by the next health care professional.
- Technological solutions are promising, but still need further development and need to fit into a well-designed medication reconciliation process.
- Active patient involvement in medication reconciliation is essential, as the patient is the only constant factor among the numerous health care professionals involved.

INTRODUCTION

In Chapter 5, the importance of professional communication was explained by illustrating the difficulties associated with medication information transfer between different health care professionals. Medication discrepancies arise frequently at all levels of care transitions, putting the patient at risk for adverse drug events (ADEs). Important transitions occur at admission to and discharge from the hospital, and so many studies have focused on medication reconciliation at these transition points to improve continuity of care. Nonetheless, other transitions, such as to nursing homes or ambulatory care clinics, are equally important. The potential causes of errors can serve as the starting point for finding solutions to the problems, especially causes related to patients or information technology (IT).

This chapter will consider various solutions for improved communication of medication information. First, we will discuss solutions within the current, generally paper-based systems. Next, IT-based solutions will be discussed, subdivided into solutions for communication between professionals and solutions involving the patient. At the end of this chapter, there are two expert commentaries, both written by patients, who describe how the academic research presented resonates with their personal experience of living with long-term illnesses and experiencing medication use firsthand.

SOLUTIONS WITHIN THE CURRENT SYSTEM

Hospital Admission

On admission to a hospital, the current system of medication information transfer mostly relies on referral notes from general practitioners (GPs) and patients bringing either their medication itself and/or their medication lists. As mentioned in Chapter 5, the GP's information may be incomplete, due to multiple prescribers from different health care settings. The information gathered from the patient's own medication may also be incorrect, as patients do not always bring along all of their medication (Varkey et al. 2007), or may bring medication that is no longer in use or for which the dosage regimen has changed (Gleason et al. 2010). Medication lists brought by patients can also be incomplete or outdated (Green et al. 2010; Meyer et al. 2012). Including additional sources of information may, therefore, be a solution for better medication information transfer.

The community pharmacy records for the patient could be helpful, provided the patient generally uses the same community pharmacy. The patient may be given a role as the carrier of this medication record, after verification that the record is complete and accurate. This is feasible in cases of planned admissions. For unplanned admissions, the medication records may be transferred in other ways, such as by fax. Interpretation of community pharmacy medication records, however, is not always easy for other health care professionals, because they are, in fact, dispensing records. They often exclude over-the-counter medication, may not reflect recent adjustments made by prescribers, and may list medications that are no longer used by the patient. Hospital pharmacists or pharmacy technicians are better trained than doctors and nurses in the correct interpretation of these medication records and of the pitfalls in other sources used for assessing an accurate and complete medication overview of the patient. Therefore, improvement strategies for information transfer at hospital admission could involve a combination of expanding the sources of information to include the community pharmacy records and using dedicated hospital pharmacy personnel.

Several studies on the effects of this combined strategy have been performed. All show that it results in a reduction in unintentional medication discrepancies. For example, employing pharmacy technicians in preoperative screening resulted in a decrease in the proportion of patients with one or more medication discrepancies from 18.6% to 5.4% (van den Bemt et al. 2009a). The Dutch multicenter World Health Organization (WHO) High 5s study on standardization in medication reconciliation in acutely admitted elderly patients showed that a pharmacy-led reconciliation program resulted in a significant reduction of the proportion of patients with one or more unintended medication discrepancies from 63% to 22%, while a mixed strategy (with shared responsibilities between pharmacy and other health care professionals) showed no effect (van den Bemt et al. 2013).

Several systematic reviews drew the same conclusion (Kwan et al. 2013; Mueller et al. 2012; Spinewine et al. 2013), although good-quality studies are still needed to assess the effects of complete and accurate medication reconciliation, especially on clinically relevant outcomes such as ADEs, duration of hospitalization, or readmission rates. Furthermore, additional studies are needed to investigate whether dedicated and well-trained pharmacy personnel perform better than dedicated and well-trained doctors or nurses. The positive effect of pharmacy interventions may very well arise from the fact that, apart from training, pharmacy personnel may have dedicated time to perform detailed medication reconciliations, unlike doctors and nurses.

Involvement of the patient (and/or relatives) in the medication reconciliation process is also important. Patients are most likely to know which medicines from different prescribers they use and which medicines they have purchased over the counter (Urban et al. 2014), as patients and/or their carers are the only constant factors in the health care continuum. This is why obtaining information from patients forms an essential part of the standard operating procedure for medication reconciliation in the WHO High 5s project (Barber et al. 2014; Leotsakos et al. 2014). However, when looking at the medication reconciliation practices of four acute hospitals in the United Kingdom, Urban and colleagues (2014) found that patients were only involved in 50% of all medication reconciliations performed.

In conclusion, improvement of the transfer of medication information at hospital admission, within the current system, can be achieved by using as many sources of information as possible, by employing pharmacists or pharmacy technicians, and by involving the patient and/or carers.

Hospital Discharge

Pharmacists and pharmacy technicians also have a role in medication reconciliation at hospital discharge, as confirmed in a systematic review focused on handovers at discharge (Hesselink et al. 2012). The same applies to the role of the patient: medication reconciliations involving patients identify many more discrepancies than those without patient involvement (Karapinar-Çarkıt et al. 2009).

However, the process of hospital discharge differs from that at admission in a number of important ways. First, the role of the patient is different—on admission, the patient is mainly used as a source of information, whereas at discharge, the patient is mainly provided with information. While in the hospital, medication may be stopped and new medication added, and the patient needs to be informed about these changes and how to use the new medication using the teach-back method. The health care professional also needs to discuss whether newly prescribed short-term medications, such as laxatives, analgesics, and hypnotics, are still needed at discharge. Second, instead of gathering as much information as possible, the hospital now becomes the source of information. Staff need to communicate information to the next health care professionals involved in the patient's care, such as the GP, nursing home physician, or community pharmacist. Traditionally, discharge letters or summaries are used for communication between secondary and primary care, but many GPs experience a delay in receiving these (Karapinar et al. 2010). Furthermore, the information regarding medication is often limited or even missing in 2%–40% of discharge summaries, as was shown by Kripalani and colleagues (2007) and in two small Australian (Belleli et al. 2013) and British studies (Hammad et al. 2014). When medication information is present, it often contains inaccuracies, as was shown in a study of the quality of discharge summaries prepared by first-year internal medicine residents (Legault et al. 2012). One may argue that these inaccuracies are due to the inexperience of the residents, but in another study, the number of medication errors in discharge summaries was compared between doctors at various levels of medical training and no difference in error rates was found (Callen et al. 2010). The authors of that study blamed the inaccuracies on the fact that medication information in discharge summaries was transcribed from other sources, which is an error-prone behavior, even when electronic discharge summaries were used. In the United Kingdom, a large audit of discharge information included in 3444 discharge summaries revealed that only 67% of summaries provided medication information and even less (49%) provided information on therapy change (Hammad et al. 2014).

Finally, in order for the discharge communication to be effective, the information needs to be documented by the next health care professional in their patient records. As discussed in Chapter 5, only 22% of adverse drug reactions (ADRs) mentioned in discharge letters were documented by the GPs in their records (van der Linden et al. 2006). Lalonde and colleagues (2008) and Karapinar-Çarkıt and colleagues (2014) showed that, even when a detailed medication discharge plan was communicated to the community pharmacy, the prevalence of medication discrepancies remained high, with no difference when compared to usual care. A potential explanation may be the fact that community pharmacies do not document medication that has been discontinued, but only document information that is necessary for dispensing medication.

Improvements within the current system of hospital discharge communication need to focus on patient education and involvement, a timely and accurate communication of discharge medication information (preferably within 24 hours, or at least before the first contact with primary care), and the documentation of this information by the next health care professional. Patient education and involvement is covered in detail in Chapter 22. The latter two topics are discussed below in more detail.

Communication by Discharge Letters

Instead of manually transcribing medication information into discharge letters, hospital information systems (provided medication is electronically prescribed) should enable automatic electronic extraction of medication data into the discharge letter. Often, this is not feasible, so another solution would be to print the electronic extract (including reasons for medication changes), which can be faxed or e-mailed to the GP and community pharmacist. GPs mostly do not perceive it as a problem if discharge medication is communicated separately from the discharge letter, as long as they receive the information quickly and preferably by e-mail (Karapinar et al. 2010). Separating this information from the (often delayed) discharge letter could enable communication within 24 hours of discharge as part of the pharmacy-based medication reconciliation process. Finally, separating information on discharge medication may fit better with the current work processes, where medication information is documented in another part of the GP's electronic patient record than the medical information.

Documentation by the Next Health Care Professional

Instruction manuals on how to document the discharge medication information in community pharmacy systems have been tested in a small Dutch study (Karapinar-Çarkıt et al. 2014), but had little or no effect on the completeness of documentation, which was generally low. Another Dutch study also concluded that a transitional pharmaceutical care intervention alone did not improve the documentation of medication changes by either GPs or community pharmacists (Lefeber et al. 2014). Both studies confirm what was shown for the documentation of ADRs by GPs (van der Linden et al. 2006) and the documentation of medication by community pharmacists (Lalonde et al. 2008). As such, it seems that, in improving the continuity of care post-discharge, the documentation of information is the most difficult step to change. Documentation, of course, takes time, and without reimbursement for that time, it may be difficult to change current practice. Especially in this field, electronic communication between health care professionals may be helpful, and this is dealt with in the next part of this chapter.

INFORMATION TECHNOLOGY FOR COMMUNICATION BETWEEN HEALTH CARE PROFESSIONALS

By using shared electronic health records (EHRs) or medication records, health care professionals have access to the same medication information. Theoretically, this should result in a reduction of medication transfer errors (Cortelyou-Ward et al. 2012). Much depends, however, on how well the electronic medication record is kept up to date and actively used (Tully et al. 2013). In other words, before IT solutions are adopted, the process of medication reconciliation needs to be redesigned (according to the suggestions made earlier).

For an up-to-date electronic medication record, the following conditions are needed:

- All prescribers of medication for the patient need to have access to the shared EHR.
- It should be possible to include medication information that is provided by the patient.
- The documentation efforts need to be easily incorporated into routine clinical practice and no extra efforts should be necessary.

The second and third conditions cannot be met if a stand-alone electronic system is used for the sole purpose of medication reconciliation. The ideal situation would be either to choose one system to be used by everyone or to link together the information systems used by the different health care professionals, so that each professional can continue to use his or her own system. By linking them, all medication information would come together in one electronic medication record. Needless to say, this requires many technical and privacy issues to be resolved before it can be realized.

The ideal system has not yet been invented, which is reflected in the limited number of studies of interventions in this field. In 2007, Salemi and Singleton published a report on the use of computerized pharmacy data for medication reconciliation at hospital admission. Although this early study relied on printouts rather than electronically transferred data, a reduction in medication discrepancies was nevertheless seen after implementation. A few years later, another study investigated the effects of an electronic medication reconciliation application on unintentional, potentially harmful medication discrepancies between preadmission medication and either admission or discharge medication (Schnipper et al. 2009). The application used in that study integrated the medication information from several electronic sources (including two outpatient electronic medical record systems and discharge orders from two study hospitals). From the integrated medication list, it was possible to move medication to the inpatient medication record (to continue the medication), to add medication, or to change medication. The electronic tool was combined with process redesign, making physicians responsible for taking preadmission medication histories, pharmacists responsible for confirming the reconciliation process at admission, and nurses responsible for reconciliation at discharge. This combined intervention resulted in a decrease in potential ADEs of 1.44 to 1.05 per patient. The benefit was mostly achieved by one hospital, which succeeded in better integration of the medication reconciliation tool into the computerized prescriber order entry application, but numerous other differences between the two hospitals could have accounted for the different results (Schnipper et al. 2009). The proportion of potential ADEs remained high after the intervention, indicating that the medication reconciliation tool and process were still not perfect.

Even more disappointing results were reported in a study by Boockvar and colleagues (2010), comparing the communication of medication information between nursing homes and hospitals for patients with and without an EHR. No differences in medication discrepancies and ADEs between the groups could be demonstrated, possibly due to the lack of dedicated tools for medication reconciliation within the EHR.

The EHR can thus be seen as only one of the potential sources for medication information still requiring the usual medication reconciliation process to be performed, as was found in studies using either a shared EHR (Moore et al. 2011) or an integrated EHR linking a health record to a pharmacy record (Linsky and Simon 2013).

The problem that EHRs still contain only part of the medication information may be overcome by using one nationwide computerized medication record, as was piloted in Austria in the "e-Medikation" project (Ammenwerth et al. 2014). Although physicians and pharmacists confirmed the potential for patient safety improvement, the pilot project also revealed that many technical and patient consent issues needed resolving before the system would be sufficiently user-friendly.

All these studies clearly show that the requirements mentioned in the beginning of this part of the chapter have still not been met, and thus electronic medication information transfer is only in its infancy. Therefore, the contribution of interprofessional communication using IT to ensure safe medication information transfer is still limited. However, the experiences can be used to build more user-friendly, better-integrated systems that, one day, will be able to contribute to patient safety within the context of a well-designed medication reconciliation process. Recently, designs for improved medication reconciliation applications have been published, such as by Cadwallader and colleagues (2013). Their design displayed information from multiple sources, indicating discrepancies between the sources and sorting the lists into displays that were useful for clinicians. The application was designed in close collaboration with health care professionals, enabling developers to tune the system to the professionals' needs. An important part of their design was provision for patient input, recognized as a very important source for medication information, as mentioned earlier in this chapter. Several IT applications focus on this patient input, and therefore, we dedicate the last part of this chapter to such patient involvement.

INFORMATION TECHNOLOGY INVOLVING THE PATIENT

This chapter has repeatedly stressed the importance of active patient involvement in the processes of medication reconciliation, as the patient is the only constant factor during all transitions in care, whereas there are numerous health care professionals involved. The patient should always be included in medication reconciliation, but obtaining information can also be facilitated using IT. This would enable the patient to provide health care professionals with information other than through face-to-face contact. Although this may never fully eliminate the need for a personal interview, it will at least contribute to its efficiency. Several IT solutions involving patients have been explored: secure messaging, consumer-based kiosk technology, personal health records, and tablet computers will be discussed next.

SECURE MESSAGING

Secure e-mail provides safe electronic communication within a platform with restricted access. It has been used for sending electronic reminders to patients, either for appointments or to enhance adherence to therapeutic interventions. Another way to use secure e-mail technology is to use it to ask patients to bring their medication list for scheduled hospital appointments or admissions. This was assessed in a small study in which a secure e-mail was sent to 62 patients prior to a scheduled clinic appointment, instructing them to review and update their medication list on their personal EHR and bring a printout of the list with them (Burke-Bebee et al. 2012). Only 29 patients actually kept their appointment and were included in the study, and only five patients were asked by their physician about the e-mail. Two acknowledged having received it, and with such low patient numbers, it was not possible to draw any conclusions on the effects of secure e-mails on medication information transfer.

Heyworth and colleagues (2013) have further explored the use of secure messaging in a qualitative study among primary care professionals. Structured interviews with 15 health care professionals revealed that they were all in favor of using secure messaging and believed that secure messaging and virtual medication reconciliation could reduce ADEs. The same group also studied the implementation of a "Secure Messaging for Medication Reconciliation Tool" within a web portal in a pilot project of 60 patients being discharged from a hospital. After discharge, a secure e-mail with their medication list was sent to the patients, allowing them to perform medication reconciliation via the portal. As a result, 108 medication discrepancies and 23 potential ADEs were identified by the patients themselves (Heyworth et al. 2014).

Consumer-Based Kiosk Technology

One way of obtaining medication information from the patient through IT is to use consumer-based kiosk technology at the point of hospital check-in (Lesselroth et al. 2009, 2011). In U.S. studies of such kiosks, the patient used a program on a computer terminal to check in for a planned hospital admission. The program gathered medication lists from various connected Veterans Affairs (VA) electronic sources and presented the composite list to the patient. The list contained both text and pictures of the medication to help the patient verify the information. The patient checked a box alongside each medicine in order to confirm that the medication was "taken as directed"; a free-text box was available to add comments about the medication (for example, how it was taken if not taken as directed). Additionally, a free-text box was presented in order to enter medications not presented on the composite list. The information entered by the patient was then checked in an interview. The introduction of the kiosk technology led to an almost 50% reduction in time needed for medication reconciliation, while the accuracy did not change (Lesselroth et al. 2009).

Personal Electronic Health Records

As early as 2008, the contribution of patients to updating their medication list via a secure web-based patient portal to an EHR was studied (Staroselsky et al. 2008). The EHR was designed for primary care use and included the ability to store health care professional notes from primary care encounters, results of diagnostic tests, family history, medication lists, and automated physician reminders for preventive tests and procedures. Through a dedicated web-based patient portal, patients were allowed access to their own EHRs, enabling them to view their medication and allergy lists and to perform administrative tasks such as scheduling appointments or requesting referrals. Medication records could not be corrected directly by the patient, but a request to do this could be sent to the doctor.

Eighty-four patients from a primary care practice using the web-based portal were compared with 79 patients from the same practice not using the portal. The results were disappointing: no differences in the proportion of medication discrepancies between the two groups were found and patients notifying physicians of discrepancies did not result in the latter updating the medication list in the EHR. The authors stated that this could potentially be improved if more efficient ways to process the information were presented to the physician, as well as by clarifying the benefits.

A few years later, Turvey and colleagues (2012) explored the interest of patients in transferring and sharing information from personal EHRs. Two samples of patients (25,834 and 18,471, respectively) participated in a web-based survey. Almost half of patients in the first sample reported printing information from their record and 20% reported saving information on their local computer, but only 4% ever sent information from their EHR to another person. About a third of patients in the second sample reported entering medication information themselves, and 18% shared this information with their VA health care professional. Only 10% of patients shared the information with non-VA health care professionals. The authors concluded that patient participation needed to be promoted in order to improve these results.

The largest quantitative study on the effect of a personal EHR on medication accuracy was performed by Schnipper and colleagues (2012). In 11 primary care practices, 1761 patients were allocated to the intervention group and 2218 to the control group. The control group received care as usual, whereas in the intervention group, a personal EHR was used, as in the study described above by Staroselsky and colleagues (2008). In a sub-study, 267 patients in the intervention group were also asked to prepare medication lists prior to a primary care physician visit and were compared with 274 patients from the control group. These lists allowed the patients to view their medication in the personal EHR and to indicate updates to their medication and allergies. The proportion of medication discrepancies was 42% in the intervention group and 51% in the control group; potentially harmful discrepancies were reduced significantly in the intervention group (Schnipper et al. 2012).

Tablet Computers

The popularity of tablet computers has inspired medical researchers to explore this new technology for all kinds of medical applications. One example is patient engagement during their hospital stay in order to improve transitions of care, for which 30 patients were included in a pilot study (Greysen et al. 2014). Through the tablet computer, an online educational health module could be accessed, as well as a personal EHR. Patients could use the tablets for 3–5 hours, after which they were collected again and a survey was performed. Almost all patients were satisfied with the tablets. Basic orientation took little time, with 15 minutes or less being needed for 70% of participants; older patients took a little longer than younger patients in this regard. The majority of patients, despite being first-time users, could perform tasks such as accessing their medication list, verifying appointments, or sending secure messages to their primary care professional.

Summarizing all of the studies on IT involving the patient, it seems that the results are somewhat conflicting, as also suggested by a systematic review on electronic patient portals in general and not just those dedicated towards medication information (Goldzweig et al. 2013). The technologies discussed are certainly promising regarding their contribution to safe medication transfer, but still need further development, and they need to fit into a well-designed medication reconciliation process. Additionally, optimizing patient participation needs to be explored, as the use of patient portals may be dependent on specific patient characteristics, such as their level of education (Goldzweig et al. 2013). Thus, what was concluded for the use of IT by health care professionals ("the ideal system has not been invented yet") also applies to its use by patients.

PUTTING SOLUTIONS INTO PRACTICE

In this chapter, we have discussed many potential interventions aimed at improving communication about medication information at points of transfer. We have seen that practical solutions that will likely result in the greatest improvements in the short term are pharmacy-based medication reconciliation programs and programs improving discharge summaries, the documentation of medication information, and patient involvement. These are generally interventions that would involve employing additional and/or dedicated staff, making them relatively costly for many institutions. This probably explains the ongoing search to find more efficient ways of improving medication information transfer, which is directed towards IT support for the medication reconciliation process. While this certainly holds high hopes for the future, the current IT systems have not yet succeeded in optimal medication information transfer.

Whether using the short-term practical solutions to improve medication reconciliation or focusing on the future of IT, the essential role of the patient and/or carer should always be taken into account. For all solutions, it is clearly shown that the patient is the ultimate information carrier, and the most appropriate way to conclude this chapter is by stressing this universal solution to most medication reconciliation problems: invite and involve your patients!

COMMENTARY (I)

Fran Husson
Patient, United Kingdom

Reading this chapter as a patient, I was struck by how very authentic the arguments and the evidence presented sounded. Communication failures concerning my medications have taken place at all points of my medical history—hospital admission, discharge, and GP surgery—and on a repeated basis. While over the last 10 years the main hospital looking after me has made fantastic improvements in providing clear and meaningful outpatient follow-up letters to both GPs and patients, patients do not get to see if and how this information is recorded by their various care providers and are unable to check whether it is correct.

The authors quite rightly point out that "the patient is the only constant factor during all transitions in care, whereas there are numerous health care professionals involved." However, the patient is not always aware of his/her role and status in their medical (and medication) journey. This is why one hospital trust in England worked with patients such as myself to develop a handheld patient record called "My Medication Passport" to help patients become more aware of what medicines, over-the-counter remedies and supplements they take (and why), and to have a place to keep this medication list up to date. This will, in turn, help healthcare professionals deliver accurate medication reconciliation, as well as provide a focus for a patient-centered medication review. The patient, consciously or not, is the carrier of important medication information, which initiatives such as My Medication Passport can formalize. The use of IT will potentially facilitate any sharing of information in the future, but as a patient, I wonder whether we are ignoring another important factor: the patient's relationship with medication in general and their own in particular.

Because of sea-change quality improvements in pharmacology, I wonder whether patients have become complacent towards medicines and, to some extent, ignore the risks of medication interactions, side effects, and severe reactions. The expectation of many members of the public that antibiotics are required for simple coughs and colds is an example of this. I fear that unless we find ways to redress this situation, to help people start seeing medicines as strong and potentially dangerous therapeutic agents and to help them consider what they can do to help in this process, it will be difficult to achieve significant improvements in medication communication.

COMMENTARY (II)

Carolyn Gamble
Patient, Canada and United Kingdom

I can relate as a patient to the issues outlined in this chapter—from information exchanges at admission and discharge in a hospital setting to the management of multiple medications in primary care. I have multiple conditions that require safe medication management. I currently receive care from primary and secondary health care providers, including alternative dispensaries for specialist medications that I receive from a company that delivers the medication to my home.

I am a Canadian, but have been living in the United Kingdom for the past 10 years. I was initially diagnosed with a musculoskeletal disease (ankylosing spondylitis) in Ontario, Canada, while still under the care of my parents, at the young age of 17. I did not have a real sense of "responsibility" for *anything* in life yet, let alone understanding what "patient responsibility" entailed. Communication from the rheumatologist was limited. I did not speak with my family doctor about it at the time, and I did not take the medication I was prescribed, as it made me feel unwell.

I had no concept that my future would be filled with the management of multiple medications—in a different country, navigating through a healthcare system that was foreign to me.

My disease progressed rapidly when living in the United Kingdom, and I started to receive aggressive medication treatment under the care of a rheumatologist. I was away from family support and I realized quickly that *I* was the one who had to take responsibility for the management and safe use of multiple medications received from various clinicians—if I did not take responsibility, then who would?

I took it upon myself to understand what prescriptions I was taking and how to take the medicines safely. I wanted to fully understand their function and role in disease management and the associated risks. I educated myself on side effects and which over-the-counter medications were also safe to take in conjunction with my prescriptions. Overall, my health care has been excellent. However, I can think of a number of occasions that relied on me, the patient, to provide essential medication information to the clinician to prevent an adverse event. I believe that all patients (or their carers/relatives) should understand the overall medication requirements of their own condition/situation, and be clear on what actions to take.

Reading this chapter, it is clear there is no "single point of truth" where all my medication information is stored for all health care professionals to access. The current system is reliant on people manually reproducing and resending information on multiple occasions, which leaves room for human error, and patients can suffer from time delays.

The introduction of shared electronic records should help improve communication, but should never absolve patients of their personal responsibility to have an overall view of requirements. If electronic shared healthcare records were in place, I would be confident that all my healthcare information was stored in one place, available for all relevant parties to understand the overall requirements of my treatments. A shared electronic information approach involving the patient would make me have more trust and confidence that the system is accurate. This would enable me to feel more relaxed and my patient responsibility would feel less onerous.

18 Interventions for Safer Work Systems

Johanna I. Westbrook, Gordon Caldwell, and Bryony Dean Franklin

CONTENTS

KEY POINTS

- Improving error awareness, more effective teamwork, and providing adequate learning opportunities can improve the "people" component of the health care system.
- Double-checking and enabling organizations to learn from workarounds can improve the safety of the tasks that people perform.
- Effective improvements in tools and technology can be relatively "low tech," such as improving prescribing stationery or access to information.

- Within the environment, reducing interruptions and distractions for health care professionals may reduce errors in their practice.

INTRODUCTION

There have been a great many studies exploring a very wide range of approaches to improving the safety of medication use. In this chapter, we consider some examples of approaches to developing safer work systems; these are envisaged as being complementary to other solutions presented elsewhere in this book. The chapter is based around four of the five structural components of the work system described in the Systems Engineering Initiative for Patient Safety model (Carayon et al. 2006) presented in Chapter 11: people, tasks, tools and technology, and the environment. The fifth component, the organization, is outside the scope of this chapter.

PEOPLE

As highlighted in Chapter 11, people are central to any work system and interact with all the other components. Here, we focus on the specific issues of error awareness and teamwork, before briefly considering some implications about learning.

Error Awareness

Increasing health care professionals' awareness of the risks of medication use is likely to be helpful in their mitigation. As in most situations, complacency can be dangerous. Error awareness can be considered in three categories:

1. A general awareness that errors do occur, how and why they occur, their consequences, and some general principles for prevention
2. Awareness of specific errors that have occurred locally or in similar situations
3. Awareness of one's own errors, providing opportunities for self-checking, personal reflection, and change

General Awareness

All health care professionals should be aware that medication errors are common, a significant proportion is undetected, and many of those that are detected are unreported. This should be addressed in undergraduate and postgraduate education, as well as being a topic of ongoing and open discussion in the workplace.

Errors of Others

Awareness of others' errors in similar settings is likely to provide more specific opportunities for learning and action at organizational, team, and individual levels. This is part of the rationale for incident reporting systems (Chapter 15) and underpins the way in which safety cultures within the organization develop and improve (Chapter 13). On a more local level, facilitating shared learning may involve sharing details of common or serious errors, together with tips on how to avoid them, and inviting other suggestions for prevention. This may be via email bulletins, posters, team meetings, screensavers, and other media. Learning from patient safety incidents can also be incorporated into formal training programs, an example of which has been shown to deliver significant improvements in patient safety knowledge, skills, and behaviors among junior doctors (Ahmed et al. 2014).

Awareness of One's Own Errors

Unless an individual is aware of their own errors, they are unlikely to be able to reflect on the reasons for these errors or change their practice to prevent recurrence. Interviews with prescribers suggest they underestimate their risk of prescribing errors, perhaps as a result of limited feedback when errors are detected (Baysari et al. 2011a). Where there are formal second checks of any stage of the medication process (as described later in this chapter), it is important that any errors identified are fed back to those responsible. This should be in a constructive manner, encouraging reflection on how the error could have been prevented, either by the individual or by making more systemic changes.

In relation to prescribing, pharmacists in both hospital and community settings often identify prescribing errors and ensure these are rectified, but may not always feed back to the individual prescriber concerned (Bertels et al. 2013). This may be because of a lack of confidence or opportunity. With paper-based prescribing, signatures and any associated prescribers' names may also be illegible. In a small, tightly knit team, a person's initials or scrawled signature may be recognizable and errors can be fed back to the original prescriber. However, in increasingly large and diffuse teams, staff will encounter prescriptions or medication orders written by prescribers with whom they are not familiar. In addition, even when they can be identified, the doctor who made the error may no longer be on duty, and another doctor would have to make the necessary amendments to ensure patient safety. This creates challenges in verifying authority to prescribe, as well as in letting prescribers know about any errors made. The use of name stamps, in parallel with campaigns to improve prescriber identification, has shown some evidence of benefit (Reynolds et al. 2014). Electronic prescribing systems resolve some of these problems regarding prescriber identification and present new opportunities to provide feedback to prescribers (Baysari et al. 2014).

After receiving feedback, the health care professional has to decide what to do differently in order to change their behavior. Providing feedback alone has been shown to be less effective than feedback that includes both explicit targets and an action plan (Ivers et al. 2012). Implementation intentions, or "if–then" plans, have been shown to be effective in changing behavior in general (Gollwitzer and Sheeran 2006), and preliminary work has shown that workshops based on these psychological theories may be helpful in improving prescribing safety (Tully et al. 2014).

Another approach that has been used in the hospital setting is a more detailed real-time audit of prescription writing, with immediate correction of any deficiencies. Such a "check-and-correct" approach (Caldwell 2010) has been used during ward rounds, focusing on identifying problems such as illegibility, use of abbreviations, not specifying indications and durations for antimicrobials, and failure to document allergies clearly. A member of the team is allocated the role of checker and highlights any problems verbally to the rest of the team. This can also support a safety culture in which team members gain experience in speaking out when they identify poor practice. An evaluation suggests these approaches bring about some improvement in prescription writing (Lépée et al. 2012).

Being aware of the situations in which errors are likely to occur can also support personal self-checking practices, whereby an individual formally rechecks their own work. In a study of the causes of prescribing errors, several doctors suggested that, when they had written the prescription, something "had seemed wrong"; however, they could not identify the problem at the time (Dean et al. 2002a). This inner "warning bell" could serve as a cue to perform a self-check, or ask another health care professional for a second check, as discussed later in this chapter.

Teamwork

Some aspects of teamwork—focusing on interprofessional communication—have been considered in Chapter 17. There is limited existing evidence regarding the role and extent of communication in terms of medication management among clinical staff. Studies employing social network analysis techniques have demonstrated that communication between doctors and nurses about medications issues is limited, but that specific staff, such as ward pharmacists, often act as central information hubs (Creswick et al. 2009; Creswick and Westbrook 2007, 2010). Senior doctors on hospital wards have been identified as being on the periphery of communication networks regarding medication advice provision, indicating that they are rarely sought out or available to provide medication-related advice to support junior doctors (Creswick and Westbrook 2015). As such, strategies to improve the advice-giving networks between senior and junior physicians may be a fruitful area for intervention to improve medication safety. Doctors and nurses surveyed on hospital wards believe that improved communication between staff regarding medication issues could reduce errors. An Australian study showed that strong medication communication networks among ward clinical staff, measured using social network analysis, were associated with a significantly lower prescribing error rate (Creswick and Westbrook 2015).

Other important skills include providing high-quality handovers—both written and verbal—and encouraging a culture of speaking up about any concerns. A specific issue relating to prescribing in hospitals is that the person who makes the prescribing decision is often not the person who actually writes the prescription (Baysari et al. 2011d; Coombes et al. 2008b; Ross et al. 2012). This also means that decision support incorporated into electronic prescribing systems may not be seen by the decision-maker (Baysari et al. 2011d). Prescribing training should therefore target all grades of staff, empower junior doctors to seek clarification from their seniors if needed, and encourage senior doctors to be more supportive of their juniors' needs for support (Ross et al. 2012). Another approach to enhancing teamwork is to ask participants on ward rounds to sign in as part of the team. Team members commit themselves to active participation in the round and to be vigilant for and speak out about errors. This also provides an opportunity for brief introductions to confirm the roles of all present.

Implications for Learning

Educational interventions are discussed in detail in Chapter 16; here, we consider the implications of how health care professionals learn, particularly within medicine. Newly qualified doctors often report feeling insecure when writing prescriptions for hospital patients, despite their time spent studying at medical school (Coombes et al. 2008a; Han and Maxwell 2006). Their theoretical knowledge and conceptual understanding of pharmacology may not be matched by expertise in choosing and safely prescribing medications in practice. In the United Kingdom, for example, prescribing and many other skills were historically developed according to an apprenticeship model, both during medical school clinical attachments and after starting work as a junior doctor. However, this model has changed over time due to concomitant changes in both health care and education. More recently, U.K. medical schools have recognized that final-year students were not well prepared for practical prescribing, with consequent risks to patient care. Prescribing examinations, objective structured clinical examinations, and simulations have subsequently been included in final-year examinations, or very early in the first year as a junior doctor. However, it can be suggested that these assessments often resemble memory tests rather than reflecting the realities of actual practice. Many are closed-book, whereas in practice, doctors would be expected (and indeed encouraged) to use reference sources, either paper or electronic. Boxes 18.1 and 18.2 summarize reflections from two of the authors in this book on learning the skills needed for safe prescribing under both the traditional apprenticeship model and within a modern medical school.

BOX 18.1 TRADITIONAL LEARNING OF SAFE PRESCRIBING

I learned the importance of correct doses and the preparation of drugs in the 1970s, starting in school in chemistry and biochemistry experiments. An error in the dilution of an active agent could ruin an experiment. Later, in medical school physiology experiments, I saw the effects of tiny changes in drug concentrations on contractions of guinea pig ileum. Success in these experiments depended on the careful choice of agents and meticulous preparation and administration, coupled with careful observation and note-keeping; failure seemed easy to achieve. I learned that almost obsessive care is essential when dealing with any active agent. Of course, at medical school, I also studied pharmacology and therapeutics, learning groups of drugs, mechanisms of action, side effects, and interactions. These classroom studies equipped me to pass written exams, but did little to prepare me for my first day in work, prescribing for acutely unwell patients. Feeling ready for that first day on the wards was a result of the apprenticeship style of learning used in my medical school. As medical students, we were expected to participate in the work of patient care and to be taught by the rest of the medical team in return. Three tasks prepared us for future prescribing:

1. We were expected to clerk as many admissions as possible, including clarifying the list of medications that the patient was taking before admission. This helped us familiarize ourselves with common medications and doses.
2. Whenever the current paper medication chart became full, it was the student's job to draft a rewritten drug chart clearly and legibly, and have it checked and signed by a qualified doctor. In this way, we became familiar with the layout of the medication chart, as well as the commonly used medications. Poor legibility was considered a sign of disrespect.
3. Students drafted the discharge communication, including the discharge medications, again before checking and sign-off by a doctor.

We were constantly reminded of the potential side effects of medications because we were taught to always consider iatrogenic side effects as a possible differential diagnosis in a "surgical sieve" clinical thinking process. The number of effective medications in use in the 1980s was much smaller than now, and it was relatively easy to become familiar with a small armamentarium of therapeutic drugs. When we were unfamiliar with medications, we always had the British National Formulary in one of the large pockets of our white coats.

Dr. Gordon Caldwell
Worthing Hospital, Brighton, United Kingdom

BOX 18.2 MODERN DAY LEARNING OF (SAFE) PRESCRIBING

In August 2014, I was entrusted with the responsibility of prescribing for 30 patients on a respiratory ward. On my first day as a junior doctor, with no senior support, I was asked to prescribe medicines that I had not even heard of in medical school. The learning curve has been steep and, in the first few weeks of my professional career, I have grown increasingly confident and (hopefully) more capable as a prescriber. Most of my learning has occurred "on the job," as opportunities to participate in prescribing activities, such as those mentioned in Box 18.1, were limited during medical school.

My research background has heightened my awareness of prescribing errors and important issues such as antimicrobial resistance, but despite this, I have made mistakes. As far as I am aware, none of these have caused serious harm to my patients, but nonetheless, suboptimal prescribing can have a detrimental effect on patients and colleagues and the potential for a more serious adverse event to occur is ever present.

The reasons for my prescribing errors are both varied and complex. Errors generally arise when I am not able to recognize my underperformance in a given situation; a hurdle that can be overcome by receiving feedback and by being able to recall and act upon it in future situations. In addition, technology, while undoubtedly helpful in many situations, can also be a hindrance. Our discharge prescription software auto-selects the most common dose and timings for many frequently prescribed drugs and, just like the auto-correct on my overenthusiastic smartphone, can lead to slips.

Unfamiliarity with medications made me feel uneasy during the first few days of being a prescriber. However, on reflection, these drugs seem to be the ones I am least likely to prescribe incorrectly, as these are known unknowns, and I recognize the need to be cautious and seek help with these. Although today's doctors can no longer accommodate the British National Formulary in their pockets, a smartphone (provided we can spell correctly) or a friendly pharmacist will usually come to the rescue.

Dr. Lucy McLellan
Department of Educational Development and Research,
Maastricht University, Maastricht, the Netherlands

TASKS

Here we consider task-related factors, with a focus on double-checking and learning from workarounds.

Double-Checking

Double-checking of medications prior to administration is a widely used strategy designed to reduce the risk of error (Alsulami et al. 2012; McLeod et al. 2014); here, we provide an overview and evidence for its effectiveness in preventing error.

Double-checking refers to a process in which two health professionals independently check nominated components of the medication use process. While double-checking may apply to prescribing, dispensing, or the administration of medications, the majority of the literature relates to nurses preparing and administering drugs. This part of the chapter, therefore, focuses on nurses' double-checking, although the principles are likely to apply to other contexts.

Independence in the checking process is viewed as critical to its effectiveness (Baldwin and Walsh 2014; IMS Institute for Healthcare Informatics 2013). If two nurses check together, there is a risk of bias in that the second nurse will have a tendency to see what they expect to see (Alsulami et al. 2012, 2013). Variants of double-checking that are subject to bias include two nurses simultaneously checking together, one nurse watching the second nurse, or one nurse showing a second nurse (Dickinson et al. 2010).

There is often confusion among staff as to what double-checking involves, and professional guidelines and hospital policies often fail to clearly describe the process (National Patient Safety Agency 2007; Ramasamy et al. 2013). For example, the

American Society of Clinical Oncology/Oncology Nursing Society safety standards (2013) for oral chemotherapy state that, during drug preparation, "A second person … independently verifies each order for chemotherapy before preparation, including confirming: two patient identifiers…" and, for chemotherapy administration, "At least two practitioners or personnel approved by the practice/institution to prepare or administer chemotherapy, verify the accuracy of: drug name, dose, volume…" However, the standards provide no guidance regarding how this verification should occur. Surveys demonstrate inconsistencies in nurses' and pharmacists' understanding of what is meant by double-checking. Conroy and colleagues (2012), in a survey of 105 pediatric nurses and pharmacists from 69 U.K. hospitals, found 34% of respondents incorrectly explained double-checking as "One nurse performs the task, the other one checks her or his work." Only 40% correctly identified the important element of independent checking. Double-checking and verification policies and guidelines should thus be explicit in stating how, when, and with whom checking is to be undertaken.

Effectiveness of Double-Checking in Preventing Medication Errors

Despite the widespread use of double-checking (Conroy et al. 2012; McLeod et al. 2014), there is limited research evidence to demonstrate its effectiveness in reducing medication errors and harm. A systematic review (Alsulami et al. 2012) identified only one controlled study that quantitatively compared error rates with double- versus single-checking (Kruse et al. 1992). The study involved three hospital wards (two intervention and one control). In period 1, one intervention ward was randomized to a single-check procedure, while the second used a double-check procedure. In period 2, these allocations were reversed. In both periods, the control ward used a double-check procedure. Medication administration errors were identified using an audit of medication charts as well as incident reports. The researchers found a small but statistically significant difference in medication error rates (2.12 errors per 1000 medication administrations for double-checking versus 2.98 for single-checking). Other studies are based on self-reports or incident reports as the primary source of outcome data and are therefore of limited value.

In the absence of strong evidence of effectiveness, at face value, a double-checking process would appear theoretically safer than one single check. Double-checking builds redundancy into the system and is a common approach used across many industries. However, there are some suggestions that such redundancy may have the opposite effect to that which is desired. It has been argued that diffusion of responsibility may result in what has been labeled as "social shirking," whereby an individual's responsibility is diffused by the presence of another person (Sagan 2004). In support of this theory, Sagan (2004) presents evidence of crime reporting, where it has been found that if there is a single witness to a violent crime, there is a 75% chance that it will be reported, but in the presence of multiple witnesses, the likelihood is decreased by 15%. In health care, similar effects may manifest with an individual's perception of personal responsibility influencing their compliance and diligence in the double-checking process. As such, double-checking may hypothetically result in feelings of reduced responsibility and, as such, predispose to error (Alsulami et al. 2012). Linden and Kaplan (1994), following a review of blood transfusion errors, suggested that this effect may have operated, resulting in a potential increased risk of error. However, substantial evidence to support this theory in relation to double-checking in medication administration remains elusive. Qualitative studies have demonstrated that many nurses believe that double-checking is a helpful process (Alsulami et al. 2012), but equally, nurses have reported that they would be confident with single-checking (Jarman et al. 2002; Winson 1991).

Compliance with Double-Checking

Multiple internal and external contextual factors are likely to contribute to the extent to which checking procedures will be effective in achieving the desired goal of preventing errors (Shillito et al. 2010). These include an individual's knowledge of the checking process and whether they remember to complete all the necessary steps, particularly within interruptive environments. Time pressures with multiple competing priorities may also result in shortcuts to procedures (Manias et al. 2005). Deference to authority has also been reported (Armitage 2008), whereby double-checking the work of a more senior or expert staff member may result in the second checker being less likely to question the perceived expert. There is limited evidence of the extent to which staff comply with double-checking policies. In one study, Alsulami and colleagues (2014) observed 2000 drug administrations by pediatric nurses in a U.K. children's hospital. They identified 11 steps required for accurate drug double-checking (such as checking for the correct drug, dosage formulation, and drug route), with an additional four steps for intravenous medications (such as checking the volume and rate). Examination of each of these checking stages showed high compliance, with most exceeding 90%. An exception was for drug dose calculations, where independent checking occurred in only 30% of doses.

Resources Associated with Double-Checking

Substantial resource costs are associated with double-checking procedures. Two nurses are required and time is consumed by nurses seeking out a second nurse to participate in the checking process, and this will often involve interrupting a nurse, which itself may pose a safety risk (Westbrook et al. 2010a, b). Few studies have quantified the impacts of double-checking policies on nurses' time. In a 1-week time and motion study, Kruse and colleagues (1992) estimated that changing to a single-checking procedure could save around 17 hours per 1000 medication administrations.

Recommendations in Relation to Double-Checking

Checking processes are fundamental to many aspects of health care delivery, from correct patient identification to counting surgical instruments. However, the available evidence suggests that compliance with the independent double-checking of medication administration is hindered by a poor understanding and clear guidance of what is required, environmental factors such as time pressures, the interruptive nature of hospital wards, the complex hierarchical nature of clinical roles, and resource limitations. More recently, the value of checklists (Gawande 2009; Winters et al. 2009) as an effective safety strategy has highlighted the potential value of systematic checking processes more generally in preventing errors. Designing detailed checklists for double-checking has been suggested as one way to guard against some of the more frequent problems of limited guidance, missed steps, and inaccurate completion (Institute for Safe Medication Practices 2013). For example, checklists that detail the specific components of the medication order that should be compared against the drug label may support improved compliance. Thus, double-checking should be used judiciously, with a clear understanding of its challenges and limitations (Institute for Safe Medication Practices 2013). Strategies may include prioritizing certain patient groups or medications for double-checks and formally recording double-checks to ensure that both parties take responsibility for the process, while ensuring that double-checking is not relied on alone as a safety intervention.

Learning from Workarounds

Workarounds are deliberate actions that do not follow the explicit or implicit rules, assumptions, workflow regulations, or intentions of system designers (Debono et al. 2013; Koppel et al. 2008b). Many examples of workarounds have been reported for medication systems (Koppel et al. 2008b; McAlearney et al. 2007; Rack et al. 2012). They typically occur due to problems in system design, and identifying workarounds, therefore, provides an opportunity to highlight such problems and identify corresponding solutions. Chapter 10 presents the sociotechnical explanation for why these workarounds may occur. As an example, in a study of a closed-loop system comprising electronic prescribing, automated dispensing, barcode patient identification, and electronic medication administration records (Franklin et al. 2007a), patient identity was checked prior to medication administration for 17% of 1344 doses observed pre-intervention and 81% of 1291 doses afterwards. While this is an impressive improvement, it might be expected that such a closed-loop system would ensure 100% adherence to verification of the patient's identity. However, full compliance was not achieved because of informal practices such as sticking barcodes to patients' hospital furniture, which were scanned instead of the patient's wristband. Further exploration indicated that these workarounds occurred due to wristband barcodes not always scanning reliably due to creases, and confusion over the correct identification procedure for patients in single rooms where there were infection control precautions. Similar workarounds have been reported on hospital wards in the United States (Koppel et al. 2008b). Understanding the reasons for these workarounds could thus be used to develop solutions.

TOOLS AND TECHNOLOGY

Various information technology solutions are discussed in Chapters 19 through 21. We focus here on more basic but equally important tools and technologies, such as prescribing stationery and information sources.

Prescribing Stationery

There is increasing evidence that the design of stationery, such as prescription pads, inpatient medication administration records, and drug charts, can influence medication safety. We draw here on some examples from the U.K. context, although the same principles are likely to apply elsewhere. The vast majority of National

Health Service hospitals continue to use traditional paper prescription charts for hospital inpatients; these are used both for prescribing and to record the administration of medicines (Ahmed et al. 2013). Concerns have been raised about the contributory role of poorly designed prescription charts in facilitating prescribing errors (Franklin et al. 2011) and, in an experimental study specifically designed to investigate this issue, differences in prescription chart designs were associated with significant variations in prescribing error rates (Tallentire et al. 2013). While a standardized inpatient prescription chart exists for Wales, charts across the rest of the United Kingdom often have very different features. The U.K. Academy of Medical Royal Colleges released guidelines in 2011 detailing an expert panel's view of the essential components of a safe and effective chart. The idea of a standardized drug chart across the country is more controversial (Barber et al. 2013; Rawlins 2013). In a simulated context, a newly designed prescription chart incorporating behavioral insights into the user-centered design significantly reduced a number of common prescribing errors, including dosing errors and illegibility (King et al. 2014). Positive behavioral change was seen without prior education or support, suggesting that some common prescription writing errors are rectifiable simply through changes in the content and design of prescription charts. Points to consider include whether there is enough space to be able to write units such as micrograms in full, thus avoiding potentially dangerous abbreviations, as well as space for signatures, and avoiding the requirement for multiple charts for the same patient.

Information Resources

Safe prescribing, dispensing, administering, and monitoring of medication often requires access to information such as dosing guidelines, treatment protocols, and other reference sources. The sheer volume of information concerned can make it challenging to ensure that the right information is available at the point of need. Traditionally, such information was held within reference books in the pockets of health care professionals' white coats; today, it is more likely to be on the internet, intranet, or smartphones. This creates both opportunities and challenges. Opportunities include the provision of locally relevant, interactive sources of information that can be targeted to different user groups (Baysari et al. 2013a; Charani et al. 2012); challenges include the risk of fragmenting clinical practice as a result of multiple smartphone applications and other reference materials being used, as well as identifying appropriate governance arrangements (Charani et al. 2014).

ENVIRONMENT

The working environment itself can affect safety. Hospital wards are increasingly busy, noisy places, and staff are subject to repeated interruptions. Every ward should ideally have a quiet area, with good desk space and access to online and paper-based information, for staff needing to do work requiring concentration and care. Some units have introduced "prescribing stations," at which the doctor writes up medications (Booth et al. 2012). With this model of practice, no changes to medication are made during ward rounds, and instead medication is prescribed afterwards at the prescribing station, which is equipped with reference sources, calculators, pens, and a phone. Staff are not allowed to interrupt doctors when they are prescribing at the station. In other settings, the goal is to support real-time prescribing at the patient's bedside, which provides more opportunities for involving the patient in the process.

Interruptions and Distractions

The terms interruptions and distractions are often used interchangeably in the medication safety literature, but arguably should be differentiated, as they are likely to exert different cognitive demands on those affected. Where the two are differentiated, an interruption results in the individual stopping the task they are doing before it has been completed (Biron et al. 2009; Flynn et al. 1999). In contrast, a distraction is generally considered to be a stimulus from an external source that may be ignored or only requires a momentary lapse of attention from the primary task (Boehm-Davis and Remington 2009). Thus, distractions are not followed by cessation of the primary activity. Staff often identify interruptions and/or distractions as contributing to medication errors (Ulanimo et al. 2007). In a study in two Australian hospitals, nurses made significantly more errors—and more serious errors—as interruptions increased during medication preparation and administration (Westbrook et al. 2010b). In another study, this time using simulation, more medication administration errors occurred when nurses were interrupted (Prakash et al. 2014). In an ambulatory care pharmacy, interruptions and distractions were associated with dispensing errors, the majority of which involved incorrect label information (Flynn et al. 1999).

A small number of studies have tested the effectiveness of specific interventions to reduce interruptions during medication-related tasks. For example, Relihan and colleagues (2010) tested a set of interventions (staff education, checklists, red aprons, "do not disturb" signs, and patient information leaflets) in an acute medical admissions unit and halved the number of interruptions to nurses administering medications. Others have used a human factors approach and redesigned the area used to prepare medications in a pediatric ward to incorporate frosted glass barriers; evaluation suggested a significant reduction in interruptions (Colligan et al. 2012).

Far fewer studies have examined the effects of interventions on medication errors, rather than on interruptions and/or distractions. A recent systematic review identified only three studies that investigated the effects of interventions on medication errors, all of which reported reductions (Raban and Westbrook 2014). In a more recent simulation study in the ambulatory chemotherapy setting, some types of errors were reduced with a suite of interventions that included a physically distinct space for medication verification, standardized workflow, a speaking-aloud protocol, and a demarcated "no interruption zone" around the intravenous pumps (Prakash et al. 2014). While there is limited evidence of the sustainability of both such interventions and their benefits, reducing interventions and distractions seems a suitable area on which to focus. However, in clinical settings, some interruptions are appropriate and necessary (Westbrook 2014), and few studies have explored this issue or attempted to differentiate appropriate and inappropriate interruptions. The potential for unintended consequences through reducing appropriate interruptions and distractions should therefore be considered when designing interventions and their evaluation.

CONCLUSION

In this chapter, we have considered a range of approaches to designing safer work systems. While the evidence for many of these is relatively limited, judicious consideration of such approaches is likely to be beneficial, especially if they are designed using the theoretical perspectives described in Section II of this book.

COMMENTARY

Mário Borges Rosa
Pharmacist and Medication Safety Officer, Fundação Hospitalar do Estado de Minas Gerais, and President of Instituto para Praticas Seguras no Uso de Medicamentos (ISMP Brasil), Brazil

Nilson Gonçalves Malta
Pharmacist, Hospital Automation Manager, Hospital Israelita Albert Einstein, Brazil

It can be argued that Brazilian educational institutions do not provide health care professionals with adequate knowledge about patient safety. This aspect has been quite clear in my experience as a pharmacist. The pharmacy degree is neither patient-centered nor safety-oriented. My very first contact with medication safety was as an intern, but even so, I was not prepared for the real-life clinical practice I would eventually become involved with. I learned about the theories of pharmacokinetics, but little on how to put these theories into practice; I learned nothing about high-risk medicines, nor how to communicate with physicians about any concerns I had with their medication orders. I felt that there was a strong need for pharmacists to be part of the multidisciplinary team, but that undergraduate preparation was insufficient in terms of both clinical knowledge and communication skills. As a result, the pharmacist has to learn these skills on the job in the health care environment, with potential risks to the patient.

There are currently around 6000 hospitals in Brazil, of which over 50% have fewer than 50 beds. This hampers the introduction of interventions to support the safe use of medications; in general, small hospitals face financial constraints and are short-staffed. Most hospitals have attempted to adopt safer dispensing systems, and a considerable number of hospitals already rely on basic electronic prescribing systems, though few have clinical decision support. An important movement that is improving medication safety is that some health care institutions are now seeking accreditation by the Brazilian Organização Nacional de Acreditação. A National Patient Safety Program was also launched by the Brazilian Ministry of Health in 2013, which includes some of the "lower-tech" aspects relating to medication safety as described in this chapter. It mandates that hospitals should make reliable drug information readily available and that the dispensing and administration of high-alert medications should be double-checked. The program also requires that areas used for prescribing, dispensing, preparation, and administration must be adequate, with minimal interruptions. However, national audits suggest that adherence to such medication safety guidance still needs to be improved.

Another limitation evident in Brazil is the lack of teamwork among health care professionals. The difficulty in accepting changes in routines among both experienced and inexperienced professionals is clear in day-to-day hospital activities. Error is not yet perceived as a real risk, suggesting a poor safety culture. Generally, organizations do not have specific error prevention and handling policies in place.

As highlighted in this chapter, double-checking is a good example of how the understanding of a safety process is often unclear among the health care professionals concerned, and ultimately, the process is prone to personal interpretations of what is required. In my experience, the second professional usually verifies what the first has done, rather than the recommended procedure of independent verification by both professionals. New technologies are welcome to improve safety, but we still have a long way to go in making the current working practices safer. In Brazilian hospitals outside major urban centers, the use of technology to improve safety is still in its infancy. Much of this is due to high costs, the low budgets of institutions and the lack of an organizational safety culture. Many of the lower-cost interventions relating to safer work systems described in this chapter are already applied in the hospitals where we work, and we believe they can be applied in most Brazilian hospitals. The National Patient Safety Program plays an important role in encouraging the implementation of these measures. Nonetheless, it is true that budgetary concerns may limit hospital access to some solutions. A simple double-check eventually may not occur due to nurses' workloads and/or a lack of staff. Another issue is related to the dramatic variations in health care provision around the country. Richer regions are attended by high-level health care services, while there are areas that have difficulty in hiring professionals. These challenges will also need to be addressed in order to ensure safe systems of work.

19 Electronic Prescribing and Medication Administration Record Systems

Johanna I. Westbrook and Melissa T. Baysari

CONTENTS

KEY POINTS

- Electronic prescribing and electronic medication administration records have been implemented to varying degrees internationally to reduce medication errors, and there is some evidence as to their success.
- Well-designed decision support systems in particular have a positive effect, although the use of alerts can be problematic.

- There is evidence, however, that these technologies can create new types of errors.
- Such systems will change the way in which health care professionals work, including impacting on their efficiency and interactions with others.

INTRODUCTION

This chapter will provide an overview of the evidence related to the effectiveness of electronic prescribing (e-prescribing) and medication administration systems in a range of settings to reduce medication errors and adverse drug events (ADEs). At the end of the chapter, there are two expert commentaries from authors who reflect on how such systems have been introduced into their very different countries.

E-PRESCRIBING SYSTEMS

E-prescribing in both hospital and community settings has been consistently identified as an important intervention with the potential to reduce medication errors and ADEs (Ammenwerth et al. 2008; Bates et al. 1998, 1999; Nuckols et al. 2014; Reckmann et al. 2009). E-prescribing systems allow clinicians to electronically enter medication orders using drop-down menus and/or free-text typing and include varying levels of decision support to guide prescribing. Even e-prescribing systems with very limited decision support have potential advantages over handwritten prescribing, including improved standardization of orders, a full audit trail, legibility of orders, use of approved drug names and units, specification of key data fields such as route of administration, improved access to medication information, and secure storage and recall of order records (Reckmann et al. 2009; Westbrook et al. 2012). One study of two commercial e-prescribing systems with limited decision support implemented at two major hospitals showed a more than 90% reduction in prescribing errors related to incomplete, illegible, and illegal medication orders (Westbrook et al. 2012).

E-prescribing systems can be integrated or interfaced with other clinical information systems (for example, as an application within a computerized prescriber order entry [CPOE] system). Ideally, e-prescribing systems should support closed-loop medication management (Franklin et al. 2007a) and thus it is preferable for them to interface seamlessly with a hospital's pharmacy dispensing and medication administration system (Bates et al. 2001). The benefits to medication safety are likely to be related to the extent to which a closed-loop medication system is achieved.

IMPACT ON MEDICATION ERRORS

Robust evaluations of e-prescribing systems that measure changes in medication error rates are still relatively limited (Ammenwerth et al. 2008; McKibbon et al. 2012; Reckmann et al. 2009). In an extensive review of systems used in hospital and ambulatory settings, McKibbon and colleagues (2012) identified 87 randomized controlled trials (RCTs) assessing the effect of integrated medication management information technology on multiple aspects of the medication process. Most ($n = 67$, 72%)

were conducted in the United States, with a further 16 (18%) from Europe. Only one RCT measured changes in prescribing errors and reported a significant reduction in errors following e-prescribing implementation in an intensive care unit (ICU). In a broader review, McKibbon and colleagues (2011) identified eight of ten non-RCT quantitative studies, including those using a controlled before-and-after design, which measured changes in ADEs post e-prescribing implementation and reported significant medication error reductions. A meta-analysis of 16 studies assessing the effectiveness of e-prescribing systems in hospitals to reduce potential ADEs and medication errors relative to paper-based systems concluded overall a 50% reduction in potential ADEs and error rates (Nuckols et al. 2014). This is consistent with a broader systematic review (Ammenwerth et al. 2008) examining studies in both inpatient and outpatient settings, which found 23 of 25 studies showed medication error reductions. Nine studies measured changes in ADEs and six reported a significant reduction in ADEs (ranging from 35% to 98%). However, all reviews (Ammenwerth et al. 2008; Nuckols et al. 2014; Reckmann et al. 2009) have noted concerns regarding the quality of the study designs and outcome data used, as well as inconsistencies in medication error definitions applied. Few studies have examined effects in multiple wards or over different hospitals or compared e-prescribing systems. An Australian controlled before-and-after study compared changes in prescribing error rates on multiple wards at two hospitals, each of which implemented a different commercial e-prescribing system (Westbrook et al. 2012). Significant reductions of over 50% in overall prescribing error rates in all three intervention wards were found, with no significant changes in error rates on the control wards. The study also found that the most serious prescribing errors—with the potential to cause permanent harm to patients—significantly declined by 44% (from 25 serious prescribing errors per 100 admissions, 95% confidence interval [CI]: 21–29, to 14 per 100 admissions, 95% CI: 10–18) compared with control wards. The e-prescribing systems implemented had limited decision support at the time of the study. Decision support is viewed as a central element to increasing the effectiveness of e-prescribing systems (Kadmon et al. 2009; Teich et al. 2005). Thus, as decision support within these systems continues to improve, further declines in error rates would be expected.

ELECTRONIC CLINICAL DECISION SUPPORT SYSTEMS

One of the core benefits of e-prescribing is the ability to provide clinicians with information and guidance using electronic decision support at the time decisions are being made or being entered into the system. There is now good evidence that when well designed and targeted, electronic decision support can have significant and positive impacts on care outcomes in both ambulatory and acute care settings (Bright et al. 2012; Garg et al. 2005; Schedlbauer et al. 2009). E-prescribing with embedded decision support can change prescribing behaviors and improve patient outcomes (Garg et al. 2005; Schedlbauer et al. 2009) while reducing costs (Forrester et al. 2014; Vermeulen et al. 2014; Weingart et al. 2009). A U.S. RAND study estimated that decision support in e-prescribing systems could prevent 200,000 ADEs a year, saving U.S. $1 billion (Hillestad et al. 2005). Even simple decision support, such as reminders for those over 65 years old to receive influenza vaccinations, has been estimated in the United States to potentially reduce deaths by 5200–11,700 and prevent 1.0–1.8 million hospital bed-days each year (Hillestad et al. 2005). A further U.S. study in the hospital setting has shown that decision support improved rates of venous thromboembolism prophylaxis by over 30%, from 61.9% to 82.1% (Bhalla et al. 2013).

However, accompanying this evidence of the positive effects of electronic decision support is an increasing number of reports demonstrating high rates at which alerts are overridden by users, along with accounts of alert fatigue and frustration. Studies have found that clinicians override 49%–96% of drug alerts (van der Sijs et al. 2006), and an observational study showed that clinicians did not read the majority of alerts presented (Baysari et al. 2011a).

Decision support may include structured data fields providing clear guidance of the information required, access to online reference materials that may be used in response to a clinical question, order sets (defined lists of medication orders for specific conditions or specific patient groups, which can be ordered together without having to individually order each medication), alerts that draw a clinician's attention to a particular issue in relation to a medication order, or warnings to identify a potential error. Commonly used alerts include drug–allergy, drug–drug interaction, drug–pregnancy, drug–age, and drug–disease checking (Greenes 2007).

Drug–drug interaction alerts have an important role to play in the prevention of serious medication errors due to failures to take interactions into account. However, determining the decision rules for the triggering of interaction alerts can be challenging. If the threshold is set too low, there will be a large number of alerts with subsequent alert fatigue, meaning that most of them will be ignored. A Dutch study (van der Sijs et al. 2009) found that 98% of interaction alerts were ignored by clinicians. There has been considerable discussion and debate regarding the best way to optimize interaction alerts for greater effectiveness (Hines et al. 2011; Olvey et al. 2010; Peterson and Bates 2001; Phansalkar et al. 2012, 2013). A review of the quality of drug–drug interaction decision support in a range of ambulatory e-prescribing and dispensing systems identified high variability in the sensitivity and specificity of interaction alerts (Sweidan et al. 2009).

Electronic decision support is a deceptively simple concept, while its design and execution is extremely difficult and there is still much to learn about how to deliver system-wide change through electronic decision support systems (Bates et al. 2003; Baysari et al. 2011b; Bobb et al. 2007; Coiera et al. 2006; Colombet et al. 2004; Elwyn et al. 2008; Osheroff et al. 2005; Sittig et al. 2006; Weingart et al. 2009). Core attributes of effective decision support include that it should provide immediate real-time information to inform a decision when it is being made, integrate this into the user's workflow, and be as non-interruptive as possible (Bates et al. 2003). Passive decision support, such as presenting one or more default options for dosing, is likely to integrate seamlessly into existing work processes (Bates et al. 2003). In contrast, by making it a requirement for clinicians to seek out the decision support themselves, there is a risk that clinicians will not access or utilize the support available, as one of the most significant predictors of improved clinical outcomes with decision support is the automatic provision of advice (Kawamoto et al. 2005). At the same time, presenting clinicians with large numbers of automatic alerts leads to users becoming desensitized to alert presentation, hampering the effectiveness of alerts as a safety intervention. In a study that compared the types of alerts seen and responded to by two groups of doctors—one where alerts were computer triggered and the other where alerts were presented only when a doctor requested a check—it was found that doctors in the computer-triggered group saw more alerts and made more changes to orders following alert presentation, but ignored 88% of the problems identified by the computer system. Doctors in the "on-demand" group rarely requested a check and so were warned about less than 1% of the prescribing problems identified by the alert system, but they ignored only 25% of the problems identified (Tamblyn et al. 2008).

Understanding the drug databases and logic behind decision support alerting is also central to safe and effective decision support. Those implementing and using systems need to understand whether—and how—decision support can be customized. Having processes and governance structures in place for maintaining and monitoring decision support within e-prescribing systems is crucial (Sittig and Ash 2011). While the focus of much attention is on adding new decision support rules, consideration also needs to be given to processes for establishing when ineffective, redundant, or unsafe decision support rules should be removed (Baysari et al. 2013a, b). Attempts to determine which alerts to remove from an e-prescribing system (Baysari et al. 2013a, b; van der Sijs et al. 2008) have shown that the process can be difficult, because neither users nor experts agree on which alerts can be safely removed from a system.

Alert design is also believed to play a role in the clinician acceptance of alerts within e-prescribing systems. Interviews with users have shown that poor design, such as missing information, unhelpful layout of content, and failure of alerts to be distinctive, are all barriers to clinicians reading and interpreting alerts (Baysari et al. 2011b; Russ et al. 2012b). In a study that quantified the impact of alert features on interaction alert acceptance (i.e., percentage not overridden), alert display was found to be one of the strongest predictors of acceptance (Seidling et al. 2009). Display comprises variables such as the visibility, legibility, and color of alerts. Addressing these variables should be informed by a number of human factors principles for effective alert design (Horsky et al. 2012, 2013; Phansalkar et al. 2010, 2014; Zachariah et al. 2011). However, research systematically examining the impact of these different variables on alert effectiveness, particularly in real-world clinical contexts, is rare. In a simulation study, prescribers were presented with six patient scenarios and required to prescribe and respond to alerts (good versus bad design) using an e-prescribing system (Russ et al. 2014). It was found that when the alerts were well designed, prescribers were faster to read the alerts and complete the scenarios, reported lower workload scores, and made fewer prescribing errors. These findings suggest that the adoption of human factors design principles can increase an alert's capacity to improve prescribing, at least in a simulated environment.

Although impressive, the simulation results above should not be taken at face value, as research has also shown that the context of decision support is central to the impact it will have. For example, in a study on the use of an e-prescribing system on hospital ward rounds conducted by 14 medical teams in a general hospital, only 17% of alerts were read and no changes were made to medication orders in response to the alerts generated (Baysari et al. 2011d). In contrast, junior doctors working after-hours in the same hospital read 78% of alerts and changed 5% of medication orders following alert presentation (Jaensch et al. 2013). It is likely that some alerts work in some contexts but not in others, and that decision support may be more beneficial to junior doctors who are working unsupervised than to doctors prescribing as part of specialty medical teams during ward rounds.

Customizing Alerts

Customizing alerts for clinicians or groups of clinicians has been suggested as a potential strategy for reducing unnecessary alerts and alleviating alert fatigue (Isaac et al. 2009; van der Sijs et al. 2008). Presenting specific alert types to particular specialties or clinicians of certain skill levels (i.e., junior doctors) would ensure that specialists with a high level of knowledge in an area do not receive the alerts related to that area (for example, cardiologists may not need to receive alerts about cardiology drugs). A variation is that clinicians could receive a particular type of alert a certain number of times, and then this alert will no longer trigger for this individual. Customizing alerts may be problematic because computerized alerts should serve as a safety net in times of forgetfulness or time pressure, even for experts.

Other potential strategies for reducing alert numbers include tiering alerts according to severity, presenting only high-level (severe) alerts, increasing alert specificity, and tailoring alerts to patient characteristics. Organizing alerts into levels based on severity and presenting each severity level in a different way to users (for example, different colors, signal words, or symbols) would instantly cue users to the importance of the alert and may result in fewer important alerts being missed or overridden. In a study that examined this approach, a review of alert log data was performed at two U.S. hospitals. One hospital displayed all alerts equivalently, while the other presented serious drug–drug interactions as hard-stop alerts (i.e., the prescriber could not continue with the order), less severe interactions were presented as alerts that required an override reason, and the least severe interactions were presented as information-only alerts (i.e., non-interruptive warnings) (Paterno et al. 2009). The study showed that such tiering of alerts according to severity resulted in greater acceptance of severe alerts (100% versus 34%) and moderate alerts (29% versus 10%), although by design, the most severe alerts at the hospital adopting tiered alerts had to be complied with. No data were collected on ADEs, so the impact of tiering on ADEs could not be determined.

Low-priority alerts have been shown to cause user frustration and slow down the medication ordering process (van der Sijs et al. 2006). The results from Paterno and colleagues (2009), shown above, suggest that classifying alerts based on severity and presenting low-priority information in a non-interruptive way may improve prescriber acceptance of more serious alerts. A scenario-based simulation study revealed that interruptive alerts are three-times more effective at reducing prescribing errors than non-interruptive alerts (Scott et al. 2011), highlighting a need for careful consideration when deciding whether alerts should be interruptive or non-interruptive. Widespread agreement is needed on what constitutes low- or high-priority information to inform these decisions about alert design.

To increase alert specificity, several other approaches have been suggested, including ensuring drugs are classified individually, rather than by drug class, and ensuring that the mode of administration of drugs (topical, oral, intravenous, etc.), the time of administration, and the dosages of drugs are taken into consideration by the alert system (Seidling et al. 2009; Smithburger et al. 2011; Sweidan et al. 2009; van der Sijs et al. 2006). These approaches should minimize the number of clinically irrelevant alerts being presented, but there has been limited research to investigate systematically the impact of these factors on prescriber acceptance of alerts. Although more sophisticated and currently out of reach for many organizations, a more effective alert system is one that considers individual patient characteristics prior to the triggering of an alert. For example, integrating laboratory results into the alert system may ensure alerts are more patient-relevant and useful for prescribers (Zwart-van Rijkom et al. 2009). Simpler strategies based on the same principle include presenting pregnancy alerts only for patients who are known to be pregnant, rather than all female patients in the hospital, and only presenting allergy alerts for patients with a complete list of allergies documented.

SYSTEM-RELATED MEDICATION ERRORS ASSOCIATED WITH E-PRESCRIBING

E-prescribing systems can introduce new types of medication errors. These have been variously labeled as system-related, technology-induced, and "e-iatrogenic" (Westbrook et al. 2013a). These new errors arise due to the ways in which health professionals interact with the electronic system and its design. In a study of 1164 prescribing errors arising following the use of two different e-prescribing systems, Westbrook and colleagues (2013a) identified those errors that may have been directly facilitated by use of an e-prescribing system and that would not have occurred when using paper medication charts. They found that these system-related errors occurred at a rate of 78 per 100 patient admissions. The most frequent mechanism was clinicians incorrectly selecting from a drop-down menu, which accounted for 43% of all system-related prescribing errors identified. However, only 11 system-related errors were rated as serious, defined as having the potential to cause permanent patient harm. Additional features, such as setting default administration times (not always used in paper-based systems), while designed to improve the efficiency of the prescribing process, also led to new errors when clinicians forgot to change times when this was required. Comparisons between two commercial e-prescribing systems showed that both produced similar rates of system-related errors; however, the mechanisms for those errors differed and were related to differences in system design features (Westbrook et al. 2013a).

Multiple further factors have been identified as potentially contributing to new types of errors following e-prescribing system implementation. These include technical issues, such as an inability for users to view an entire medication chart at one time on a computer screen due to poor interface design, or limited integration with other clinical information systems (Koppel et al. 2005; Redwood et al. 2011; Wetterneck et al. 2011). Changes in work practices and workflows following system implementation can result in shifts in areas of responsibility and roles associated with system use. For example, e-prescribing systems require user privileges to be specified that define the information that can be accessed and the tasks that can be performed by individuals or those in a specific role (such as an intern). While this may provide certain safeguards, it can also result in important treatments being delayed and may reduce the efficiency of work (Ash et al. 2004, 2006; Koppel et al. 2005; Redwood et al. 2011; Sittig and Ash 2011). Many existing e-prescribing systems continue to have limited functionality related to some complex medications. Studies have reported hospitals initiating workaround processes to deal with these limitations, such as keeping some medication orders (such as insulin) on paper medication charts (Ahmed et al. 2013; Westbrook et al. 2013a). This dual process of electronic and paper charting may require a new task for prescribers, namely to insert an electronic alert within the e-prescribing system to notify users that a paper medication chart also exists for that patient. This process has been shown to result in "new" errors arising, with paper-based orders being missed or failing to be discontinued (Westbrook et al. 2013a).

Hybrid paper and electronic systems within the same organization are a further mechanism by which new medication errors may be introduced. E-prescribing system implementation within a given organization will often occur over a period of many months, and sometimes years, due to a range of factors, most often resource availability. During this period, a hospital may have some wards where e-prescribing is used while other wards remain paper-based (NHS Connecting for Health 2009; Westbrook et al. 2012). The hospital pharmacy and individual wards will have to deal with both and provide a point of integration. Potential safety risks and work inefficiencies are introduced as a result of patients needing to transfer between wards with and without e-prescribing systems. This will often involve the printing out of electronic medication charts that may present information in different ways to those on the hospital's standard paper medication chart. While printed, typed orders may be easier to read, and the different ways in which orders and information are presented may pose new risks of information being missed or misinterpreted. Hospitals that upgrade systems over time or change to new systems may encounter similar issues.

ELECTRONIC MEDICATION ADMINISTRATION RECORDS

Electronic medication administration records (eMARs) are electronic records of medications prescribed and administered to individual patients. eMARs are populated with medication orders generally entered by prescribers (often in conjunction with an e-prescribing system), but also in some instances by non-prescribers (i.e., pharmacists, pharmacy technicians, or nurses entering orders on behalf of a prescriber). eMARs may be integrated with barcode technology or radio frequency identification technology. Some eMARs allow for the provision of photos of patients, and this feature is of particular benefit in nursing homes, where residents may not be able to assist in their identification. Some eMARs designed for nursing homes also include images of drugs, which provide an additional cue for staff administering drugs, some of whom may not be clinically trained.

When nurses log onto an eMAR and open a patient's electronic drug chart, they can select a drug due for administration. Some systems use color coding to indicate drugs already administered and those due or overdue. Once administered, the nurse confirms the administration, or if unable to administer a dose, they are required to document the reason (for example, if the patient was absent or refused the dose). The electronic signature of the nurse is linked to the doses administered, along with the time at which the information was entered.

Perceived benefits of eMARs include the ability to track dose omissions, enforce documentation of reasons for dose omissions, and improve the timing of administrations due to clear displays that can alert to overdue doses and improve medication management (for example, by having the eMAR cue nurses to check vital signs or other test result values prior to the administration of some drugs).

Few studies have focused on assessing the direct impact of eMARs alone on medication administration errors. A study in a U.K. pediatric ICU completed three audits of an e-prescribing with eMAR system (at pre-implementation, at 1 week post-implementation, and again at 6 months post-implementation) to identify dose omissions (Warrick et al. 2011). A reduction in omitted medications was observed (8.1% at pre-eMAR versus 1.4% at 6 months post-eMAR): dose omissions documented as being for "other reasons" or left blank were eliminated, and dose omissions documented as being due to the drug being "unavailable" were reduced (Warrick et al. 2011).

A second U.K. study used an interrupted time series analysis to measure changes to an existing e-prescribing and eMAR system on dose omission rates (Coleman et al. 2013a). Four interventions were implemented sequentially and evaluated across the hospital. Over the entire 4.5-year study, dose omission rates for medications declined by 53%. The effect of each of the four interventions was also examined independently. Clinical dashboards that displayed individual ward performances on omitted doses for all staff and monthly executive team meetings with a focus on omitted doses were associated with a statistically significant reduction in dose omissions post-implementation. However, the implementation of a visual indicator to show overdue doses was not associated with a change in dose omission rates.

In a large controlled before-and-after study at two Australian hospitals, each of which implemented an eMAR along with an e-prescribing system but did not include the use of barcodes, 153 nurses were observed preparing and administering 4176 medications across three intervention and three control wards (Westbrook and Li 2013). Following e-prescribing and eMAR introduction, the overall rate of medication administration errors on the intervention wards significantly declined by 4.24 errors per 100 administrations (95% CI: 0.15–8.32, $p = 0.04$) compared to the control wards. Dose timing errors experienced the greatest reduction (3.35 per 100 administrations; 95% CI: 0.01–6.69, $p < 0.05$).

Qualitative studies (Culler et al. 2011; Hsieh et al. 2004, 2009; Mitchell et al. 2004b; Moreland et al. 2012) seeking to explore users' perceptions of eMARs have shown that most were positive after system implementation, and in two studies, users' perceptions improved over time (Culler et al. 2011; Moreland et al. 2012). Concerns voiced by users have centered around problems with the integration of the eMAR with other systems and its impact on workflow and patient safety. Elements of medication administration documentation, such as accuracy (Moreland et al. 2012) and quality of information (Mitchell et al. 2004a), were believed to have improved following eMAR implementation, while staff perceptions of the effects on teamwork and communication between health care professionals have varied between studies (Culler et al. 2011; Moreland et al. 2012).

IMPACT OF E-PRESCRIBING AND ELECTRONIC MEDICATION ADMINISTRATION RECORDS ON WORK EFFICIENCY

Impact of Systems on Doctors' and Nurses' Work

Clinicians often raise concerns that the introduction of electronic medication systems will disrupt existing patterns of work and negatively impact upon the efficiency of their work (Cresswell et al. 2014; Georgiou et al. 2009). Qualitative studies have revealed that many doctors and nurses report both improved and worsening efficiency following system implementation (Culler et al. 2011; Devine et al. 2010; Holden 2010).

A number of studies have attempted to measure the impact that e-prescribing systems—usually with eMARs—have on the work of doctors, but relatively few have investigated nurses' work on general hospital wards. It is well recognized that, for most people, typing a medication order using a computer will take longer than handwriting it. Conversely, the ability to access e-prescribing and medication information from a range of locations (both within and outside the hospital) is a clear advantage to work flexibility, speed, and ease of information access. However, what is less clear is whether the increased information access and shortcuts that e-prescribing systems provide, such as not requiring medications to be re-charted after a defined period, being able to modify existing orders without retyping the entire order (in some cases), and the use of order sets, saves clinicians' time overall. Research on early e-prescribing systems published in the 1990s (Bates et al. 1994) showed that, following CPOE implementation, time spent by doctors ordering medications increased, with medical interns significantly increasing their time spent ordering from 5.3% to 10.5% of their time ($p < 0.001$) and surgical house officers from 6.4% to 15.5% of their time ($p < 0.001$). The use of computerized order sets, however, was found to take less time than ordering the individual medications on paper.

Relatively few quantitative studies to identify the impact of modern e-prescribing and eMAR systems on clinicians' work have been published (Poissant et al. 2005). A study on one surgical ward in a U.K. hospital (Franklin et al. 2007a) reported that medication administration rounds were significantly shorter, but nurses spent a significantly greater proportion of time on medication tasks outside of these rounds following the introduction of a closed-loop medication system (comprising e-prescribing, automated ward-based dispensing, barcode patient identification, and eMARs). However, in another U.K. hospital using a different system (Mitchell et al. 2004a), the time taken to complete drug rounds was reported to increase from 69 seconds per item to 98 seconds per item ($p > 0.05$) after the implementation of an integrated e-prescribing and eMAR system. Studies of the use of eMARs in nursing homes are rare, however. In a study across five nursing homes in the United States, nurses were observed administering approximately 57 medications per hour following eMAR introduction, compared to 40 medications per hour before its introduction (Scott-Cawiezell et al. 2009).

An Australian study using a controlled before-and-after design investigated how doctors and nurses changed the ways they distributed their time following the introduction of an e-prescribing system with eMARs (Westbrook et al. 2013b). Doctors ($n = 59$) and nurses ($n = 70$) were directly observed for over 630 hours across four wards and observers recorded details of how they spent their time (what tasks, with whom, with what, and where). Doctors and nurses on the wards with the electronic medication systems spent similar proportions of their time on medication tasks, direct care, and professional communication as their colleagues on wards that did not have the electronic systems implemented. Differences between doctors and nurses were also found. Doctors on wards with the e-prescribing system spent significantly more time with other doctors and with patients compared to doctors on the control wards with no e-prescribing system. On the other hand, nurses on the wards with the electronic systems spent significantly less time with doctors compared to nurses on wards without the electronic medication systems (Westbrook et al. 2013b). This latter finding is consistent with a French study that found that doctors and nurses interacted less frequently following the introduction of an e-prescribing system (Beuscart-Zephir et al. 2005). Reasons for this reduced interaction may be that nurses were no longer required to clarify orders with doctors, as they were easy to read in the e-prescribing system. However, it could also be due to the fact that the location of the doctors' work changed following e-prescribing implementation (Westbrook et al. 2013b). Prior to the system's introduction, doctors conducted most of their paper-based prescribing at the central nurses' station, which provided opportunities for unplanned conversations. When the e-prescribing system was introduced, doctors shifted their work to computers located in separate rooms or in the corridors, which reduced the chances of interactions with nurses. Whether these reduced interactions had an impact on care is unknown. However, previous studies on these same hospital wards with these electronic systems showed significant reductions in both prescribing and medication administration error rates following system implementation, whereas the control wards experienced no significant changes in medication error rates (Westbrook et al. 2012; Westbrook and Li 2013). Thus, these results indicate that the introduction of e-prescribing and eMARs did not result in clinical staff significantly changing the proportions of the time they devoted to medication tasks, direct care, or professional communication in general. However, within these broad task areas, there were some subtle shifts in behaviors; for example, doctors with access to the e-prescribing system spent more time overall with other doctors. A study in a U.S. pediatric ICU before and after CPOE implementation (Zheng et al. 2010) observed medical residents for 68 hours and found no significant changes in the time doctors spent in direct care or ordering. Thus, as technology has developed over time and the computing skills of the clinical workforce have improved, there appears to be increasing evidence that these electronic medication systems are able to integrate into clinical workflows without significantly impacting overall work efficiency.

Impact of Systems on Pharmacists' Work

Despite the significant impact that e-prescribing and eMARs may have on the work of pharmacists, there have been few studies on this topic. One direct observational study of eight pharmacists over 37 hours reported that pharmacists on hospital wards with an e-prescribing system spent a significantly greater proportion of their time alone, less time in transit, and conducted more and faster clinical chart reviews compared to pharmacists who worked on wards with no e-prescribing system (Lo et al. 2010). A recent focus group study that aimed to explore pharmacists' perceptions of how e-prescribing and eMARs impacted on their work revealed that pharmacists felt they had become less visible in the clinical environment and had less patient contact following system implementation (Burgin et al. 2014).

Challenges in Measuring and Interpreting Changes in Work and Communication Patterns

Methods for measuring changes in clinicians' patterns of work and communication have also continued to improve; there are now recommended guidelines for the conduct of quantitative observational studies (Zheng et al. 2010), and standardized electronic measurement tools are available (Westbrook and Ampt 2009) that provide the opportunity to make meaningful comparisons between study results (Ballerman et al. 2011). When measuring changes in patterns of work, a controlled study design is strongly recommended in order to avoid falsely attributing temporal changes in work patterns to clinical system implementations. In their study of doctors' and nurses' work, Westbrook and colleagues (2013b) separately analyzed their data to identify whether they would have come to different conclusions if they had used an uncontrolled before-and-after study design, rather than a controlled study design. They showed that temporal changes had occurred in the pattern of nurses' work in the post-implementation period, with nurses in both study groups spending significantly more time on medication tasks and direct care in the post-implementation period. Such changes in practice may reflect a general increase over time in the severity of patients admitted to hospitals who required more intensive nursing care. Failure to use a control group to assess the effects of the e-prescribing and eMARs would have resulted in these temporal changes being incorrectly attributed to the introduction of the electronic medication systems, clearly demonstrating the importance of using a controlled study design wherever possible. Thus, whether interpreting results from published studies or designing studies to prospectively measure changes in work patterns or system effectiveness, consideration should be given to the influence of study design.

USING ELECTRONIC MEDICATION SYSTEMS TO REPLICATE CURRENT PRACTICES OR TO FACILITATE NEW PRACTICES

The extent to which electronic systems should replicate existing medication prescribing and administration systems versus trying to change these processes to take advantage of potential technology-related work efficiencies is somewhat contentious. A good example is the use of ordering shortcuts that may save clinicians many mouse clicks, but that may contradict long-held practices engrained in clinical education (Baysari et al. 2012a). For example, in a study to investigate the extent to which medication orders triggered unnecessary alerts, Baysari and colleagues (2012a) found that a third of therapeutic duplication alerts that clinicians received were generated because doctors did not use the e-prescribing system as intended. Consistent with paper-based medication order practices, many doctors would discontinue an order in the e-prescribing system before writing a new order for the same drug, but with a modified dose. This would trigger a therapeutic duplication alert because alerts would be generated if the first order was still active on a patient's chart or if the first order was no longer active but discontinued less than 24 hours previously. It was intended that doctors wishing to modify a dose would click on the existing order and directly modify the order, rather than ceasing and writing a new order, on average saving 11 mouse clicks. However, modifying an existing order in a paper-based system is generally not allowed, and it became apparent that despite training in the use of the e-prescribing system shortcuts, this engrained behavior was hard to change. Training programs need to pay particular attention to areas of clinical practice where the desired use of e-prescribing or eMAR systems may be inconsistent with previous practice.

Consideration should also be given to how the imminent introduction of medication technology may be seen as an opportunity to review existing practices versus a focus on embedding current practices (Westbrook and Braithwaite 2010). For example, the practice of nurses double-checking and double-signing for certain drug types could be reviewed, given the limited evidence of the effectiveness of such behaviors to improve medication safety (Alsulami et al. 2012), rather than assuming that this practice should be embedded within eMAR systems. Some argue that systems should largely replicate users' current practices in order to support easy adoption. However, this should be balanced against taking advantage of the opportunities information technology presents in terms of making work practices more efficient and/or safer (Baysari et al. 2012b; Russ et al. 2012a).

PLANNING FOR E-PRESCRIBING AND ELECTRONIC MEDICATION ADMINISTRATION RECORDS

The introduction of electronic medication systems is complex and there is a consistent stream of evidence to indicate that this complexity is often underestimated by health care organizations. Given the highly collaborative nature of medication processes, the introduction of these systems can fundamentally change the flow of work, communication practices, and where and when work is physically undertaken for a range of health professionals within an organization. Some changes are quite radical, while others are subtle and may go almost unnoticed until the consequences become apparent. The introduction of clinical information systems can make existing poor practices, which are largely hidden within a paper-based system, more explicit. As a result, the criticisms that are often targeted at an e-prescribing system may in fact result from requiring clinical staff to undertake medication tasks in the manner in which they should always have been performed.

CONCLUSION

The motivation for system introduction is to support the improved quality and safety of medication practices, but to achieve these goals, great attention needs to be placed on the sociotechnical context in which the systems are to be integrated, as described in Chapter 10 (Aarts et al. 2004; Cornford 2004). Producing high-quality, robust evidence on the effectiveness of systems in addressing medication safety issues is critical to engaging clinical staff and changing cultural attitudes to the system (Day et al. 2011). Health care organizations must also ensure that they have the capacity to respond to the issues identified. This should include appropriate governance structures that acknowledge that e-prescribing systems are dynamic interventions that must be monitored and will evolve over time. It is only through identifying and minimizing unintended consequences of system implementation (including new error types) that the potential benefits of the technology will be realized.

COMMENTARY (I)

Derar H Abdel-Qader

Clinical Adjunct Professor, School of Pharmacy, The University of Jordan; Regional Director for Research, Training and Development in Middle East and North Africa (MENA), The International Group for Educational Consultancy, Jordan

The implementation of e-prescribing and eMAR systems is in its infancy in the Middle East, although major medical centers and public hospitals have started to implement such systems, either as standalone systems or integrated with electronic health records. Surprisingly, only a few Middle Eastern articles have mentioned the importance of implementing e-prescribing in clinical practice, though prescribing errors with handwritten prescriptions have been documented. The evidence-based advantages reported in this chapter present a strong case that medical centers in the Middle East could benefit from e-prescribing, if they have not yet implemented it. The key advantages described by current e-prescribing users in the Middle East mirror those in the literature and include more legible and accurate prescriptions, clinical decision support embedded in the system, and the availability of an audit trail.

On the other hand, lessons learned in this chapter help reveal crucial limitations as to how health care information technology is used in the Middle East. For instance, the e-prescribing system has been seen by most of its users in the Middle East as a monolithic block in terms of its lack of adaptability to different hospital specialties, poor interfacing with other systems across different tiers of care, and difficultly in upgrading to newer versions. Unfortunately, electronic information exchange between medical centers also does not happen, meaning that patients' clinical data may be either incorrect or incomplete. Our experience has taught us that any electronic system is built up and improved incrementally, rather than set up as a ready-to-go-live device. System builders and hospital administrators should support the deployment of the e-prescribing system by efficiently providing both soft and hard infrastructures.

Contrary to what they expected prior to implementation, clinicians often complain afterwards that they spend a long time entering information at the computer workstations, instead of providing direct patient care. I have noticed that users also have a strong belief that the documented information cannot help them provide better patient care; its use is perceived as being mainly for managerial and legal purposes. All this encourages clinicians to create workarounds. Clinicians should be advised that e-prescribing does not make the prescribing process faster; it can change the way, for example, doctors make decisions about prescribing by providing predefined medication orders and other related information. They need to adapt to the new transformational technology of e-prescribing and behave differently. Researchers should use the large amount of data collected by e-prescribing systems to understand the processes and outcomes of health care better, and then improve the clinicians' interactions with the computer. Policy-makers have to present the outcome data regularly to e-prescribing system users to motivate them to continue improving their performance.

The "best" e-prescribing and eMAR systems are not born complete, but nurtured over time. After all, even the best systems are not perfect; their supporters must acknowledge their limitations and learn to work with—rather than work around—the systems and their associated problems.

COMMENTARY (II)

Marc Oertle
Chief Medical Information Officer, Spital STS AG, Spital Thun, Krankenhausstrasse, Thun, Switzerland

This chapter gives an excellent summary of the most important aspects of CPOE and eMAR systems based on a wide range of literature. However, the richness and granularity of information concerning clinical decision support systems in the literature does not accurately reflect everyday experiences. Implementing CPOE systems often creates more problems with integrating the new technology into new and/or existing workflows or processes. Regarding these problems, the fact that CPOE represents a transformational technology and not an efficiency technology cannot be emphasized enough. One example of the consequences of a transformational technology might be the effect of order sets that we face in our institution. Junior doctors, in particular, are losing the know-how to deliver good medication practice when the sophisticated clinical decision support systems and order sets define the correct drugs at the correct dosage and correct route of administration in a given clinical context.

In our institution, the cornerstones of successful implementation involved fitting our work processes (sometimes even defining these processes completely afresh, thanks to the new technology) and integrating our entire clinical documentation and automation systems around the new CPOE technology. Although sophisticated clinical decision support systems have been implemented, they rather represent the cherry on top of the cake instead of a necessity. First, we built a stable, well-performing and trustworthy CPOE system, and only then did we add clinical decision support on top. In addition, standardization (of formularies, prescribing habits, and so on) is an extremely important step in building and maintaining a CPOE system. Standardizing data, databases, software, and functionalities is crucial as processes change, guidelines are incorporated into the CPOE system, institutions merge, care pathways arise, and transitions of care have to be managed. Once a clinical decision support system has been implemented, keeping it up to date and able to handle changing practices within all different clinical specialties is a challenging task.

Prior to implementing or changing the system, processes must be designed for the system's maintenance, including how to deal with downtime (both scheduled and non-scheduled), how to seamlessly coordinate CPOE with, for example, automated dispensing systems, and how to update or upgrade the system without losing functionality or data accuracy in current or previous medication records.

As a result, the so-called sociotechnical aspects (covered in detail in Chapter 10) deserve major attention. In our institution, for example, not only did new processes have to be managed, but also the new systems meant that workloads were heavily shifted towards doctors. Work that was previously done by multiple types of staff became the sole responsibility of doctors. The impact of many of these issues can be attenuated by a deep commitment of the medical and administrative leadership in a given institution and by making the best use of medical informatics staff.

Perhaps the most important point we should always keep in mind is that the medication use process is extremely multilayered and multidimensional, concerning patients with multiple diseases and based on a huge variety of information around the medications themselves and on the experiences and skills of health professionals. We cannot manage or support this process with linear, single-minded pieces of software or single-dimensional alerts without taking this sociotechnical context into consideration.

20 Technological Approaches for Medication Administration

Johanna I. Westbrook and Melissa T. Baysari

CONTENTS

KEY POINTS

- Automated dispensing systems in patient care areas, particularly in hospitals and nursing homes, can support the safe and efficient administration of medications.
- Smart pumps can reduce administration errors with the high-risk intravenous administration of medication.
- Barcode medication administration systems can help ensure that the right patient gets the right medication.
- Integrating these technologies into a closed-loop process is likely to produce the greatest benefit compared to standalone systems.

INTRODUCTION

Internationally, there is increasing investment in medication-related technologies designed to improve the safety and appropriateness of medication therapy. This chapter will discuss a range of technologies designed to support the safe and efficient administration of medications to patients in hospitals and in care or nursing home settings. The introduction of individual technologies may each produce benefits, yet increasingly, evidence indicates that technological interventions that are integrated and support a closed-loop medication process will produce the greatest reductions in medication errors and harm.

Regardless of the technological innovation to be implemented, careful consideration needs to be given to understanding the potential benefits expected, the mechanisms by which the intervention will achieve those benefits, and current evidence that demonstrates these effects. For example, smart pump technology is expected to deliver reduced medication administration errors (MAEs) through mechanisms such as reducing cognitive demands on nursing staff (for example, by reducing the need for calculations) and alerting nurses when they try to administer a drug outside recommended infusion rates. Later in this chapter, we outline some of the evidence regarding the successes of this technology, as well as automated dispensing systems (ADS) and barcode medication administration (BCMA) systems, in achieving safer medication administration. Consideration also needs to be given to the context in which technologies are to be used and how workflows may be disrupted or enhanced. Finally, with all such technologies, it is likely that they will introduce new problems, and thus monitoring for both expected and unexpected consequences is crucial to ensuring new technologies are both safe and effective. At the end of this chapter, there is an expert commentary about how such approaches have been introduced in the United Kingdom.

AUTOMATED DISPENSING SYSTEMS

ADS are variously described as automated dispensing cabinets, automated dispensing devices, automated distribution cabinets, automated dispensing machines, and dispensing robots. These computer-controlled devices are designed to securely store, dispense, and track medications, and as a result, reduce the occurrence of medication errors. Improvements in workflow and cost efficiencies are also expected as a result of reduced staff time requirements, improved storage capacity and stock control, and the more appropriate allocation of staff to tasks according to skill levels (Franklin et al. 2008; Fung and Leung 2009).

ADS may be based within pharmacies for the dispensing of medications or based in hospital wards or other patient care areas for the storage and supply of medications to individual patients. Pharmacy-based ADS are discussed in Chapter 21; here we focus on ADS in patient care areas. The adoption of ADS varies substantially by country. A 2011 survey in 100 English National Health Service Trusts (McLeod et al. 2014) found that only 7% used an electronic drug cabinet in some wards. In contrast, a 2011 study of 562 U.S. hospitals (Pedersen et al. 2012) found that 89% used ward-based automated dispensing cabinets.

Ward-based ADS are electronic medication storage devices that usually comprise a cabinet and/or trolley. They can be used to replace some or all of the manual drug distribution processes, including traditional floor stock and individual patient dispensing systems, and provide nursing staff with ready access to patients' medications at the point of care. These ADS may be profiled, which means that a patient's orders are entered into the system either manually or via an electronic medication management system (James et al. 2013). Medications are stored within individual patient- or product-specific drawers. Typically, when a patient's details are entered, the appropriate drawer opens, enabling selection of the correct medication (James et al. 2013). Non-profiled ADS provide secure access to medications via staff entering a password or the use of biometric identification, then providing access to all of the medications in the cabinet. Some ADS provide access to individual unit doses (Schwarz and Brodowy 1995), while others provide access to a multi-dose pack from which the nurse has to then select the appropriate number of doses (Franklin et al. 2007a).

Promoted advantages of ward-based ADS include error reductions, faster times to the administration of first doses of newly prescribed medications, improvements in the efficiency of the medication management process due to reduced time spent by nurses searching for medications and counting controlled (narcotic) drugs, the ability to reduce interruptions if ADS are in a quiet location, improved tracking of medications, and reduced data entry. ADS interfaced with barcode scanning and other clinical or administrative information systems further facilitate functions such as restocking, discharge medication preparation, and billing. However, the evidence base to support these claims is relatively limited, with few rigorous evaluations in different settings.

Several observational studies have assessed the impact of ward-based ADS on medication errors. Decreases in missed medication doses (Borel and Rascati 1995; Schwarz and Brodowy 1995) and the rate of wrong dosage administrations (Borel and Rascati 1995) across wards using different brands of ADS have been reported. Implementation of ADS in an intensive care unit (ICU) in a French hospital resulted in an overall reduction in errors related to the selection, preparation, and administration of medications (Chapuis et al. 2010). In an uncontrolled before-and-after study in a U.K. hospital, a closed-loop system comprising electronic prescribing (e-prescribing), ward-based ADS, and barcode patient identification was reported to reduce MAEs from 7.0% of 1473 non-intravenous (IV) doses pre-intervention to 4.3% of 1139 doses post-intervention (Franklin et al. 2007a).

ADS also provide the opportunity to deliver alerts or to monitor drug use. For example, Sikka and colleagues (2012) reported the use of prompts in an emergency department system that asked nurses to confirm that blood cultures were taken prior to dispensing antibiotics. In a Canadian hospital, physicians were requested to enter an indication for the use of aprotinin (an expensive medication used to reduce bleeding and the need for blood transfusions during complex surgery) whenever they removed it from the ADS (Fung and Leung 2009). Reports generated from the system were then used to monitor the appropriateness of aprotinin use and to provide feedback to physicians. The hospital reported a 50% reduction in inappropriate use. ADS also support the tracking of medication use in real time. For example, administered doses of "as-required" medications can be tracked within a patient's profile in the system, allowing for up-to-date information and monitoring (Fung and Leung 2009). A number of studies evaluating staff attitudes towards ward ADS have been published (Chapuis et al. 2010; Fanikos et al. 2007; Sirois et al. 2013), and a before-and-after study suggests that staff attitudes tend to improve after ADS implementation (Chapuis et al. 2010).

SMART PUMPS

"Smart pumps," also known as "intelligent infusion devices," are IV infusion pumps that incorporate dose error reduction software to alert users when administration rates are programmed outside the recommended parameters for that medication. A drug library is developed that contains selected medications and fluids with pre-programmed elements, including dose, rate, or concentration parameters. If a clinician programs an infusion outside the pre-set limits, an alert is triggered. For high-risk drugs, "hard limits" may be set, preventing the clinician from proceeding with the administration at all. In these cases, the pump will need to be re-programmed within the set limits. In contrast, "soft limits" can be overridden by a clinician (with or without providing a reason) and the administration can proceed. Drug libraries can be tailored for different ward contexts, but development and maintenance is often resource intensive and appropriate governance structures for their review and monitoring are required (Phelps 2011). The accompanying commentary at the end of this chapter considers these issues further.

Wireless smart pumps have the advantage of enabling the fast downloading of information from and to pumps across the hospital (for example, to facilitate the updating of drug libraries). Interfacing with barcodes and e-prescribing and electronic medication administration record (eMAR) systems supports a closed-loop medication process and is likely to provide the greatest safety outcomes (Institute for Safe Medication Practices 2009). A feature to permit the remote adjustment of smart pump infusions via the internet in response to changes in patients' needs has also been suggested (Macaire et al. 2014).

Given the high error rates found in IV medication administrations (Taxis and Barber 2003b, 2004; Wahr et al. 2014; Westbrook et al. 2011), smart pumps provide a potentially valuable technology for supporting safer medication administration. They potentially reduce the risk of calculation errors and may prevent errors from reaching the patient. The data stored within smart pumps can also be used to guide quality-improvement activities (Institute for Safe Medication Practices 2009; Mansfield and Jarrett 2013). All data about alerts generated, including the time, date, and drug details, are automatically recorded, and reports of these data can be generated. Importantly, these data can be used to modify the drug libraries and provide feedback in close to real time to clinicians about potential errors (for example, instances when a nurse attempted to program an infusion outside recommended limits, was alerted and then re-entered a correct order) or actual errors (for example, when an infusion outside the recommended limits proceeded to administration).

Several studies have reported significant decreases in serious IV medication error rates following smart pump introduction (Fanikos et al. 2007; Manrique-Rodriguez et al. 2015; Pang et al. 2011; Wood and Burnette 2012). However, differences in study designs, settings, smart pumps, and outcomes assessed make direct comparisons difficult. Compliance and training are central factors associated with positive outcomes (Breland 2010; Herring et al. 2012; Manrique-Rodriguez et al. 2015). Rothschild and colleagues (2005) evaluated the impact of smart pumps implemented in cardiac surgery ICUs and step-down wards in a U.S. hospital in 2002. No significant change in medication errors was found following smart pump introduction. This was attributed to the fact that their smart pumps had no hard limits and nurses were able to bypass the use of the drug library, and did so for 25% of all infusions. Such results highlight the need for training and close monitoring of use. However, nearly a decade later, a study at the same hospital reported that the drug library was bypassed in only 8% of infusions, primarily for electrolyte solutions (Ohashi et al. 2013). Ohashi and colleagues (2013) reflected on the fact that there had been considerable efforts to improve the effective and safe use of smart pumps, including review of the drug dictionary by a multi-disciplinary committee on a routine basis, incident reporting and review, and the investigation of human factors issues.

Following the implementation of smart infusion pumps in a pediatric ICU in a Spanish hospital, 92% of infusions complied with the drug library. Review of hard-stop alerts suggested that 44 high-risk errors were prevented as clinicians re-programmed these orders in response to the alerts (Manrique-Rodriguez et al. 2015). While the use of hard stops within smart pumps has been shown to be effective, they should be used sparingly, as their overuse is likely to encourage the adoption of workarounds and negatively impact upon workflows and efficiency.

Interviews with staff have described some of the potential challenges of working with smart pumps, which include limited battery life, weight, and problems with the drug library and managing tubing (McAlearney et al. 2007). However, surveys have shown high levels of staff compliance and satisfaction with smart pump technologies that persist with use (Carayon et al. 2010; Manrique-Rodriguez et al. 2015; Mason et al. 2014).

Smart pumps only target a specific range of error types (such as dose calculation or infusion concentration or rate errors), and many errors associated with IV infusions will persist regardless of their use, particularly if smart pumps are not integrated with other clinical information systems. Training that targets both improved decision-making by clinicians and the use of the technology to support work practices is recommended (Trbovich et al. 2014).

BARCODE MEDICATION ADMINISTRATION SYSTEMS

BCMA systems are designed to enable verification of the "five rights" of medication management and therefore to prevent wrong patient, wrong dose, wrong time, wrong drug, and wrong route errors (Young et al. 2010). Patients are issued with identification tags—typically wristbands—which incorporate a barcode that can then be scanned and matched electronically against the patient's medication chart (or electronic equivalent) and barcodes on medications. If a mismatch is detected, the nurse is alerted, typically via a visual or auditory warning on the computer screen. The system also helps to ensure accurate and complete documentation of the medication administration process (Barry et al. 1989).

Several studies, but not all, have demonstrated significant reductions in MAE rates following the introduction of BCMA (Bonkowski et al. 2013, 2014; Hassink et al. 2013; Paoletti et al. 2007; Poon et al. 2010). In a well-known evaluation, a direct observational method was used to identify and classify MAEs before and after the implementation of a BCMA system with an electronic medical record (Poon et al. 2010). Results for timing and non-timing MAEs were reported separately. Overall, the non-timing MAE rate declined from 11.5% to 6.8% of doses administered and the timing error rate decreased from 16.7% to 12.2% of doses administered following BCMA introduction (Poon et al. 2010). Similar studies in other settings report significant benefits: a study in a U.S. emergency department reported a reduction in the administration error rate from 6.3% to 1.2% of administrations (Bonkowski et al. 2013), a study in a U.S. transplant inpatient unit reported a reduction in the administration error rate from 4.8% to 1.5% of administrations (Bonkowski et al. 2014), and a study in a community hospital in the Netherlands reported a decline in the administration error rate from 7.2% to 3.6% of all administrations (Hassink et al. 2013). In all of these studies, BCMA was introduced simultaneously with another intervention (typically an electronic medical record) and no control groups were included, making it difficult to determine whether the observed changes were in fact the result of BCMA implementation.

In one study, where BCMA appeared to be introduced as a single intervention in a U.S. medical ICU, the administration error rate declined significantly from 19.7% of 775 administrations to 8.7% of 690 administrations following BCMA introduction. This result was primarily due to a significant reduction in wrong time errors (from 18.8% to 7.5%), while there were no significant changes in other error types before and after BCMA introduction (DeYoung et al. 2009). This last finding is interesting, because whether administration errors are reduced post-BCMA may be at least partly related to the inclusion or exclusion of timing errors in calculating error rates and how this error type is defined (such as 90 minutes either side of the intended administration time). Some studies reporting benefits of BCMA have excluded wrong time errors (Bonkowski et al. 2014), but not all (DeYoung et al. 2009). In studies that have included this error type, wrong time errors represent the majority of MAEs observed (Hassink et al. 2013; Morriss et al. 2009; Paoletti et al. 2007). Conversely, some studies report an increase in MAE rates following BCMA introduction, primarily driven by an increase in wrong time errors (Helmons et al. 2009; Morriss et al. 2009). Morriss and colleagues (2009) reported that the introduction of BCMA in a neonatal ICU was associated with an increase in the MAE rate from 69.5 to 79.7 per 1000 doses, principally because of a 117% increase in wrong time errors. Helmons and colleagues (2009) reported that the introduction of BCMA in two medical/surgical wards and two ICUs was not associated with changes in error rates, but when wrong time errors were excluded from the analysis, the error rate on medical/surgical wards decreased by almost 58% following the introduction of BCMA. There is no indication that wrong time errors are the result of BCMA negatively affecting workflows (Morriss et al. 2009). Instead, it has been suggested that the increase (or failure to observe a decrease) in this error type may be due to the increased precision of the recorded administration times in BCMA (DeYoung et al. 2009).

Another U.S. study that reported an increased rate of MAEs following BCMA introduction utilized log data exported from a BCMA system in order to determine the number of errors detected and averted by BCMA introduction (Low and Belcher 2002). In this study, which was conducted in two medical–surgical inpatient units, MAE rates were derived from standard incident report forms pre-BCMA and from BCMA logs post-system implementation. An 18% increase in MAEs was reported, but these findings are difficult to interpret because different sources of error data were used in each study period (Low and Belcher 2002).

Errors identified by BCMA (i.e., alerts triggered) and events averted (i.e., alerts resulting in a revised administration) have been used as measures of the impact of BCMA on medication errors (Lawton and Shields 2005), but studies that have assessed the appropriateness of the alerts generated in these systems suggest that the majority are not clinically relevant (FitzHenry et al. 2011; Radecki et al. 2012; Sakowski et al. 2005). In a review of alerts generated by a BCMA system, it was revealed that 70% of the events that were classed as errors and triggered an alert were in fact appropriate administrations (Sakowski et al. 2005). When errors detected by a BCMA were presented as de-identified scenarios to a panel of clinicians, fewer than 10% were rated as moderate or severe MAEs (Sakowski et al. 2005). Similarly, when the alerts triggered during warfarin administrations with a BCMA system were reviewed by pharmacists to assess appropriateness, only 4.3% ($n = 99$) were judged to be clinically meaningful (FitzHenry et al. 2011). In an interesting study, three data sources (incident reports, direct observation, and BCMA data logs) were utilized to identify MAEs before and after the implementation of BCMA (Sakowski et al. 2005). The BCMA warned nurses of medications not currently in a patient's eMAR, medications for the wrong patient, abnormal test results, and allergies. It was concluded that BCMA logs grossly overestimated MAE rates, with the number of events classed as "averted medication errors" far exceeding both voluntarily reported and directly observed medication errors (Seibert et al. 2014). This may be at least partly because the majority of BCMA-averted medication errors related to abnormal laboratory results, an error type not identified by the other sources.

Few studies have examined the use of BCMA in nursing homes. One used BCMA data to determine the impact of BCMA on MAEs (Szczepura et al. 2011). BCMA log data were extracted and reviewed for 13 U.K. nursing and residential homes over a 3-month period. Discrepancies identified by the BCMA systems between prescribed and administered medications were classed as "averted" MAEs. The BCMA found discrepancies in 1.2% of medication administrations, with the most common error type being attempting to give a 4-hourly medication too early (Szczepura et al. 2011). The potential role of BCMA in this setting is also highlighted in the commentary at the end of the chapter.

A high rate of alerts generated by BCMA has been identified as one of the factors contributing to their misuse (Koppel et al. 2008a). Studies assessing BCMA effectiveness report that the impact of the technology on error rates is hampered by the system not always being used as intended (Poon et al. 2010). Much research has investigated nurse workarounds following the introduction of BCMA (Koppel et al. 2008b; Patterson et al. 2006; Rack et al. 2012; van Onzenoort et al. 2008). For example, in an ethnographic study of BCMA use by nurses, 28 nurses were directly observed for 79 hours during medication administration rounds in three Veterans Administration hospitals (Patterson et al. 2006). Various workaround practices were identified, including nurses bypassing the use of the scanner by typing in patients' identification numbers, scanning medications for multiple patients before medications were administered to the first patient, and scanning multiple patients' wristbands at once. In a systematic evaluation of BCMA workarounds, a mixed-methods approach was used to identify workarounds, probable causes, and potential errors resulting from each workaround (Koppel et al. 2008b). Fifteen types of BCMA workarounds were identified, including the omission of process steps (such as failing to scan the medication barcode), steps performed out of sequence (such as documenting administration before medication is administered), and unauthorized BCMA process steps (such as disabling audio alarms on the device). Probable causes included a range of factors: technology related (such as multiple scans needed to read an individual barcode), task related (such as the scanning procedure being too slow), organizational (such as a medication order not being in the system), patient related (such as patient activity interfering with BCMA use), and environmental (such as the location preventing appropriate BCMA use). Overall, it appears that workarounds are the result of limitations in BCMA design, implementation, and integration into workflows (Koppel et al. 2008b).

Observational and qualitative work with nurses has also revealed that the implementation of BCMA systems affects nurses' work practices (Bargren and Lu 2009; Novak and Lorenzi 2008; Novak et al. 2013), time spent completing work (Dwibedi et al. 2011; Poon et al. 2008; Tsai et al. 2010), and nurses' problem-solving (Holden et al. 2013b). Several studies have demonstrated that the introduction of BCMA frees up nurse time and allows them to spend more time in direct care with patients. For example, in a time-and-motion study of nurses' work before and after the implementation of BCMA in a U.S. tertiary hospital, it was found that BCMA did not increase the proportion of time nurses spent in medication administration activities, but did increase the proportion of time spent in direct care (from 26.1% to 29.9%) (Poon et al. 2008). Similarly, in a time-and-motion study in an ICU, the mean time spent in direct patient care activities increased following the introduction of BCMA, from 47.4 seconds to 182.3 seconds per medication administration (Dwibedi et al. 2011).

CONCLUSION

This chapter has described a range of technologies designed to support the safe and efficient administration of medications in different settings. ADS, BCMA systems, and "smart" IV pumps are all potentially useful tools, although a careful understanding of context and workflow is needed. Integrating these technologies to produce a closed-loop medication process is also likely to be of the greatest benefit. However, as with all technologies, new problems can arise, and monitoring for these is essential to ensure overall safety benefits.

COMMENTARY

Dr. Yogini Jani

Medication Safety Officer and Lead Pharmacist for Medication Safety, University College London Hospitals NHS Foundation Trust, London, United Kingdom

Safe medication administration practices rely heavily on an individual's knowledge, skill, and ability to identify and rectify any problems that may exist before the patient receives a medication. Administration is a critical stage to ensure the "five rights" of medication practice are achieved: the right patient gets the right medicine, at the right dose, by the right route, and at the right time. The cognitive load on the individual health care professional is considerable; the individual needs to complete the mechanistic task of administration, as well as more analytical and investigative processes to assure appropriateness and accuracy.

In the United Kingdom, medication administration in hospitals is rarely done in isolation or without interruption; there are often other demands on the nurse's attention, which can adversely affect safe and accurate administration. In care home settings, medication administration may be the responsibility of workers who have been trained, but are not from a nursing background. Therefore, some of the technological approaches described in this chapter, such as BCMA systems that check and alert any mismatch between medication, patient, and the prescription, could be beneficial in either setting.

Safe and efficient medication administration also relies on accurate dispensing. In most U.K. hospital settings, medicines are provided in their original packaging on a "ward stock" basis, with only non-ward stock items dispensed and labeled for individual patients. In contrast, U.K. care homes often use individualized packaged medication to enable administration by workers who are not trained nurses. Centralized ADS are increasingly the norm in hospital pharmacy settings, with their potential to increase the efficiency of the dispensing process and eliminate medication selection errors. However, the use of ADS in the clinical area or for the provision of individualized packaged medication has been limited to date. This is partly due to the limited evidence for the cost-effectiveness of such systems in the context of a centrally funded health service where medication costs for individual patients are not itemized.

Selecting and introducing new technologies also requires consideration of existing workflows, as these systems often require changes in the workflows and practices of the immediate users, as well as those involved in upstream stages of the medication processes. For example, the planned implementation of smart infusion pumps in the neonatal unit in our hospital required a review and standardization of infusion solutions, in contrast to the previous practice of prescribing infusions based on the individual prescriber's preference and familiarity with certain medicines, or based on patient-specific characteristics, such as weight or age. An unexpected change as a result of this standardization was the opportunity to provide some infusions as "ready-to-administer" formulations from pharmacy, using a centralized IV additive service, thus further reducing the risk of errors.

This chapter presents the evidence for the technological approaches that are available to support safe, efficient, and effective medication practice, but also highlights the need for their evaluation, as new technologies may bring about unexpected changes and effects.

21 Innovations in Dispensing

Matthew Reynolds, K. Lynette James, and Johanna I. Westbrook

CONTENTS

KEY POINTS

- This chapter considers a range of innovations relating to the transmission of prescriptions, the dispensing label, drug selection, and the use of pharmacy automated dispensing systems, among others.
- There is evidence that both appropriately configured barcode technology and the use of automated dispensing systems can reduce dispensing errors within the hospital pharmacy setting.

- Technological solutions should not be seen as a panacea, and any negative unintended consequences should be identified and addressed.

INTRODUCTION

This chapter describes innovations in pharmacy dispensing practices, with a focus on patient safety. As well as physical and procedural interventions, we consider in some detail the potential benefits of pharmacy automated dispensing systems (ADS). Whether or not something is considered an innovation depends on the baseline system: some of the innovations detailed here are already embedded firmly into practice in some settings, whereas they may be aspirational in others. The interventions described are by no means an exhaustive list, and not all will be suitable for every setting. We present them as examples for consideration, together with some evidence for their benefits.

As outlined in Chapter 3, the term "dispensing" covers the range of processes used to prepare medication for consumption by a patient in accordance with a prescription or medication order. Dispensing forms part of the medication use pathway in primary, secondary, and tertiary care settings; unsurprisingly, the processes involved vary between countries, health care systems, and settings. However, the shared goal is to ensure that prescribed medicine is dispensed safely and to a high-quality standard.

We consider innovations relating to the transmission of prescriptions, the dispensing label, drug selection (including pharmacy ADS), and other relevant safety initiatives, including identifying and rectifying drug-related problems (DRPs), as part of the dispensing process. At the end of the chapter, there is a commentary from a pharmacist working in Portugal referring to how these innovations relate to the context in which she works.

TRANSMISSION OF PRESCRIPTIONS TO THE PHARMACY FOR DISPENSING

Electronic Prescriptions

As health care moves towards the digitalization of medical records, pharmacies will increasingly have to dispense from electronic prescriptions. Different types of "electronic" prescriptions exist. Many prescriptions are created with the help of some sort of electronic support. In many settings, prescriptions are routinely typed on a computer and a printed copy is produced for dispensing, analogous to using a word processor instead of handwriting. However, more advanced electronic prescribing (e-prescribing) systems exist, as discussed in Chapter 19. Where outpatient prescriptions are also transmitted electronically, such as via the U.K.'s Electronic Prescription Service, these can be used both to populate patients' medication records within the pharmacy and to generate dispensing labels automatically, removing the requirement for manual data entry in the pharmacy. There are, therefore, potential advantages in terms of preventing labeling errors, although these benefits may not automatically be realized without collaboration between prescribers, pharmacists, and system suppliers (Franklin et al. 2014b).

IDENTIFYING AND RECTIFYING DRUG RELATED PROBLEMS AS PART OF THE DISPENSING PROCESS

Many potential definitions and classification systems exist for DRPs (van Mil et al. 2004). Here, we consider DRPs to include any problems associated with prescription legality and validity, supply problems, and issues relating to patient preference, in addition to clinical problems in relation to the appropriateness of the medication prescribed.

An example of an innovative practice to address DRPs is the New Medicines Service in England. Community pharmacists proactively offer this service, funded by the National Health Service, to patients who have recently started taking medicines for a certain specified long-term condition. Patients make three appointments with their community pharmacist over the first 4 weeks of their new drug treatment, providing the opportunity to ask questions and discuss any problems that they are experiencing. The pharmacist can then take action—for instance, recommending changes of formulation for patients with swallowing difficulties—to try and ensure that the prescribed regimen is clinically appropriate and acceptable for the patient. Importantly, patients can also be referred to the New Medicines Service by hospital staff or by their general practitioners.

Pharmacies often keep electronic records of all prescriptions dispensed for an individual patient. This is often called the patient medication record (PMR), but may also be referred to as the pharmacy information management system or pharmacy computer system. Some PMRs also capture "over-the-counter" sales on the same database. PMR databases are now being targeted as a resource capable of supporting patient safety initiatives, as described in Chapter 17. Decision support software is also being developed and incorporated into pharmacies' computer systems, using the information contained within the PMR to highlight potentially unsafe circumstances (such as drug–drug interactions or cautions related to pregnancy or age) to allow remedial action to be taken. The impact of such decision support systems has been examined in a systematic review (Ojeleye et al. 2013). The authors found that some systems are prone to multiple false alerts, in which users are unnecessarily alerted to safe dispensing, resulting in disruption to workflows and potentially increasing the chance that users will ignore important alerts. Ojeleye and colleagues (2013) recommend that alerts should only appear when there is a maximum chance of them resulting in the user taking some action. Using information specific to a patient may help increase the specificity of alerts. Setting the sensitivity and specificity of alerts is a key challenge, as it is for the e-prescribing systems discussed in Chapter 19.

INNOVATIONS IN LABELING

In many countries, it is a legal requirement that dispensed medicines have a dispensing label applied at the point of dispensing that includes a core set of information. Guidance also exists from the World Health Organization (WHO) for health care workers in countries where information on labels is not mandated (Management Sciences for Health 2012).

Computer-Generated Labels

In comparison to handwritten labels, computer-generated and printed labels reduce the manual labor needed to dispense an item and have other potential patient safety benefits. Additional warnings and other information can be automatically included for specific products, and the software can be updated to automatically reflect any changes in labeling requirements. Labels are usually legible, and their content can be standardized. However, as many of these benefits rely on up-to-date software, any software error can give rise to a systemic error and consequently affect all labeled items (Franklin et al. 2013). In order to deal with infrastructure failures such as power cuts or software crashes, disaster-management plans also need to be in place.

Typographical Innovations

"Tall man" lettering has been introduced to emphasize the differences in commonly confused drug names. Part of a drug's name is written in capital letters to highlight the difference in spelling from other easily confused drugs. For instance, the two commonly confused medicines hydroxyzine and hydralazine would be presented as hydrOXYzine and hydrALAzine. Tall man lettering may be used by the manufacturer on the original medication packs, but can also be incorporated into dispensing labels applied within the pharmacy. The U.S. Food and Drug Administration (2013b) has requested that manufacturers use tall man lettering to distinguish between 16 commonly confused pairs of lookalike generic drug names; other recommended lists also exist (Institute for Safe Medication Practices 2014a).

There are also other typographical innovations designed to reduce errors. Parts of the name may be presented in white font on a black background, for example, or certain font styles used to better differentiate between upper- and lower-case letters (Gabriele 2006). While there has been relatively little testing of such typographical innovations in real-life environments (Filik et al. 2010; Gabriele 2006; Or and Chan 2014), they are inexpensive and unlikely to have significant disadvantages.

Labeling for Individuals with Different Needs

Individuals' ability to read instructions on dispensed medicines varies, both throughout the world and within individual health care systems. An increasing number of regulations and techniques have been introduced to enable people with different needs to read and comprehend instructions for using a medicine. For medicines used within member states of the European Union, Council Directive 2001/83/EC (European Parliament and Council of European Union 2009) requires the name of the medicine to be displayed in Braille on the package. In parts of the world with lower literacy levels, WHO suggests pictograms may be useful to indicate when a dose should be taken (Management Sciences for Health 2012). The pictograms should be presented alongside rather than instead of written instructions, as the instructions may be useful for a family member. Recently, some states in the United States (for example, New York State Education Department 2004) have passed laws requiring community pharmacies to label medications with languages other than English for patients with limited English proficiency.

DRUG SELECTION

Purchasing for Safety

One way organizations can promote safe systems is by ensuring that they purchase products whose packages differ from other similar products. If alternatives do not exist, staff can be informed of the potentially confusing products (for example, by using warnings on the dispensing shelves or via the computer system used for labeling and dispensing).

Barcodes

Barcodes are already widely used to support patient safety in health care. However, in many settings, the use of barcodes to support dispensing lags behind the use of barcodes in administration. Few studies exist that show the impact of barcodes on dispensing-related safety outcomes (Leung et al. 2015), but evidence exists of reductions in dispensing errors when supported by appropriately configured barcode technology in the hospital setting (Poon et al. 2006). As with any system, barcode systems are not infallible. On top of potentially dangerous workarounds, such as those described in Chapter 20 (for example, Koppel et al. 2008b), physical limitations exist. Barcodes can be obscured by over-labeling or damaged by movement or moisture. Unavailable or uncharged scanners may also encourage staff to take shortcuts.

At present, the most common types of barcode are linear (one-dimensional) barcodes, which can store a limited amount of data. Typically, this is a unique numerical code that is linked to the manufacturer, medication name, strength, form, and pack size. A matrix (two-dimensional) barcode can store much more information, such as batch numbers and expiry dates. The forthcoming European Falsified Medicines Directive will require at least some categories of medication to have a unique identifier on each individual pack. While mainly intended to prevent the use of counterfeit medication, this initiative has potential patient safety benefits in enabling the verification of medication identities and expiry dates at the point of dispensing and facilitating product recalls.

Automated Dispensing Systems

The use of ADS is growing. In the United States, 4.5% of all hospitals reported having a fully automated pharmacy distribution system in 1999 (Ringold et al. 2000); this had risen to 10.8% by 2011 (Pedersen et al. 2012). There are numerous types of ADS, sometimes known as "dispensing robots." James and colleagues (2013) identify three types: pharmacy-based original-pack dispensers; repackaging systems; and ward-based automated dispensers. We discuss the first two here; ward-based systems are discussed in Chapter 20.

Original-pack ADS are now used in many U.K. hospital pharmacies (for example, Swanson 2004). These systems automate medication storage, stock selection and, occasionally, product labeling within the pharmacy. Medication is usually stored within the ADS based on recognition of the product barcode, and retrieved via an interface with the pharmacy labeling and stock control software. During label generation, a signal is transmitted from the labeling software system to the ADS, initiating stock selection. The ADS selects the requested medication and transfers the product to the delivery station via a conveyor belt or chute. Some ADS also have an integrated labeling device that affixes the appropriate dispensing label to the product. The loading system affects the degree of staff input required to operate the ADS. Some ADS employ a hopper loading system, whereby the robot itself picks boxes from the hopper, reads the barcodes, and automatically loads them into an appropriate place. Other ADS use manual loading systems, in which an operator must scan each individual box and place it onto a conveyor for loading.

ADS can also be used to produce both unit dose and multi-dose monitored dosage systems (Neuenschwander 1996). Some ADS contain single or multiple hand-filled canisters, each containing a particular medication used to fill the unit dose or multi-dose blister cards. Other ADS function as repackaging systems by automating the removal of medications from the manufacturers' original pack, then repackaging the medication into unit dose packs or multi-dose blister cards.

There are many arguments in favor of ADS: increased dispensing safety, shorter drug-retrieval times, reduced shelf space requirements, and cost savings through stockholding and stock rotation. We next explore these aspects, focusing on those relating to patient safety.

Dispensing Errors

One of the perceived advantages of ADS is that many of the factors contributing to human error do not apply to robots: ADS work to a predefined set of rules from which they cannot deviate. They cannot become flustered by high workloads or become complacent at quieter times. One contributor to "picking errors," in which the wrong pack of medication is selected for labeling, is confusion between look-alike or sound-alike drugs. These are pairs of medicines with names that look or sound very similar. Many techniques to reduce individuals' propensity to confuse look-alike or sound-alike drugs have been suggested (Ciociano and Bagnasco 2014), but ADS are likely to be particularly helpful in this respect. However, ADS are not a panacea for all dispensing problems—they will always pick the medication selected by the operator, for example, even if this is not the medication required.

The majority of research on the impact of original-pack ADS on dispensing errors originates from the U.K. James and colleagues (2013) conducted a detailed evaluation of the impact of ADS on self-reported internal ("near-miss") dispensing errors in a 600-bed hospital. The study concluded that the rate of internal dispensing errors decreased from 0.64% of dispensed items to 0.28% post-automation. Similar findings have been reported by others (Fitzpatrick et al. 2005; Franklin et al. 2008).

Flynn and Barker (2006) evaluated the impact of multi-dose ADS on dispensing errors in two U.S. community pharmacies. This system automated selection of the appropriate medication, counting, filling, and labeling of medication bottles. An observer inspected dispensed medications and noted any errors. At one of the study pharmacies, the ADS significantly reduced the overall rate of dispensing errors from 2.7% of prescriptions to 1.8%. However, there was no significant difference at the second pharmacy.

Several researchers have also evaluated ADS that automate the selection, packaging, and labeling of unit doses. Klein and colleagues (1994) compared the accuracy of filling unit dose carts manually and with the ADS. Independent checking of the carts revealed that automation was associated with fewer filling errors (0.65% of doses) compared to manual cart filling (0.84%). Previous research has also reported manual cart filling to be 92.62% accurate, while an ADS was 99.98% accurate (Kratz and Thygesen 1992).

Use of pharmacy-based ADS within community pharmacies to deliver medications to nursing homes has also been reported to improve the efficiency of dispensing, allowing more regular deliveries of medications, supporting more rapid responses to changes in residents' medications. Nursing staff also reported the need to undertake fewer verification tasks. A Canadian study across multiple care homes examined the number of voluntarily reported medication administration errors before and after the introduction of an automated pharmacy packaging device combined with mobile dispensing carts. The interventions were associated with a significant increase in medication administration errors; however, this increase was attributed to the new technology making errors more apparent to staff and thus more likely to be reported (Baril et al. 2014).

Stock Rotation

Good stock rotation helps ensure that dispensed medicines are within date and therefore suitable for use. Stock rotation can be based on the "first in, first out" system, whereby medicines received first are used first, but should ideally be based on the actual expiry date. Regardless, such systems rely on dispensers' vigilance and adherence to procedures. Some ADS can be programmed with information on expiry dates so that stock with the earliest date can be preferentially dispensed; some can also record batch numbers, which can be used to help respond effectively to medicine recalls. One study (Brinklow 2006) reported a reduction in expired stock from 0.5% of the drug budget to 0.3% after the introduction of an ADS, and substantial savings have been reported elsewhere (BBC News 2010).

Dispensing Time and Patient Waiting Times

A further driver for ADS is the potential to reduce the staff time needed to fulfill prescriptions. A U.K. study (James et al. 2013) showed that introducing original-pack ADS increased the number of items dispensed by pharmacy staff from 9.2 to 31.2 items per staff member per hour. In an earlier U.K. study, Franklin and colleagues (2008) found that ADS reduced the time taken to select the medicine from a median of 49 seconds per item to 32 seconds in one hospital pharmacy, and from 19 seconds to 0 seconds per item in another. The median of 0 seconds was because, with the ADS, selected items were delivered before the labeling process had been completed. However, no subsequent reduction in patient waiting times was identified, indicating that the overall dispensing time is a function of a wider range of activities than just selecting the medicines. Conversely, Fitzpatrick and colleagues (2005) showed a substantial decrease in prescription turnaround times after implementing ADS. Finally, a carousel system was shown to increase the turnaround time of "stat" (i.e., immediately administered) medicines in a U.S. hospital (Temple and Ludwig 2010).

Reducing the amount of time staff spend dispensing medicines does not necessarily translate into safer care, however. Patient safety could benefit only if staff were able to spend more time engaging in other activities related to patient safety, which is not always the case (Temple and Ludwig 2010). ADS also require staff members to operate or load them (Fitzpatrick et al. 2005), and this must be factored into any evaluation of time savings.

Disadvantages of Automated Dispensing

As well as the capital and maintenance costs, there may be unintended negative consequences of automated dispensing. There is also the possibility that automating systems may result in the de-skilling of staff members; this has not yet been studied.

OTHER SAFETY INITIATIVES

Second Checks

In many settings, pharmacists (or other pharmacy staff) perform second checks on dispensed items. Non-pharmacist second-checkers usually have to first demonstrate their checking accuracy by passing examinations and maintaining accreditations through periodic assessment. The general principles of double-checking are discussed further in Chapter 18.

It is also possible to organize workflows in such a way that the chances of detecting errors are maximized. For instance, WHO advocates that one person receives and checks the prescription, a second dispenses the medicine, and a third hands the medicine to the patient (Management Sciences for Health 2012). All three then perform checks in isolation from each another. The effectiveness of this system (and indeed any double-checking system) is underpinned by staff following the standard operating procedures, checking independently and not assuming the item they are checking is correct.

Supporting Safe Use by the Patient

Interventions aimed at supporting safe use by the patient include providing child-resistant packaging, opaque bottles with tight-fitting caps to help maintain medicine stability, and oral syringes to make it easier to measure and administer precise volumes of liquid.

THE FUTURE FOR DISPENSING

Telepharmacy may prove to be the largest change to pharmacy and dispensing practices in the next few years. The internet allows huge amounts of information to be sent rapidly and securely. Building on this, "off-the-shelf" technologies, such as teleconferencing and videoconferencing, could enable pharmacists to review prescriptions remotely and provide pharmaceutical care consultations to patients outside of traditional health care settings. Bespoke innovations such as remote dispensing machines mean that medications can also be dispensed remotely. Remote dispensing machines typically comprise secure storage under the remote control of a pharmacist, akin to a vending machine, which retrieves the prescribed medicines and produces a patient-specific label and affixes it to the box. The pharmacist reviews the prescription remotely and then electronically transmits dispensing instructions to the remote dispensing machine. An authorized person (often a health care professional, but sometimes the patient) can then remove the dispensed item from the cabinet. These systems are often employed where it is not viable to station a pharmacist or build a traditional pharmacy. Remote dispensing machines can also support a 24-hour dispensing service and often go hand-in-hand with telepharmacy services (Clifton et al. 2003) so that patients can be counseled by pharmacists on the medicines they receive.

In many countries, pharmacies are now able to supply medicines online. Such pharmacies fulfill prescriptions from a central location and deliver them by post or courier to the patient's registered address. In response to concerns about illegitimate online pharmacies and the supply of fake medicines, many regulatory authorities have taken action. For instance, the General Pharmaceutical Council in Great Britain allows pharmacies to register as an internet pharmacy if they fulfill certain conditions relating to the safe and appropriate supply of medicines, and the U.S. Food and Drug Administration's BeSafeRx campaign aims to educate consumers on how to safely use online pharmacies (U.S. Food and Drug Administration 2013a).

CONCLUSION

A range of interventions to support safe dispensing have been suggested. While regulation can drive innovation, in many cases, it is the relentless pursuit of improvement that propels the development of new and innovative working procedures. Some innovations, particularly ones based on modern technologies, will only be possible in settings in which the infrastructure is available to support them. Technological solutions should not be seen as a panacea. Time and commitment are required to develop functional solutions, and care should be taken to look for and ameliorate any unintended negative consequences.

COMMENTARY

Mara Pereira Guerreiro
Lecturer (Instituto Superior de Ciências da Saúde Egas Moniz and Escola Superior de Enfermagem de Lisboa) and Consultant (4Choice Health Consultancy), Lisbon, Portugal

This chapter discusses a number of innovations in dispensing that may benefit patient safety, ranging from the electronic transmission of prescriptions to the less sophisticated "tall man" letters.

In Portugal, there has been a strong government push towards the adoption of e-prescribing, both in hospital and community settings. Many hospitals have implemented e-prescribing, either throughout the entire site or just in some wards, sometimes integrated with electronic health records and/or clinical decision support (CDS) systems. Although the integration of such systems may prove highly beneficial, electronic prescriptions per se have the potential to avoid illegibility issues and unsafe abbreviations, as well as fostering completeness of information. These aspects have implications for patient safety, both in terms of the accuracy of dispensed items and pharmacists' ability to detect DRPs. Additionally, in the hospital setting, e-prescribing may eliminate the previous need for transcribing, provided that there is integration between prescribing and pharmacy systems. As in other countries, Portuguese hospitals have often implemented information technologies in a stepwise manner, which occasionally raises interoperability problems.

Portuguese community pharmacies routinely receive printed copies of prescriptions produced by e-prescribing software; handwritten prescriptions are only accepted in particular circumstances. Following pilot studies involving the Portuguese National Health Service and community pharmacies, the goal is to progress to a completely paperless e-prescribing system.

Ensuring dispensing accuracy is fundamental, although the verification of clinical appropriateness and the identification of other DRPs are also required to ensure patient safety at this stage of the medication use process. It is notable that Portuguese community pharmacies have used a computer system featuring PMRs and CDS for more than a decade. The PMRs are populated by both the prescription and nonprescription medicines dispensed, while also enabling documentation of patient-reported medical conditions and clinical data obtained in the pharmacy (such as weight, blood pressure, blood sugar, and cholesterol levels). The CDS system (linked to PMRs or, in their absence, focusing on dispensed items) produces tiered, color-coded alerts for drug–drug interactions, duplications, contra-indications, drug allergies, and previous adverse drug reactions; moreover, it provides default dosage information. Serious alerts are interruptive and require a justification from users; in all cases, the system provides information on issues that are flagged up and actions the pharmacist can take. At present, PMRs and CDS systems are available in nearly all community pharmacies. Despite their tremendous potential for improving patient safety, their adoption and implementation have not yet been studied, and their impacts on health outcomes, if any, remain unknown.

In terms of sophistication, the use of tall man letters is probably on the other side of the spectrum. The Portuguese medicines regulatory agency does not require tall man letters to be used on medication packaging by market authorization holders. Nonetheless, the agency has updated its guidance for the preapproval design of medicine names to minimize confusion with marketed products. Although not strictly linked with dispensing, this is an important step, as it addresses a key contributing factor for mix-ups involving look-alike or sound-alike medicines. Its impact will probably depend on the rigor of the preapproval tests used to ascertain susceptibility to confusion.

The use of tall man letters in dispensing labels produced by hospital and community pharmacies is discretionary and perhaps still uncommon. In these settings, an approach that also seems to be unexplored is employing tall man letters to differentiate look-alike or sound-alike medicines when selecting products from drop-down menus in software that is linked to ADS. It is likely that improving safety in dispensing practice will need a mixture of high-tech and low-tech solutions in the future.

22 Patient Involvement in Medication Safety

Sara Garfield and Anam Parand

CONTENTS

KEY POINTS

- There is growing awareness of the important role that patients and their carers can play in supporting medication safety in both community and hospital settings, although most empirical research has taken place in hospitals.
- A range of factors have been shown to affect patients' adoption of medication safety behaviors.
- The existing evidence suggests that the best way to promote proactive patient involvement is through direct encouragement by health care professionals; there is less evidence to support marketing campaigns or other interventions that do not involve direct professional contact.

INTRODUCTION

One approach to reducing the incidence of medication errors is to facilitate the greater involvement of patients with medication safety. It is increasingly recognized that it is essential to involve patients with their medication (Department of Health 2010; Royal Pharmaceutical Society of Great Britain 1997), including involvement in enhancing safety (Department of Health 2010; Lawton and Armitage 2012). Such involvement of patients can increase their satisfaction, improve their health outcomes, and reduce the likelihood of avoidable harm (Coulter and Ellins 2006). Patients and their carers are arguably likely to know a great deal about their usual medication. They are thus an important (and often final) defense against errors relating to their medication.

In this chapter, we will discuss the extent to which patients and health care professionals engage in and are willing to support patient involvement in medication safety. We will then present the facilitators and barriers to that involvement before moving onto an overview of intervention studies that have been carried out in order to encourage patient involvement in safety. The chapter concludes with a commentary from a patient on the points raised in this chapter, together with her views on the patient's role.

As there has been much more research in secondary rather than primary care settings on the topic of patients' active involvement in preventing medication errors, most of the examples we present in this chapter are from secondary care.

PATIENT-RELATED MEDICATION SAFETY BEHAVIORS

Typically, patient behaviors relating to medication safety include (but are not limited to) viewing their medication records, prompting staff to avoid errors, providing information to aid handovers between health care interfaces or professions, raising queries with prescribers, pharmacists, or nursing staff, and self-administering medication. It could be argued that being the *passive* recipients of information and adhering to prescribers' instructions are also safety behaviors. Although there has been extensive research on these areas, this chapter will focus on the more active ways in which patients can contribute to medication safety.

There is some evidence that patients do adopt such active behaviors in both primary and secondary care settings. Observational studies have shown that patients sometimes intervene to prevent medication errors in the hospital setting (Bolster and Manias 2010). In cross-sectional surveys in primary and secondary care, the majority of patients have reported being involved by asking questions about their medicines (Mohsin-Shaikh et al. 2014; Schwappach and Wernli 2010a) or carrying out some activities for monitoring their safety, such as checking medication names (Nau and Erickson 2005; Schwappach and Wernli 2010a). However, far fewer patients have been involved in self-administering their medications in hospitals (Mohsin-Shaikh et al. 2014) or reviewing their inpatient medication records (Cumbler et al. 2010; Mohsin-Shaikh et al. 2014), although the majority of patients have expressed a desire to do the latter. A systematic review of patient access to online primary care medical records (including records of their medication) found that such access may result in improvements in patient safety, primarily through patients identifying adverse drug reactions (ADRs) and errors in medication lists (de Lusignan et al. 2014).

There is a gap between inpatients' reported desire to be involved in medication safety and their reports of actually carrying out such behaviors. Mohsin-Shaikh and colleagues (2014) found that patients were significantly more likely to report that they would like to ask questions about their medicines in the hospital and review their inpatient drug charts than they were to report having actually done so. It is therefore important to understand what the barriers might be to inpatient involvement in medication safety in order to address this gap. Patients' concerns about speaking up may be warranted; a qualitative study has shown how speaking up to ensure medication was given at the correct time can lead to a breakdown of the relationship between the patient and the health care professional (Entwistle et al. 2010).

There is also some evidence that health care professionals are more positive towards patient involvement than are patients themselves. Davis and colleagues (2012) found that doctors, but not nurses, were significantly more likely to be willing to support other patients asking challenging questions than they were willing to ask these questions as patients themselves. In addition, Mohsin-Shaikh and colleagues (2014) found that health care professionals were more likely to report that they would support involvement than patients were to report that they would like to be involved, although the methodology used meant this could not be tested for statistical significance.

FACTORS CONTRIBUTING TO PATIENT INVOLVEMENT IN MEDICATION SAFETY

Vincent's adaptation of Reason's accident causation model (Vincent 2010) has been described in Chapter 8. In this section, we will use an adaptation of the model to categorize the factors perceived by patients and health care professionals as contributing to the likelihood of the former being involved in their medication safety.

The majority of studies in this area have been quantitative surveys of patients. Some quantitative studies have also investigated the views of health care professionals. Some of the studies have been carried out in specific patient groups, such as those undergoing chemotherapy (Schwappach and Wernli 2010a, b, c, 2011) or university employees (Hibbard et al. 2005), or have been limited to one hospital or organization (Cumbler et al. 2010; Davis et al. 2011, 2012; Mohsin-Shaikh et al. 2014). However, combined with other studies that include wider patient groups, they build a picture of the factors affecting patient involvement.

PATIENT FACTORS

Demographic Factors

Female gender (Mohsin-Shaikh et al. 2014; Schwappach and Wernli 2010b), younger age (Mohsin-Shaikh et al. 2014; Waterman et al. 2006), and non-Caucasian ethnic groups (Waterman et al. 2006) have been shown to be associated with increased patient desire and reported experience of participating in medication safety-related behaviors, although there are some conflicting findings in the literature. In an exploratory survey of 100 patients in one U.K. hospital, women were found to be more likely than men to report engaging in medication safety-related behaviors, such as looking at their drug charts, asking questions about their medications, prompting the administration of a forgotten medication, and alerting the health care professional to a wrong medication being given (Mohsin-Shaikh et al. 2014). In addition, patients under the age of 65 years were more likely to have engaged—and have the desire to engage—in such behaviors than were patients over the age of 65 years (Mohsin-Shaikh et al. 2014). Similarly, Waterman and colleagues (2006) found that in a sample of 2078 hospitalized patients in the United States, those aged between 40 and 65 years were more likely to report that they were comfortable asking about the purpose of a medication than were those over the age of 65 years. Waterman and colleagues (2006) also found that being of a non-Caucasian ethnic group was associated with an increased likelihood of asking about the purpose of a medication. In contrast to the findings of these two studies, a multiple regression analysis showed no association between age and engagement in safety-related behaviors in a study of 479 patients being administered chemotherapy in Switzerland (Schwappach and Wernli 2010b). However, in agreement with Mohsin-Shaikh and colleagues (2014), female gender was associated with such behaviors. In addition, Schwappach and Wernli (2011) found that education was not associated with engagement in safety-related behaviors in this group of patients.

Illness-Related Factors

Perhaps not surprisingly, the health condition of the patient has been perceived as a barrier to their participation in medication safety behaviors, both by health care professionals and by patients themselves. Mohsin-Shaikh and colleagues (2014) found that health care professionals believed that having a cognitive impairment impeded patients from participating in their medication safety. Schwappach and Wernli (2011) found that their health condition was also perceived by the patients themselves to be a barrier to their engagement in error prevention. A U.K. qualitative study, using interviews and focus groups, found that hospital inpatients reported that they were too unwell or not fully aware of what they were being given, which was a barrier to checking their medication in a hospital (Watt et al. 2009).

Patient's Knowledge and Experience

Increased knowledge of medication has been found to be associated with a higher level of engagement with medication safety-related behaviors. In a qualitative study of 30 chemotherapy patients, Schwappach and Wernli (2010c) found that patients were aware of their limited capability to detect errors. Some patients explained that at the start of therapy they had very limited knowledge, but as therapy continued, increased experience meant that it would be more likely that they would be able to detect errors. In a quantitative survey in the United States, 94% of 50 patients felt that reviewing hospital medication to increase their knowledge would have the potential to reduce errors (Cumbler et al. 2010).

In addition, knowledge and previous experience of errors have been shown to be associated with increased engagement in medication safety. Hibbard and colleagues (2005) conducted a survey of a convenience sample of U.S. university staff members without an employment history in health care. They investigated the likely engagement of these staff members in a number of patient safety-related behaviors, including those related to medication (for example, making sure the doctors knew about every medication being taken, making sure the doctors knew about allergies and previous ADRs, and confirming that the right medication and dose were being given). They found that those who had read about medical errors were more likely to report that they would take preventive action against errors. In addition, simply participating in the survey potentially made these people more likely to take preventive action (they were asked the same question at the beginning and end, and there was a significant difference in response). Another survey of U.S. university employees (not specifically health care professionals) found they were more likely to report that they engaged in medication safety behaviors, such as checking the name of the medication on the label, if they also reported that they had previously experienced a dispensing error (Nau and Erickson 2005). Similarly, other researchers have found that patients having experienced errors were more likely to report that they had challenged others regarding the safe administration of their chemotherapy (Schwappach and Wernli 2010a, b). Schwappach and Wernli (2010b) found that patients reported being more attentive to the drug administration processes during a follow-up interview than they had been at baseline. Many said they had learned from the initial interview about which issues to monitor and how to respond. In contrast, further work conducted by Schwappach and Wernli (2011) found that having experienced an error decreased the patients' intentions to perform safety-related behaviors when they were given hypothetical situations.

There is some evidence that health care professionals report more willingness to participate in their medication safety as inpatients in hospital than do "lay" patients (i.e., patients who are not health care professionals). Davis and colleagues (2011, 2012) carried out the same survey in a U.K. population of lay inpatients and health care professionals working in the hospital, asking about their willingness to participate in safety-related behaviors as inpatients, including asking whether the medication being prescribed is the correct medication for them and bringing in a list of medications and allergies. The results appeared to suggest that when they were patients in a hospital, health care professionals may be more willing to participate across all safety-related behaviors than lay patients.

Psychological Factors

Patient belief in their ability to prevent errors has been associated with involvement in medication safety. Hibbard and colleagues (2005) produced a "path analysis," which demonstrated that increased self-efficacy (i.e., patients' confidence in their ability to take effective action) was associated with an increased reported likelihood of taking preventive actions to reduce errors. Similarly, Schwappach and Wernli (2010a, b) found that increased perceived behavioral control (i.e., patients' belief in their own ability to prevent errors) was a strongly significant predictor of their intentions to participate in chemotherapy error prevention. In addition, Entwistle and colleagues (2010) found that U.K. patients from primary and secondary care settings who felt that they had the personal ability to assess problems were more likely to speak up if they thought something in their care was problematic, including issues related to their medication.

Patients were more likely to take preventive action to reduce errors if they felt that this behavior was expected of them by family and health care professionals and was a perceived social norm (Schwappach and Wernli 2010b, 2011). Entwistle and colleagues (2010) found that if patients thought there would be negative consequences from the health care professional involved in their care or had experienced a negative reaction to expressing their concerns previously, they may be less likely to speak up again out of fear of negative comments. Similarly, Watt and colleagues (2009) found that patients expressed the view that those patients who felt that they were in a vulnerable position if they raised a query might be hindered from checking their medications in a hospital. Schwappach and Wernli (2010a) found that instrumental attitudes (i.e., patients' cognitive beliefs regarding the outcomes of their behavior) were also a strong predictor of engagement and intention to engage in action to prevent errors related to their chemotherapy.

In contrast, patients' trust in health care professionals may act as a barrier to them becoming involved in safety-related behavior. Watt and colleagues (2009) found that, in the main, patients seemed to subscribe to the view that they should be able to rely on health care professionals to check medicines on their behalf in a hospital. Schwappach and Wernli (2010c) found that a global trust in staff safety practices was associated with a lower frequency of engagement in some safety strategies. It would seem that having blind faith (i.e., extremely high and perhaps unrealistic expectations) in staff effectiveness at avoiding errors is linked to more passive patient behavior. However, trust that a health care professional would be happy to engage in patients' concerns is linked to more active behavior.

Patients' perceptions of the risk of error and the associated risk of harm have also been shown to affect the likelihood of taking preventive action. Schwappach and Wernli (2011) found that most patients had the perception that there was little risk of error in chemotherapy administration itself and that only minor harm was associated with any errors that might occur. Schwappach and Wernli (2010c) found that perceptions of the risk of error were positively linked to patient safety behaviors. Entwistle and colleagues (2010) found that when patients were considering whether to speak up, they carried out their own risk assessment of how likely, grave, and imminent the potential for harm was and how serious the shortfall of standards of care was.

Task Factors

Studies suggest that patients are more willing to engage in safety-related tasks that are seen as established and non-challenging to health care professionals. Patients seem less comfortable speaking up if it is a more "confrontational" act, such as challenging health care professionals about their practices, than they are to ask for further information, such as about unknown drugs (Marella et al. 2007). Hibbard and colleagues (2005) found that patients were more likely to believe that long-standing, established actions, such as making sure that the doctor knows any allergies or ADRs that they have to medications, would be more effective than less well-established actions that require questioning clinical personnel, such as confirming whether they were being given the right medication. Waterman and colleagues (2006) found that patients were more likely to report being comfortable asking a nurse the purpose of their medication than telling staff that an error had occurred. Schwappach and Wernli (2010c) found that patients were significantly more likely to engage in traditional, role-conforming behaviors, such as reporting ADRs, than proactive behaviors, such as asking a nurse to read aloud a medication label. Patients were also significantly more likely to engage in proactive behaviors, such as asking a nurse to read aloud a medication label, than challenging behaviors, such as asking a nurse whether a drug or infusion was correct. Davis and colleagues (2011) found that patients were more willing to bring a list of allergies and any medications that they were taking into a hospital than to ask a doctor or nurse whether their inpatient medication was the correct medication for them. In addition, Davis and colleagues (2012) found that health care professionals were more likely to support patients being involved in bringing in medication and allergy lists than patients asking whether medications were correct for them. For all of these studies, apart from Schwappach and Wernli (2010c), it is not clear from the published papers whether the differences observed were significant, but there would seem to be cumulative validity to these findings.

There is also evidence that the nature of some medications given in hospitals may act as a barrier to patients being involved in identifying errors. Watt and colleagues (2009) found that patients thought that lack of familiarity with the appearance of the different brands of medications given in hospitals might hinder inpatients from checking their medications (i.e., the tablets being given looked different, but were in fact the usual treatment). They also found that patients thought it was impractical or impossible, as a patient, to check every injection, infusion, and intravenous drug that was administered. Perhaps not surprisingly, Schwappach and Wernli (2010c) found that patients were more likely to believe that they could detect a disconnected infusion tube compared to detecting confusion over infusion bags or the provision of oral chemotherapy agents being at too high a dose.

Individual Health Care Professional Factors

The level of encouragement provided by health care professionals has also been associated with patient engagement in medication safety. As noted above, patients' perceptions that health care professionals expect and would like them to be involved in their medication safety encourages participation. Davis and colleagues (2011, 2012) specifically found that patients reported that they would be more willing to engage in safety behaviors (including medication safety behaviors) if they were actively encouraged to do so by health care professionals. Therefore, health care professionals who support and encourage patients to engage in their medication safety could expect to increase patient involvement. In an Australian qualitative study (observations and interviews) of interactions between nurses and patients, Bolster and Manias (2010) noted that there was variation in how patient centered different nurses were, and to what extent they encouraged participation.

The extent to which health care professionals report that they would support patients engaging in medication safety behaviors has been related to their professional discipline. Davis and colleagues (2012) found that significantly more nurses than doctors would support patients in asking whether a medication was correct for them. Similarly, in a vignette study, Davis and colleagues (2015) found that nurses were significantly more likely than doctors to approve of and support patients intervening in their medication safety and to think that their intervention could have positive effects on the health care professional–patient relationship. Mohsin-Shaikh and colleagues (2014) found that pharmacists and nurses were significantly more likely to report that they supported patients asking questions about their medicines and self-administering their own medicines than doctors.

The data are also suggestive (although non-significant) of there being a greater likelihood of female rather than male health care professionals supporting patient involvement (Davis et al. 2012). However, another study found no association between gender and willingness to support patient involvement (Mohsin-Shaikh et al. 2014). This is an area that requires further investigation.

Health care professionals' negative perceptions and beliefs towards patient engagement have also been identified. Martin and colleagues (2013) carried out a qualitative study using observations and interviews of patients with cancer, hospital doctors, general practitioners (GPs), and nurses. They found that some health care professionals perceived there being disadvantages in drawing patients' attention to safety hazards, including the risk of causing unnecessary anxiety, discomfort, and perhaps distrust. They believed that the majority of adverse events would be resolved without affecting the patients and that ensuing mistrust might negatively affect the patient experience and treatment outcome.

Team Factors

In contrast to the individual health care professional factors described above, there has been less research on how health care professionals working in teams affect patient engagement. However, Bolster and Manias (2010) noted that no communication was observed between nurses and pharmacists during any of their ward observations and that there were missed opportunities for pharmacists and nurses to communicate with each other in order to ensure that the patient was adequately informed and educated. This study was carried out in one hospital ward with a collaboratively developed philosophy of patient-centered care, so it is also possible that these barriers are more commonly found on wards without this philosophy.

Environmental Factors

The busy environment and time constraints under which health care professionals work has been identified as a barrier to patient engagement with medication by both patients and health care professionals. Chemotherapy patients perceived staff time pressure as one of the most important barriers to patients engaging with staff to reduce errors (Schwappach and Wernli 2011). Bolster and Manias (2010) found that the nurses they interviewed identified a lack of time as a factor that limited the interactions that they were able to have with patients. Some patients in the study also commented on the busyness of the nurses and the implications of this on them getting their medicines late.

Organizational Factors

We have not identified any studies that described organizational factors that contributed to patient involvement in their medication safety. Some of the aforementioned findings, such as staff time pressures and health care professionals' attitudes and beliefs towards patient involvement, may well link to organizational factors. These, for example, may have arisen from the organizations' safety cultures, financial resources, or constraints, as described in Chapter 13.

In summary, research in this area has mainly focused on patient-, individual health care professional- and task-related factors that may affect patients' engagement in their own safety. Few studies have discussed team-related, environmental, and organizational factors, and these are therefore areas for future research.

INTERVENTIONS TO ENCOURAGE PATIENT AND CARER INVOLVEMENT

Patients

A small number of studies have attempted to encourage patient involvement. Strategies to encourage patients to speak up about their medications in order to improve their safety have been mostly implemented within secondary care (Nau and Erickson 2005; Stevenson et al. 2004; Weingart et al. 2004). Table 22.1 presents examples of implemented interventions used to influence patient involvement in medication safety. Some of the interventions show some promise, but their effectiveness in influencing patients—and their link to improved safety—is unclear.

TABLE 22.1
Examples of Strategies Used to Encourage Patient Involvement in Medication Safety

Strategies	Examples of Where Used
Online medication safety prompts	Weingart et al. (2008)
Reminders of personalized medication lists and drug safety information	Weingart et al. (2004, 2008)
Mass advertising (for example, brochures, posters at pharmacies, medication cards, and prescription folders)	Joint Commission (2008), World Health Organization and EuroPharm Forum (1993)
Telephone services involving health care professionals calling patients regularly during their drug treatment	Kelly et al. (1999), Schectman et al. (1994)
Patients asked to write down their questions	Barnett et al. (2000)
Pharmacist visits to support patients to ask questions	American Hospital Association (2001)
Health care professionals providing patients with questions to ask	Agency for Healthcare Research and Quality (2002), Joint Commission (2008)

For example, a robust randomized controlled trial of a simple intervention (Weingart et al. 2004) showed some subjective improvements in patients and their families asking more questions and increasing medication safety. Patients allocated to the intervention group received a personalized medication list every 3 days, with a glossary of technical terms and medication safety information (for example, on possible allergic reactions), while a control group received the medication safety information alone. In total, 24% of 17 nurses reported that patients and their families from the intervention group asked more questions than usual about their medications (over half [59%] were aware of the intervention). Close to a third of the nurses (29%) reported that patient's speaking up resulted in the prevention of at least one medication error. However, no objective between-group differences were found in the prevention of harm due to adverse drug events (Weingart et al. 2004).

A systematic literature review by Hall and colleagues (2010) showed that objective outcomes (such as patient safety incidents) were often not reported in studies on the usefulness of informing behaviors by patients (Atkin et al. 1998; Hall et al. 2010; Neafsey et al. 2002; Varkey et al. 2007), and only two of the six intervention studies designed to promote proactive patient involvement in medication safety by "informing the management plan" (for example, patients informing health care professionals about their drug allergies) showed any objective effectiveness (Kennedy 1990; Schnipper et al. 2006).

Other research provides partial evidence to illustrate the achievements of encouraging patient involvement. Weingart and colleagues (2008) investigated patients who received an electronic message via a web portal 10 days after a new or changed prescription, reminding them of their medication and asking them whether they had any queries or problems. They found that 13% of those who opened their message responded to it, mostly by querying the drug's effectiveness, administration, dosages, and side effects. However, while they compared patients who did and did not respond to the message, there was no baseline or reference group for comparing against the usual communication patterns in patients who did not receive such messages.

On a larger scale, international bodies such as the World Health Organization (WHO) and the U.S. Agency for Healthcare Research & Quality have initiated campaigns and provided guidance for what questions patients should ask about medications to stay safe (World Health Organization and EuroPharm Forum 1993) and the Agency for Healthcare Research and Quality (2002) in the United States. The U.S. Joint Commission (2008), for example, introduced a "Speak Up" campaign for patients, comprising information given via brochures and advertisements on how to be proactive in preventing medication errors and in patient involvement in medications generally. Box 22.1 presents some of this guidance on how patients can speak up to help avoid mistakes with medicines at hospitals/clinics and with doctors or pharmacists (Joint Commission 2008). However, even large campaigns have failed to show successes in their attempts to involve patients in speaking up about medications. The WHO/EuroPharm Forum "Questions to Ask about your Medicine" campaign (World Health Organization and EuroPharm Forum 1993), which was advertised through posters at pharmacies, medication cards, and prescription folders across eight countries, advised patients to ask five questions about their medications, such as "Whom do I contact if problems occur?" Nonetheless, an evaluation of this campaign in Finland showed no increase in the number of patients asking questions of their pharmacist (Airaksinen et al. 1998).

BOX 22.1 RECOMMENDATIONS ADAPTED (SHORTENED) FROM THE JOINT COMMISSION "SPEAK UP CAMPAIGN: HELP AVOID MISTAKES WITH YOUR MEDICINES" (2008)

Help avoid mistakes with your medicines at the hospital/clinic:

- Make sure your health care professionals/caregivers check your wristband and ask your name before giving you medicine.
- Tell a caregiver if you think you are about to get the wrong medicine.
- Know what time you should get a medicine. If you don't get it then, speak up.
- Tell your caregiver if you don't feel well after taking a medicine.
- Read intravenous fluids bag. Ask the caregiver how long it should take for the liquid to run out. Tell the caregiver if it's dripping too fast or too slow.
- Get and check carefully a list of all your medicines. If you're not well enough, ask a friend/relative to help.

Ask your doctor or pharmacist:

- How will this new medicine help you?
- Are there other names for this medicine?
- Is there any written information about the medicine?
- Can you take this medicine with your allergy? Remind health care professionals about your allergies.
- Is it safe to take this medicine with your other medicines?
- Are there any side effects of the medicine?
- Who can you contact if you have side effects or a bad reaction? When can they be reached?
- Are there specific instructions for your medicines? (E.g., take with food.)
- Can you stop taking the medicine as soon as you feel better?
- Do you need to swallow or chew the medicine or can you cut/crush the medicine?
- Is it safe to drink alcohol with the medicine?

There is perhaps more encouraging evidence associated with individual health care professionals helping patients to speak up than has been found with such marketing campaigns or interventions not involving direct professional contact. In a review of the literature on patient–health care professional communication about medicines, Stevenson and colleagues (2004) reported three interventions showing that patients talking about medications with nurses led to more medication-related discussions with doctors (Hanna 1993; Mills et al. 1999; Wilder-Smith and Schuler 1992): two telephone service interventions involving calls from health care professionals showed improvements in the number of patients contacting the clinic (Kelly et al. 1999; Schectman et al. 1994), and one study showed that patients ask more questions of their pharmacist when instructed by health care professionals to write down their questions in advance (Barnett et al. 2000).

Carers

Informal carers have an essential role to play in speaking up on behalf of their family or friends, particularly in pediatrics and mental health, or for those that are too unwell to speak up for themselves. In fact, it may be easier for relatives to speak up than the patient themselves; questioning health care professionals on behalf of someone else, particularly close family members, was considered to be easier than speaking up as a patient oneself (Watt et al. 2009). Some qualitative studies have shown that patients believed that bringing family members to outpatient clinic visits improved their care (Dowell et al. 2005). The guidance provided for carers, however, seems to focus on parental involvement, rather than a wider range of carer roles. The Joint Commission (2011) released guidance on what parents could do to ensure that it is safe for their child to take a new medicine, such as asking health care professionals whether they can cut or crush pills or put them in food if their child has trouble swallowing them. Further empirical research is required to investigate how carers of people of all different ages and vulnerabilities can successfully accomplish their role of speaking up for the person receiving care.

Most empirical research has been conducted in the hospital setting. Aside from the national campaigns, there is little evidence on the proactive involvement of outpatients or those in the primary care setting. Researchers may avoid studying patient involvement in medication safety at home due to its conflation with patient self-management of medications (Berger et al. 2014). Yet medication errors are one of the most prevalent errors within the home setting (Masotti et al. 2010). Such errors include communication failures and administration errors (Meredith et al. 2001; Walsh et al. 2013). As more and more people are being cared for at home, research could help to identify errors that could be prevented through speaking up, and help to identify how best to encourage such behavior from both patients and their carers.

LESSONS LEARNED FROM OTHER PATIENT SAFETY INTERVENTIONS

While there are few studies of interventions specifically aimed at proactive patient involvement in medication safety, there is a wider body of literature relating to interventions encouraging patient involvement in other areas of safety. For example, patients speaking up about their caregivers' hand hygiene has been widely investigated, largely due to its relevance across all care settings and its clear link to reducing infection (Pittet et al. 2000). Campaigns have used a variety of different measures, often in combination, to improve patients' abilities to speak up about the hand hygiene of their health care professionals in hospital. For example, the WHO "Hand Hygiene Campaign" used posters, leaflets, and information stands to inform and encourage patients on their potential involvement (World Health Organization 2009a). These strategies are similar to those that have been used in the context of medication safety.

A recent systematic literature review showed that while videos and leaflets were the most commonly used interventions, encouragement to the patient directly from the health care professionals themselves was the most effective strategy for improving hand hygiene (Davis et al. 2015). In six studies using video interventions and eight using personal encouragement (for example, the health care professional inviting or instructing patient participation), all improved the patients' willingness to speak up. Two of three studies of written interventions (for example, using leaflets or booklets) showed improved willingness to speak up. However, the majority of studies did not examine *actual* behaviors or use robust controlled methodologies; only one study used an objective measure of patient behavior (Lent et al. 2009).

Overall, it appears that we do not know enough about interventions on proactive patient involvement. Campaigns may not always be effective and the literature suffers from a lack of examination of actual behaviors, controlled methodologies, and information outside of secondary care. The existing evidence from both medication safety-related interventions and lessons learnt elsewhere indicates that the best way to promote proactive patient involvement is through direct encouragement by health care professionals.

CONCLUSION

Considering all of the evidence that we have presented in this chapter, a good intervention would likely benefit from positive attitudes and encouraging behavior from health care professionals suggesting that medication safety-related behaviors by patients are both welcomed and expected. This includes a positive reception to those patients raising questions or concerns and engendering a relationship based on trust with the patient. Carers should similarly be encouraged to challenge and ask questions. Safety information could also increase involvement through leaflets or videos, personalized medication lists, and asking for questions in advance of medication therapy. Particular attention could be paid to those patients who are less likely to be involved in their own medications, such as older adults, males, those with poor health or cognitive impairment, and those that have less experience and/or knowledge of medications, medication errors, and health care. Although more challenging, health care professionals could also attempt to confront negative beliefs in the usefulness of patient involvement behaviors among patients and health care professionals alike. When developing interventions to support health care professionals in encouraging patient involvement, it is likely that the team, environment, and organization in which the health care professionals work will all need to be considered.

COMMENTARY

Ailsa Donnelly
Patient, United Kingdom

As a patient and carer, most of my medication "experiences" occur in primary care, with drugs prescribed by a GP and dispensed by a community pharmacist. What therefore struck me very forcibly in this chapter was that, even though the majority of medicines are dispensed in primary care to be taken at home, nearly all the research on patient involvement in medication safety has been carried out in hospitals, much of it on cancer patients undergoing chemotherapy. This is a very different scenario from an individual at home juggling several boxes of pills trying to work out which, when, and how many of each to take, and indicates a huge research gap with clear implications for medication safety.

In terms of the actual research findings, many of them fall into the "intuitively obvious" category; for example, patients who have previously found an error or who are made aware of the potential for error are more likely to check their medication. Patients who are more confident are encouraged to ask questions and those who work as healthcare professionals are also more likely to check their medications. Worryingly, but understandably, some patients have blind faith that the professionals will get it right, while others are reluctant to speak up in case it affects their subsequent care. More worryingly still, this fear sometimes appears justified, and has certainly been the experience of some patients I have known.

Perhaps because the focus is so much on secondary care, most of the research has involved doctors and nurses, and there is little mention of the role of pharmacists—a major difference from primary care. As a patient, however, I found the idea that any doctor or nurse would be reluctant to inform patients about safety hazards because they perceived that this would lead to mistrust both extremely alarming and very patronizing.

It was excellent to see the inclusion of a section on carers, and an acknowledgement that it is often easier to speak up on behalf of someone you are looking after than for yourself.

The chapter ended with an interesting review and comparison of interventions and campaigns aimed at making patients more "safety aware" across a number of areas, using varied methods such as videos, text messaging, and written information. Patients need to realize that medication safety is a joint responsibility and that they too have a crucial role to play. The authors concluded that the best way to promote patient involvement is by active encouragement from health care professionals (i.e., a one-on-one relationship in which patients feel it is safe, expected, and normal to ask questions). In hospitals, this role may be performed by the nurse, but in primary care, the main point of contact for patients and carers is likely to be the pharmacy where the medication is dispensed. For this reason, community pharmacists should be trained and supported to carry out this vital role.

23 Conclusion

Mary P. Tully and Bryony Dean Franklin

This book has summarized a wealth of research conducted in the search for safety in medication use. Section I outlined the extent of the problems in several key parts of the medication use process. Subsequent sections have presented approaches that could be taken to investigate these problems and interventions that could be introduced to ameliorate them.

Two words that spring to mind when reading this research are *heterogeneity* and *complexity*. There is enormous variation in the research studies conducted in each area—most obviously variations in definitions, in methods, and in contexts. This makes it very difficult to conduct meta-analyses to answer overarching questions such as, "By how much will electronic prescribing reduce prescribing errors?" The contexts within which studies are conducted are particularly important to understanding whether an intervention that had an impact in one particular hospital or clinic, for example, will do likewise in another. Usually, some adaptation of the intervention will also be needed to make the intervention run smoothly in another hospital or clinic, but does that mean that it is the same intervention or another one that has been introduced? In addition, an appropriate research design should be used for evaluation. Time series studies or stepped wedge designs may be needed to account for changing contexts while maintaining rigor in the research design. Studies should also include measurement of the fidelity with which interventions have been implemented; if it transpires that an intervention is only partially implemented in actual practice, this is important to understanding study findings.

The complexity of the health care system means that affecting one element of the medication use process can affect the whole. However, this does not necessarily mean that a positive impact on a single component will have a positive impact on other parts, or indeed the system as a whole. Chapter 10, for example, highlights examples of the negative impacts of electronic prescribing on some groups of staff, despite its positive impact on reducing some types of prescribing errors. The negative consequences of the intervention can also manifest themselves as workarounds, where staff do things in unanticipated ways in order to enhance efficiency or achieve other goals. There have been studies that have even found workarounds in secondary care as a result of interventions in primary care (Tully et al. 2013). These "knock-on" effects need to be explicitly sought when interventions are introduced, so that the intervention can be tailored, if possible, to avoid the need for workarounds.

It is essential that interventions are evaluated in multiple contexts, which are then described in detail in any dissemination of an evaluation's findings, so that we can understand what works where and how. It is not sufficient that an intervention is evaluated initially when it is first introduced and then never again. It is important that we consider how interventions can be made sustainable beyond the duration of a research study. It takes commitment from senior leadership within an organization to ensure that interventions, especially behavioral changes, become embedded in routine practice. During this time, it could be argued that measurement data should be collected at regular intervals to assess whether initial promising changes have been sustained, or conversely, so that any initial "settling in" periods have passed.

Safety in medication use is not the responsibility of a single discipline, such as pharmacy or medicine. The authors of the chapters in this book, and the authors of the research cited throughout, are from multiple disciplines. In health care practice, doctors, nurses, pharmacists, patients, and carers are all involved every day in medication use. Therefore, when interventions need to be introduced, all these groups of people, as a minimum, need to be involved.

Earlier in this book, Chapter 11 described the cyclical process of measurement and change that is used in improvement science. In addition, use of some of the safety culture measurements described in Chapter 13, such as the Manchester Patient Safety Assessment Framework (Kirk et al. 2007), can itself lead to improvements in safety culture. This framework is delivered in a workshop discussion and can encourage the health care professionals attending to reflect on how they approach safety and what they can do differently (Parker 2009). Thus, there is a good argument for further investigating the impact that regular measurement can have on patient safety, as has been suggested in England when using the Safety Thermometer for the routine measurement of harm (Power et al. 2014). The social aspects regarding the creation and use of such data should not be underestimated; data are often recorded as part of a process conducted in the workplace with choices made by human beings, not automata (Dixon-Woods et al. 2012). In a seminal study in this area, health care professionals recorded the incidence of hospital-acquired infections, and the authors concluded that "variability arose not because of wily workers deliberately concealing, obscuring or deceiving, but because counting was as much a social practice as a technical practice" (Dixon-Woods et al. 2012).

We would like to end by reiterating the importance of involving patients in medication safety. Their involvement in ensuring their own safety has been mentioned throughout this book, and with good reason. Patients themselves are the one constant throughout all interactions with health care professionals. More recently, patients and the public are also becoming more involved in the research process. Indeed, in many countries, such as in the United Kingdom, this is required for all clinical research receiving government funding. Such patient and public involvement involves lay people being involved throughout the entire research process, from helping to define the research question to helping disseminate the findings. There is evidence from other areas of research that such involvement broadens attitudes and ensures that research is relevant to society and wider social concerns. However, there has been little research published to date that reports on public involvement in research on medication use (Sutton and Weiss 2008), as opposed to clinical trials research, where it is more common.

In conclusion, we encourage everyone working in the field of safe medication use to evaluate their interventions and disseminate their findings broadly. There is still much to do to make the process as safe as possible. If reducing medication errors was easy, we would have done it by now!

References

Aarts, J., Doorewaard, H., and Berg, M. 2004. Understanding implementation: The case of a computerized physician order entry system in a large Dutch university medical center. *J Am Med Inform Assoc* 11: 207–216.

Aarts, J., and Koppel, R. 2009. Implementation of computerized physician order entry in seven countries. *Health Aff (Millwood)* 28: 404–414.

Abdel-Qader, D.H., Harper, L., Cantrill, J.A., and Tully, M.P. 2010. Pharmacists' interventions in prescribing errors at hospital discharge: An observational study in the context of an electronic prescribing system in a UK teaching hospital. *Drug Saf* 33: 1027–1044.

Adedoye, D. 2007. *Dispensing Errors in an Automated Dispensary: Their Incidence and Detection*. London: King's College Hospital NHS Foundation Trust.

Agency for Healthcare Research and Quality. 2002. Quick Tips—When getting a prescription. http://www.innovations.ahrq.gov/qualitytools/quick-tips-when-getting-prescription (Accessed May 26, 2015).

Ahmed, M., Arora, S., Tiew, S. et al. 2014. Building a safer foundation: The lessons learnt patient safety training programme. *BMJ Qual Saf* 23: 78–86.

Ahmed, Z., McLeod, M.C., Barber, N., Jacklin, A., and Franklin, B.D. 2013. The use and functionality of electronic prescribing systems in English acute NHS trusts: A cross-sectional survey. *PLoS One* 8: e80378.

Airaksinen, M., Ahonen, R., and Enlund, H. 1998. The "questions to ask about your medicines" campaign. An evaluation of pharmacists and the public's response. *Med Care* 36: 422–427.

Ajzen, I. 1991. The theory of planned behavior. *Organ Behav Hum December* 50: 179–211.

Alhadeff-Jones, M. 2008. Three generations of complexity theories: Nuances and ambiguities. *Educ Philos Theory* 40: 66–82.

Allan, E.L., and Barker, K.N. 1990. Fundamentals of medication error research. *Am J Health Syst Pharm* 47: 555–571.

Allan, E.L., Barker, K.N., Malloy, M.J., and Heller, W.M. 1995. Dispensing errors and counseling in community practice. *Am Pharm* NS35: 25–33.

Alldred, D.P., Barber, N.D., Buckle, P. et al. 2009. Care homes' use of medicines study: Medication errors in nursing and residential care homes—Prevalence, consequences, causes and solutions. http://www.birmingham.ac.uk/Documents/college-mds/haps/projects/cfhep/psrp/finalreports/PS025CHUMS-FinalReportwithappendices.pdf (Accessed December 18, 2013).

Alldred, D.P., Standage, C., Fletcher, O. et al. 2011. The influence of formulation and medicine delivery system on medication administration errors in care homes for older people. *BMJ Qual Saf* 20: 397–401.

Alldred, D.P., Standage, C., Zermansky, A.G. et al. 2008. Development and validation of criteria to indentify medication-monitoring errors in care home residents. *Int J Pharm Pract* 16: 317–323.

Allport, G. 1935. Attitudes. In *A Handbook of Social Pyschology*, ed C. Murchison, 789–844. Worcester, MA: Clark University Press.

Alper, E., Rosenberg, E.I., O'Brien, K.E., Fischer, M., and Durning, S.J. 2009. Patient safety education at U.S. and Canadian medical schools: Results from the 2006 Clerkship Directors in Internal Medicine survey. *Acad Med* 84: 1672–1676.

Alsulami, Z., Choonara, I., and Conroy, S. 2014. Paediatric nurses' adherence to the double-checking process during medication administration in a children's hospital: An observational study. *J Adv Nurs* 70: 1404–1413.

Alsulami, Z., Conroy, S., and Choonara, I. 2012. Double checking the administration of medicines: What is the evidence? A systematic review. *Arch Dis Child* 97: 833–837.

Alsulami, Z., Conroy, S., and Choonara, I. 2013. Medication errors in the Middle East countries: A systematic review of the literature. *Eur J Clin Pharmacol* 69: 995–1008.

Amalberti, R., Auroy, Y., Berwick, D., and Barach, P. 2005. Five system barriers to achieving ultrasafe health care. *Ann Intern Med* 142: 756–764.

American Hospital Association. 2001. Medication safety issue brief. Asking consumers for help. Part 3. *Hosp Health Netw* 75: 57–58.

American Society of Clinical Oncology, and Oncology Nursing Society. 2013. 2013 Updated American Society of Clinical Oncology/Oncology Nursing Society Chemotherapy Administration Safety Standards including Standards for the Safe Administration and Management of Oral Chemotherapy. *J Oncol Pract* 9: 5s–13s.

American Society of Health-System Pharmacists. 2014a. ASHP guidelines on compounding sterile preparations. *Am J Health Syst Pharm* 71: 145–166.

American Society of Health-System Pharmacists. 2014b. Pharmacy technician accreditation. http://www.ashp.org/menu/Accreditation/TechnicianAccreditation (Accessed June 20, 2014).

American Society of Hospital Pharmacists. 1980. ASHP technical assistance bulletin on hospital drug distribution and control. *Am J Hosp Pharm* 37: 1097–1103.

American Society of Hospital Pharmacists. 1993. ASHP guidelines on preventing medication errors in hospitals. *Am J Hosp Pharm* 50: 305–314.

American Society of Hospital Pharmacists. 1994. ASHP technical assistance bulletin on compounding non-sterile products in pharmacies. *Am J Hosp Pharm* 51: 1441–1448.

Amin, S.G. 2001. Control charts 101: A guide to health care applications. *Qual Manag Health Care* 9: 1–27.

Ammenwerth, E., Duftschmid, G., Gall, W. et al. 2014. A nationwide computerized patient medication history: Evaluation of the Austrian pilot project "e-Medikation." *Int J Med Inform* 83: 655–669.

Ammenwerth, E., Schnell-Inderst, P., Machan, C., and Siebert, U. 2008. The effect of electronic prescribing on medication errors and adverse drug events: A systematic review. *J Am Med Inform Assoc* 15: 585–600.

Anacleto, T.A., Perini, E., Rosa, M.B., and Cesar, C.C. 2007. Drug-dispensing errors in the hospital pharmacy. *Clinics (Sao Paulo)* 62: 243–250.

Anderson, S. 2001. The historical context of pharmacy. In *Pharmacy Practice*, eds K. Taylor and G. Harding. London: Taylor & Francis.

Anon. 1982. ASHP standard definition of a medication error. *Am J Hosp Pharm* 39: 321.

Anon. 2000. Boots pharmacist and trainee cleared of baby's manslaughter but fined for dispensing a defective medicine. *Pharm J* 267: 390–392.

Anon. 2005. 33% of fatal med errors involve insulin therapy. *Healthcare Benchmarks Qual Improv* 12: 31–32.

Anto, B., Barlow, D., Oborne, C.A., and Whittlesea, C. 2011. Incorrect drug selection at the point of dispensing: A study of potential predisposing factors. *Int J Pharm Pract* 19: 51–60.

Armitage, G. 2008. Double checking medicines: Defence against error or contributory factor? *J Eval Clin Pract* 14: 513–519.

Armitage, G., Newell, R., and Wright, J. 2010. Improving the quality of drug error reporting. *J Eval Clin Pract* 16: 1189–1197.

Ash, J.S., Berg, M., and Coiera, E. 2004. Some unintended consequences of information technology in health care: The nature of patient care information system-related errors. *J Am Med Inform Assoc* 11: 104–112.

Ash, J.S., Sittig, D.F., Campbell, K., Guappone, K.P., and Dykstra, R.H. 2006. An unintended consequence of CPOE implementation: Shifts in power, control, and autonomy. *AMIA Annu Symp Proc* 11–15, Washington, DC, USA.

Ashcroft, D.M., and Parker, D. 2009. Development of the pharmacy safety climate questionnaire: A principal components analysis. *Qual Saf Health Care* 18: 28–31.

Ashcroft, D.M., Morecroft, C., Parker, D., and Noyce, P.R. 2005a. Safety culture assessment in community pharmacy: Development, face validity, and feasibility of the Manchester Patient Safety Assessment Framework. *Qual Saf Health Care* 14: 417–421.

Ashcroft, D.M., Morecroft, C., Parker, D., and Noyce, P.R. 2006. Likelihood of reporting adverse events in community pharmacy: An experimental study. *Qual Saf Health Care* 15: 48–52.

Ashcroft, D.M., Quinlan, P., and Blenkinsopp, A. 2005b. Prospective study of the incidence, nature and causes of dispensing errors in community pharmacies. *Pharmacoepidemiol Drug Saf* 14: 327–332.

Association for the Advancement of Medical Instrumentation. 2014. Small-bore connectors. http://www.aami.org/hottopics/connectors/ (Accessed October 7, 2014).

Atkin, P.A., Stringer, R.S., Duffy, J.B. et al. 1998. The influence of information provided by patients on the accuracy of medication records. *Med J Aust* 169: 85–88.

Australian Commission on Safety and Quality in Health Care. 2012. *Medication Safety Action Guide*. Australian Commission on Safety and Quality in Health Care, New South Wales, Australia.

Australian Patient Safety Foundation. 2010. Incident reporting. http://www.apsf.net.au/inc_report.php (Accessed August 24, 2014).

Avery, A.J., Ghaleb, M., Barber, N. et al. 2013. The prevalence and nature of prescribing and monitoring errors in English general practice: A retrospective case note review. *Br J Gen Pract* 63: 413–414.

Avorn, J., and Gurwitz, J.H. 1995. Drug use in the nursing home. *Ann Intern Med* 123: 195–204.

Azarnoff, D.L., Lee, J.C., Lee, C., Chandler, J., and Karlin, D. 2007. Quality of extemporaneously compounded nitroglycerin ointment. *Dis Colon Rectum* 50: 509–516.

Bagian, J.P., Gosbee, J., Lee, C.Z. et al. 2002. The Veterans Affairs root cause analysis system in action. *Jt Comm J Qual Improv* 28: 531–545.

Bagian, J.P., Lee, C., Gosbee, J. et al. 2001. Developing and deploying a patient safety program in a large health care delivery system: You can't fix what you don't know about. *Jt Comm J Qual Improv* 27: 522–532.

Baldwin, K., and Walsh, V. 2014. Independent double-checks for high-alert medications: Essential practice. *Nursing* 44: 65–67.

Balka, E., Kahnamoui, N., and Nutland, K. 2007. Who is in charge of patient safety? Work practice, work processes and utopian views of automatic drug dispensing systems. *Int J Med Inform* 76(Suppl 1): 48–57.

Ballerman, M.A., Shaw, N.T., Mayes, D.C., Gibney, R.T.N., and Westbrook, J.I. 2011. Validation of the work observational method by activity timing (WOMBAT) method of conducting time-motion observations in critical care settings: An observational study. *BMC Med Inform Decis Mak* 11: 32.

Bandura, A. 1986. *Social Foundations of Thought and Action: A Social Cognitive Theory*. Englewood Cliffs, NJ: Prentice-Hall.

Banning, D. 1995. Help to report near-misses. A study of the rate of medication errors in the dispensary. *Pharm Prac* 5: 461–464.

Barber, N., Jacklin, A., and Franklin, B.D. 2013. Essay: Of snarks, boojums and national drug charts. *J R Soc Med* 106: 6–8.

Barber, N., Rawlins, M., and Franklin, B.D. 2003. Reducing prescribing error: Competence, control, and culture. *Qual Saf Health Care* 12(Suppl 1): i29–i32.

Barber, N., Safdar, A., and Franklin, B.D. 2005. Can human error theory explain non-adherence? *Pharm World Sci* 27: 300–304.

Barber, N.D., Alldred, D.P., Raynor, D.K. et al. 2009. Care homes' use of medicines study: Prevalence, causes and potential harm of medication errors in care homes for older people. *Qual Saf Health Care* 18: 341–346.

Barber, S., Thakkar, K., Marvin, V., Franklin, B.D., and Bell, D. 2014. Evaluation of My Medication Passport: A patient-completed aide-memoire designed by patients, for patients, to help towards medicines optimisation. *BMJ Open* 4: e005608.

Bargren, M., and Lu, D.F. 2009. An evaluation process for an electronic bar code medication administration information system in an acute care unit. *Urol Nurs* 29: 355–67, 391.

Baril, C., Gascon, V., St-Pierre, L., and Lagace, D. 2014. Technology and medication errors: Impact in nursing homes. *Int J Health Care Qual Assur* 27: 244–258.

Barker, K.N., Flynn, E.A., Pepper, G.A., Bates, D.W., and Mikeal, R.L. 2002. Medication errors observed in 36 health care facilities. *Arch Intern Med* 162: 1897–1903.

Barker, K.N., and McConnell, W.E. 1962. The problems of detecting errors in hospitals. *Am J Hosp Pharm* 19: 361–369.

Barker, S. 2003. Dispensing errors: Recording, analysis and human error. *Pharm Manage* 19: 11–16.

Barnett, C.W., Nykamp, D., and Ellington, A.M. 2000. Patient-guided counseling in the community pharmacy setting. *J Am Pharm Assoc* 40: 765–772.

Barr, H., and Low, H. 2012. *Interprofessional Education in Pre-Registration Courses. A CAIPE Guide for Commissioners and Regulators of Education*. Foreham: CAIPE.

Barry, G.A., Bass, G.E., Jr., Eddlemon, J.K., and Lambert, L.L. 1989. Bar-code technology for documenting administration of large-volume intravenous solutions. *Am J Hosp Pharm* 46: 282–287.

Bastien, J.M.C., and Scapin, D.L. 1993. *Ergonomic Criteria for the Evaluation of Human-Computer Interfaces N° RT-0156*. 1–79. Rocquencourt, France: INRIA.

Bateman, R., and Donyai, P. 2010. Errors associated with the preparation of aseptic products in UK hospital pharmacies: Lessons from the national aseptic error reporting scheme. *Qual Saf Health Care* 19: e29.

Bates, D.W., Boyle, D.L., and Teich, J.M. 1994. Impact of computerized physician order entry on physician time. *Proc Symp Comput Appl Med Care* 996, Washington, DC.

Bates, D.W., Boyle, D.L., Vander Vliet, M.B., Schneider, J., and Leape, L.L. 1995a. Relationship between medication errors and adverse drug events. *J Gen Intern Med* 10: 199–205.

Bates, D.W., Cohen, M., Leape, L.L. et al. 2001. Reducing the frequency of errors in medicine using information technology. *J Am Med Inform Assoc* 8: 299–308.

Bates, D.W., Cullen, D.J., Laird, N. et al. 1995b. Incidence of adverse drug events and potential adverse drug events. *JAMA* 274: 29–34.

Bates, D.W., Kuperman, G.J., Wang, S. et al. 2003. Ten commandments for effective clinical decision support: Making the practice of evidence-based medicine a reality. *J Am Med Inform Assoc* 10: 523–530.

Bates, D.W., Leape, L.L., Cullen, D.J. et al. 1998. Effect of computerized physician order entry and a team intervention on prevention of serious medication errors. *JAMA* 280: 1311–1316.

Bates, D.W., Teich, J.M., Lee, J. et al. 1999. The impact of computerised physician order entry on medication error prevention. *J Am Med Inform Assoc* 6: 313–321.

Baysari, M., Westbrook, J.I., Richardson, K., and Day, R.O. 2011a. Influence of computerized decision support on prescribing during ward-rounds: Are the decision-makers targeted? *J Am Med Inform Assoc* 18: 754–759.

Baysari, M.T., Day, R.O., and Westbrook, J.I. 2012b. Consistency or efficiency? A dilemma for designers. *J Am Med Inform Assoc* 19: 1119–1120.

Baysari, M.T., Lehnbom, E.C., Richardson, K. et al. 2013a. Electronic medication information sources: Understanding the needs and preferences of health professionals. *J Pharm Pract Res* 43: 288–291.

Baysari, M.T., Oliver, K., Egan, B. et al. 2014. Audit and feedback of antibiotic use: Utilising electronic prescription data. *Appl Clin Inform* 4: 583–595.

Baysari, M.T., Reckmann, M.H., Li, L., Day, R.O., and Westbrook, J.I. 2012a. Failure to utilize functions of an electronic prescribing system and the subsequent generation of 'technically preventable' computerized alerts. *J Am Med Inform Assoc* 19: 1003–1010.

Baysari, M.T., Westbrook, J., Braithwaite, J., and Day, R.O. 2011b. The role of computerized decision support in reducing errors in selecting medicines for prescription: Narrative review. *Drug Saf* 34: 289–298.

Baysari, M.T., Westbrook, J.I., and Daymond, T. 2011c. Understanding doctors' perceptions of their prescribing competency and the value they ascribe to an electronic prescribing system. *Stud Health Technol Inform* 178: 1–6.

Baysari, M.T., Westbrook, J.I., Egan, B., and Day, R.O. 2013b. Identification of strategies to reduce computerized alerts in an electronic prescribing system using a Delphi approach. *Stud Health Technol Inform* 192: 8–12.

Baysari, M.T., Westbrook, J.I., Richardson, K.L., and Day, R.O. 2011d. The influence of computerized decision support on prescribing during ward-rounds: Are the decision-makers targeted? *J Am Med Inform Assoc* 18: 754–759.

BBC News. 2010. Hospital robots cut hospital pharmacy bill. http://www.bbc.co.uk/news/uk-scotland-tayside-central-11552610 (Accessed November 16, 2014).

Beaney, A. 2005. *Quality Assurance of Aseptic Preparation Services*. London: Pharmaceutical Press.

Becker, M.D., Johnson, M.H., and Longe, R.L. 1978. Errors remaining in unit dose carts after checking by pharmacists versus pharmacy technicians. *Am J Hosp Pharm* 35: 432–434.

Becker, M.H. 1974. The health belief model and personal health behavior. *Health Educ Monogr* 2: 324–508.

Beecher, H.K., and Todd, D.P. 1954. A study of the deaths associated with anesthesia and surgery: Based on a study of 599,548 anesthesias in ten institutions 1948–1952, inclusive. *Ann Surg* 140: 2–35.

Bell, C., Brenner, S., Cunraj, N. et al. 2011. Association of ICU or hospital admission with unintentional discontinuation of medications for chronic diseases. *JAMA* 306: 840–847.

Belleli, E., Naccarella, L., and Pirotta, M. 2013. Communication at the interface between hospitals and primary care—A general practice audit of hospital discharge summaries. *Aust Fam Physician* 42: 886–890.

Benneyan, J.C., Lloyd, R.C., and Plsek, P.E. 2003. Statistical process control as a tool for research and healthcare improvement. *Qual Saf Health Care* 12: 458–464.

Berdot, S., Gillaizeau, F., Caruba, T. et al. 2013. Drug administration errors in hospital inpatients: A systematic review. *PLoS One* 8: e68856.

Berg, M., Aarts, J., and van der Lei, J. 2003. ICT in health care: Sociotechnical approaches. *Methods Inf Med* 42: 297–301.

Berger, Z., Flickinger, T.E., Pfoh, E., Martinez, K.A., and Dy, S.M. 2014. Promoting engagement by patients and families to reduce adverse events in acute care settings: A systematic review. *BMJ Qual Saf* 23: 548–555.

Bertels, J., Almoudaris, A.M., Cortoos, P.J., Jacklin, A., and Franklin, B.D. 2013. Feedback on prescribing errors to junior doctors: Exploring views, problems and preferred methods. *Int J Clin Pharm* 35: 332–338.

Berwick, D., and National Advisory Group on the Safety of Patients in England. 2013. *A Promise to Learn, a Commitment to Act: Improving the Safety of Patients in England*. London: Williams Lea.

Berwick, D., and Nolan, T. 2003. High reliability health care. http://www.ihi.org/resources/Pages/Presentations/HighReliabilityHealthCarePresentation.aspx (Accessed March 17, 2014).

Beso, A., Franklin, B.D., and Barber, N. 2005. The frequency and potential causes of dispensing errors in a hospital pharmacy. *Pharm World Sci* 27: 182–190.

Beuscart-Zephir, M.C., Elkin, P., Pelayo, S., and Beuscart, R. 2007. The human factors engineering approach to biomedical informatics projects: State of the art, results, benefits and challenges. *Yearb Med Inform* 1: 109–127.

Beuscart-Zephir, M.C., Pelayo, S., Anceaux, F. et al. 2005. Impact of CPOE on doctor–nurse cooperation for the medication ordering and administration process. *Int J Med Inf* 74: 629–641.

Beuscart-Zephir, M.C., Pelayo, S., and Bernonville, S. 2010. Example of a human factors engineering approach to a medication administration work system: Potential impact on patient safety. *Int J Med Inform* 79: e43–e57.

Beyea, S.C., Hicks, R.W., and Becker, S.C. 2003. Medication errors in the OR—A secondary analysis of Medmarx. *AORN J* 77: 122, 125–129, 132–134.

Bhalla, R., Berger, M.A., Reissman, S.H. et al. 2013. Improving hospital venous thromboembolism prophylaxis with electronic decision support. *J Hosp Med* 8: 115–120.

Bierly, P.E., and Spender, J.C. 1995. Culture and high reliability organizations: The case of the nuclear submarine. *J Manag* 21: 639–656.

Bigham, M.T., Amato, R.T., Bondurrant, P. et al. 2009. Ventilator-associated pneumonia in the pediatric intensive care unit: Characterizing the problem and implementing a sustainable solution. *J Pediatr* 154: 582–587.

Billett, S. 2001. Knowing in practice: Re-conceptualising vocational expertise. *Learn Instr* 11: 431–452.

Biron, A.D., Loiselle, C.G., and Lavoie-Tremblay, M. 2009. Work interruptions and their contribution to medication administration errors: An evidence review. *Worldviews Evid Based Nurs* 6: 70–86.

Bleakley, A. 2010. Blunting Occam's razor: Aligning medical education with studies of complexity. *J Eval Clin Pract* 16: 849–855.

Bleakley, A. 2011. Becoming a medical professional. In *Becoming a Professional*, ed L. Scanlon, 129–151. New York, NY: Springer.

Bleakley, A., Bligh, J., and Browne, J. 2013. *Medical Education for the Future: Identity, Power and Location*. New York, NY: Springer.

Blumenthal, D. 2010. Launching HITECH. *N Engl J Med* 362: 382–385.

Bobb, A., Gleason, K., Husch, M. et al. 2004. The epidemiology of prescribing errors: The potential impact of computerized prescriber order entry. *Arch Intern Med* 164: 785–792.

Bobb, A.M., Payne, T.H., and Gross, P.A. 2007. Viewpoint: Controversies surrounding use of order sets for clinical decision support in computerized provider order entry. *J Am Med Inform Assoc* 14: 41–47.

Boehm-Davis, D.A., and Remington, R. 2009. Reducing the disruptive effects of interruption: A cognitive framework for analysing the costs and benefits of intervention strategies. *Accident Analysis and Prevention* 41: 1124–1129.

Bolster, D., and Manias, E. 2010. Person-centred interactions between nurses and patients during medication activities in an acute hospital setting: Qualitative observation and interview study. *Int J Nurs Stud* 47: 154–165.

Bond, C.A., and Raehl, C.L. 2001. Pharmacists' assessment of dispensing errors: Risk factors, practice sites, professional functions, and satisfaction. *Pharmacotherapy* 21: 614–626.

Bonkowski, J., Carnes, C., Melucci, J. et al. 2013. Effect of barcode-assisted medication administration on emergency department medication errors. *Acad Emerg Med* 20: 801–806.

Bonkowski, J., Weber, R.J., Melucci, J. et al. 2014. Improving medication administration safety in solid organ transplant patients through barcode-assisted medication administration. *Am J Med Qual* 29: 236–241.

Boockvar, K.S., Livote, E.E., Goldstein, N. et al. 2010. Electronic health records and adverse drug events after patient transfer. *Qual Saf Health Care* 19: e16.

Booth, R., Sturgess, E., Taberner-Stokes, A., and Peters, M. 2012. Zero tolerance prescribing: A strategy to reduce prescribing errors on the paediatric intensive care unit. *Intensive Care Med* 38: 1858–1867.

Borel, J.M., and Rascati, K.L. 1995. Effect of an automated, nursing unit-based drug-dispensing device on medication errors. *Am J Health Syst Pharm* 52: 1875–1879.

Bourrier, M. 1996. Organizing maintenance work at two American nuclear power plants. *J Contingencies Crisis Manage* 4: 104–112.

Bousquet, A., and Curtis, S. 2011. Beyond models and metaphors: Complexity theory, systems thinking and international relations. *Cambr Rev Int Aff* 24: 43–62.

Bower, A.C. 1990. Dispensing error rates in hospital pharmacy. *Pharm J* 244: R22–R23.

Boyle, T.A., Bishop, A.C., Duggan, K. et al. 2014a. Keeping the "continuous" in continuous quality improvement: Exploring perceived outcomes of CQI program use in community pharmacy. *Res Social Adm Pharm* 10: 45–57.

Boyle, T.A., Bishop, A.C., Hillier, C. et al. 2014b. Regulatory authority approaches to deploying quality improvement standards to community pharmacies: Insights from the SafetyNET-Rx program. *J Pharm Pract* 27: 138–149.

Boyle, T.A., Mackinnon, N.J., Mahaffey, T., Duggan, K., and Dow, N. 2012. Challenges of standardized continuous quality improvement programs in community pharmacies: The case of SafetyNET-Rx. *Res Social Adm Pharm* 8: 499–508.

Brady, A.M., Malone, A.M., and Fleming, S. 2009. A literature review of the individual and systems factors that contribute to medication errors in nursing practice. *J Nurs Manag* 17: 679–697.

Bramstedt, K.A., Moolla, A., and Rehfield, P.L. 2012. Use of standardized patients to teach medical students about living organ donation. *Prog Transplant* 22: 86–90.

Breland, B.D. 2010. Continuous quality improvement using intelligent infusion pump data analysis. *Am J Health Syst Pharm* 67: 1446–1455.

Bright, T.J., Wong, A., Dhurjati, R. et al. 2012. Effect of clinical decision-support systems: A systematic review. *Ann Intern Med* 157: 29–43.

Brinklow, N.A. 2006. A report assessing the impact of an Automated Dispensing System (ADS) at Kings College Hospital NHS Trust. http://media.dh.gov.uk/network/121/files/2011/03/robot-KCH.pdf (Accessed August 28, 2014).

Britto, M.T., Vockell, A.L., Munafo, J.K. et al. 2014. Improving outcomes for underserved adolescents with asthma. *Pediatrics* 133: e418–e427.

Buchanan, T.L., Barker, K.N., Gibson, J.T., Jiang, B.C., and Pearson, R.E. 1991. Illumination and errors in dispensing. *Am J Hosp Pharm* 48: 2137–2145.

Buetow, S.A., Sibbald, B., Cantrill, J.A., and Halliwell, S. 1996. Prevalence of potentially inappropriate long term prescribing in general practice in the United Kingdom, 1980–95: Aystematic literature review. *BMJ* 313: 1371–1374.

Burgin, A., O'Rourke, R., and Tully, M.P. 2014. Learning to work with electronic patient records and prescription charts: Experiences and perceptions of hospital pharmacists. *Res Social Adm Pharm* 10: 741–755.

Burke-Bebee, S., Wilson, M., and Buckley, K.M. 2012. Building health information technology capacity: They may come but will they use it? *Comput Inform Nurs* 30: 547–553.

Cadwallader, J., Spry, K., Morea, J. et al. 2013. Design of a medication reconciliation application: Facilitating clinician-focused decision making with data from multiple sources. *Appl Clin Inform* 4: 110–125.

Calabrese, A.D., Erstad, B.L., Brandl, K. et al. 2001. Medication administration errors in adult patients in the ICU. *Intensive Care Med* 27: 1592–1598.

Caldwell, G. 2010. Real time 'check and correct' of drug charts on ward rounds; a process for improving doctors' habits in inpatient prescribing. *Pharmacy Management* 26: 3–9.

Callen, J., McIntosh, J., and Li, J. 2010. Accuracy of medication documentation in hospital discharge summaries: A retrospective analysis of medication transcription errors in manual and electronic discharge summaries. *Int J Med Inform* 79: 58–64.

Candlish, C., Worsley, A.J., and Zaman, S. 2003. Do pharmacists extemporaneously dispense or do they use specials manufacturers? *Int J Pharm Pract* 11: R47.

Cant, R.P., and Cooper, S.J. 2010. Simulation-based learning in nurse education: Systematic review. *J Adv Nurs* 66: 3–15.

Carayon, P. 2010. Human factors in patient safety as an innovation. *Appl Ergon* 41: 657–665.

Carayon, P., Faye, H., Hundt, A.S., Karsh, B.-T., and Wetterneck, T. 2011. Patient safety and proactive risk assessment. In *Handbook of Healthcare Delivery Systems*, ed Y. Yuehwern, 12-11–12-15. Boca Raton, FL: Taylor & Francis.

Carayon, P., Schoofs, H.A., Karsh, B.T. et al. 2006. Work system design for patient safety: The SEIPS model. *Qual Saf Health Care* 15(Suppl 1): i50–i58.

Carayon, P., Schoofs H.A., and Wetterneck, T. 2010. Nurses' acceptance of smart IV pump technology. *Int J Med Inf* 79: 401–411.

Carayon, P., Wetterneck, T.B., Cartmill, R. et al. 2014a. Characterising the complexity of medication safety using a human factors approach: An observational study in two intensive care units. *BMJ Qual Saf* 23: 56–65.

Carayon, P., Wetterneck, T.B., and Hundt, A.S. 2007. Evaluation of nurse interaction with bar code medication administration technology in the work environment. *J Patient Saf* 3: 34–42.

Carayon, P., Wetterneck, T.B., Hundt, A.S., Rough, S., and Schroeder, M. 2008. Continuous technology implementation in health care: The case of advanced IV infusion pump technology. In *Corporate Sustainability as a Challenge for Comprehensive Management*, ed K. Zink, 139–151. New York, NY: Springer.

Carayon, P., Wetterneck, T.B., Rivera-Rodriguez, A.J. et al. 2014b. Human factors systems approach to healthcare quality and patient safety. *Appl Ergon* 45: 14–25.

Carayon, P., Xie, A., Cox, E. et al. 2014c. Participatory ergonomics for healthcare process redesign: The example of family-centered rounds. In *Human Factors in Organizational Design and Management XI*, eds O. Broberg, N. Fallentin, P. Hasle, P.L. Jensen, A. Kabel, M.E. Larsen and T. Weller, 25–30. Chicago, Illinois, USA.

Catchpole, K., Bell, M.D., and Johnson, S. 2008. Safety in anaesthesia: A study of 12,606 reported incidents from the UK National Reporting and Learning System. *Anaesthesia* 63: 340–346.

Catchpole, K., and Wiegmann, D. 2012. Understanding safety and performance in the cardiac operating room: From "sharp end" to "blunt end." *BMJ Qual Saf* 21: 807–809.

Cavell, G. 2009. How to use trigger drugs to help identify adverse medication events. *Clin Pharm* 1: 484–485.

Centre for Advancement of Interprofessional Education. 2014. Centre for Advancement of Interprofessional Education (CAIPE). http://caipe.org.uk/ (Accessed October 6, 2014).

Chapanis, A., and Safrin, M.A. 1960. Of misses and medicines. *J Chronic Dis* 12: 403–408.

Chapuis, C., Roustit, M., Bal, G. et al. 2010. Automated drug dispensing system reduces medication errors in an intensive care setting. *Crit Care Med* 38: 2275–2281.

Charani, E., Castro-Sánchez, E., Moore, L.S.P., and Holmes, A. 2014. Do smartphone applications in healthcare require a governance and legal framework? It depends on the application! *BMC Med* 12: 29.

Charani, E., Kyratsis, Y., Lawson, W. et al. 2012. An analysis of the development and implementation of a smartphone application for the delivery of antimicrobial prescribing policy: Lessons learnt. *J Antimicrob Chemother* 68: 960–967.

Chaudhry, B., Wang, J., Wu, S. et al. 2006. Systematic review: Impact of health information technology on quality, efficiency, and costs of medical care. *Ann Intern Med* 144: 742–752.

Chen, H., Reeves, J.H., Fincham, J.E. et al. 2006. Off-label use of antidepressant, anticonvulsant, and antipsychotic medications among Georgia Medicaid enrollees in 2001. *J Clin Psychiatry* 67: 972–982.

Chen, Y., Neil, K.E., Avery, A.J., Dewey, M.E., and Johnson, C. 2005. Prescribing errors and other problems reported by community pharmacists. *Therap Clin Risk Manage* 1: 333–342.

Cheung, K.C., van der Veen, W., Bouvy, M.L. et al. 2014. Classification of medication incidents associated with information technology. *J Am Med Inform Assoc* 21: e63–e70.

Chief Medical Officer. 2000. *An Organisation with a Memory. Report of an Expert Group on Learning from Adverse Events in the NHS*. London: The Stationery Office.

Chief Medical Officer. 2001. *Building a Safer NHS for Patients: Implementing an Organisation with a Memory*. London: The Stationery Office.

Child Health Corporation of America. 2009. Trigger tools. Pediatric Intensive Care Unit (PICU) trigger tool. http://chca.com/triggers/index.html (Accessed March 5, 2013).

Chua, S.S., Wong, I.C., Edmondson, H. et al. 2003. A feasibility study for recording of dispensing errors and near misses in four UK primary care pharmacies. *Drug Saf* 26: 803–813.

Cina, J.L., Gandhi, T.K., Churchill, W. et al. 2006. How many hospital pharmacy medication dispensing errors go undetected? *Jt Comm J Qual Patient Saf* 32: 73–80.

Ciociano, N., and Bagnasco, L. 2014. Look alike/sound alike drugs: A literature review on causes and solutions. *Int J Clin Pharm* 36: 233–242.

Claesson, C., Burman, K., Nilsson, J., and Vinge, E. 1995. Prescription errors detected by Swedish pharmacists. *Int J Pharm Pract* 3: 151–156.

Clarke, S. 2010. An integrative model of safety climate: Linking psychological climate and work attitudes to individual safety outcomes using meta-analysis. *J Occup Organ Psych* 83: 553–578.

Clarke, S.G. 2000. Safety culture: Underspecified and overrated? *Int J Manage Rev* 2: 65–90.

Clifford, S., Barber, N., and Horne, R. 2008. Understanding different beliefs held by adherers, unintentional nonadherers, and intentional nonadherers: Application of the Necessity-Concerns Framework. *J Psychosom Res* 64: 41–46.

Clifton, G.D., Byer, H., and Heaton, K. 2003. Provision of pharmacy services to underserved populations via remote dispensing and two-way videoconferencing. *Am J Health Syst Pharm* 60: 2577–2582.

Climente-Marti, M., Garcia-Manon, E.R., Artero-Mora, A., and Jimenez-Torres, NV. 2010. Potential risk of medication discrepancies and reconciliation errors at admission and discharge from an inpatient medical service. *Ann Pharmacother* 44: 1747–1754.

Coiera, E., Westbrook, J.I., and Wyatt, J.C. 2006. The safety and quality of decision support systems. *Methods Inf Med* 45: S20–S25.

Coleman, J.J., Hodson, J., Brooks, H.L., and Rosser, D. 2013a. Missed medication doses in hospitalised patients: A descriptive account of quality improvement measures and time series analysis. *Int J Qual Health Care* 25: 564–572.

Coleman, J.J., van der Sijs, H., Haefeli, W.E. et al. 2013b. On the alert: Future priorities for alerts in clinical decision support for computerized physician order entry identified from a European workshop. *BMC Med Inform Decis Mak* 13: 111.

Colligan, L., Guerlain, S., and Steck, S.E. 2012. Designing for distractions: A human factors approach to decreasing interruptions at a centralised medication station. *BMJ Qual Saf* 21: 939–947.

Colombet, I., Bura-Riviere, A., Chatila, R., Chatellier, G., and Durieux, P. 2004. Personalized versus non-personalized computerized decision support system to increase therapeutic quality control of oral anticoagulant therapy: An alternating time series analysis. *BMC Health Serv Res* 4: 27.

Conner, M., and Norman, P. 2005. *Predicting Health Behavior*. Berkshire: McGraw-Hill International.

Conroy, S., Davar, Z., and Jones, S. 2012. Use of checking systems in medicines administration with children and young people. *Nurs Child Young People* 24: 20–24.

Coombes, I.D., Mitchell, C.A., and Stowasser, D.A. 2008a. Safe medication practice: Attitudes of medical students about to begin their intern year. *Med Educ* 42: 427–431.

Coombes, I.D., Stowasser, D.A., Coombes, J.A., and Mitchell, C. 2008b. Why do interns make prescribing errors? A qualitative study. *Med J Aust* 188: 89–94.

Cooper, D. 2001. Improving safety culture: A practical guide. http://www.behavioural-safety.com/articles/Improving_safety_culture_a_practical_guide.pdf (Accessed September 10, 2014).

CORESS. 2010. A confidential reporting system for surgery. http://www.coress.org.uk/ (Accessed August 24, 2014).

Cornford, T. 2004. A sociotechnical perspective on ICT and drugs: Prescribing practice and medication error. In *IT in Health Care: Sociotechnical Approaches. Second International Conference*. Portland, OR: ITHC.

Cornish, P.L., Knowles, S.R., Marchesano, R. et al. 2005. Unintended medication discrepancies at the time of hospital admission. *Arch Intern Med* 165: 424–429.

Cornu, P., Steurbaut, S., Leysen, T. et al. 2012. Effect of medication reconciliation at hospital admission on medication discrepancies during hospitalization and at discharge for geriatric patients. *Ann Pharmacother* 46: 484–494.

Cortelyou-Ward, K., Swain, A., and Yeung, T. 2012. Mitigating error vulnerability at the transition of care through the use of health IT applications. *J Med Syst* 36: 3825–3831.

Coulter, A., and Ellins, J. 2006. *Patient-Focused Interventions: A Review of the Evidence*. London: Health Foundation.

Cousins, D., Rosario, C., and Scarpello, J. 2011. Insulin, hospitals and harm: A review of patient safety incidents reported to the National Patient Safety Agency. *Clin Med* 11: 28–30.

Cowley, E., Williams, R., and Cousins, D. 2001. Medication errors in children: A descriptive summary of medication error reports submitted to the United States Pharmacopoeia. *Curr Therap Res* 62: 627–640.

Cox, S.J., and Flin, R. 1998. Safety culture: Philosopher's stone or man of straw? *Work and Stress* 12: 189–201.

Craftman, A.G., von Strauss, E., Rudberg, S.L., and Westerbotn, M. 2013. District nurses' perceptions of the concept of delegating administration of medication to home care aides working in the municipality: A discrepancy between legal regulations and practice. *J Clin Nurs* 22: 569–578.

Crandall, W., Kappelman, M.D., Colletti, R.B. et al. 2011. ImproveCareNow: The development of a pediatric inflammatory bowel disease improvement network. *Inflamm Bowel Dis* 17: 450–457.

Cresswell, K.M., Bates, D.W., Williams, R. et al. 2014. Evaluation of medium-term consequences of implementing commercial computerized physician order entry and clinical decision support prescribing systems in two 'early adopter' hospitals. *J Am Med Inform Assoc* 21: 194–202.

Cresswell, K.M., Panesar, S.S., Salvilla, S.A. et al. 2013. Global research priorities to better understand the burden of iatrogenic harm in primary care: An international Delphi exercise. *PLoS Med* 10: e1001554.
Creswick, N., and Westbrook, J.I. 2007. The medication advice-seeking network of staff in an Australian hospital renal ward. *Stud Health Technol Inform* 130: 217–231.
Creswick, N., and Westbrook, J.I. 2010. Social network analysis of medication advice-seeking interactions among staff in an Australian hospital. *Int J Med Inform* 79: e116–e125.
Creswick, N., and Westbrook, J.I. 2015. Who do hospital physicians and nurses go to for advice about medications? A social network analysis and examination of prescribing error rates. *J Patient Saf* (in press).
Creswick, N., Westbrook, J.I., and Braithwaite, J. 2009. Understanding communication networks in the emergency department. *BMC Health Serv Res* 9: 247.
Croskerry, P. 2003. The importance of cognitive errors in diagnosis and strategies to minimize them. *Acad Med* 78: 775–780.
Cullen, D.J., Sweitzer, B.J., Bates, D.W. et al. 1997a. Preventable adverse drug events in hospitalized patients: A comparative study of intensive care and general care units. *Crit Care Med* 25: 1289–1297.
Cullen, D.J., Sweitzer, B.J., Bates, D.W. et al. 1997b. Preventable adverse drug events in hospitalized patients: A comparative study of intensive care and general care units. *Crit Care Med* 25: 1289–1297.
Culler, S.D., Jose, J., Kohler, S., and Rask, K. 2011. Nurses' perceptions and experiences with the implementation of a medication administration system. *Comput Inform Nurs* 29: 280–288.
Cumbler, E., Wald, H., and Kutner, J. 2010. Lack of patient knowledge regarding hospital medications. *J Hosp Med* 5: 83–86.
Dart, R.C., and Rumack, B.H. 2012. Intravenous acetaminophen in the United States: Iatrogenic dosing errors. *Pediatrics* 129: 349–353.
Darzi, A. 2008. *High Quality Care For All—NHS Next Stage Review Final Report. (Cm7432).* London: The Stationery Office.
Davis, B., and Sumara, D. 2009. Complexity as a theory of education. *Transnat Curr Inquiry* 5: 33–44.
Davis, R., Parand, A., Pinto, A., and Buetow, S. 2015. A systematic review of the effectiveness of strategies to encourage patients to remind healthcare professionals about their hand hygiene. *J Hosp Infect* 89: 141–162.
Davis, R.E., Sevdalis, N., and Vincent, C.A. 2011. Patient involvement in patient safety: How willing are patients to participate? *BMJ Qual Saf* 20: 108–114.
Davis, R.E., Sevdalis, N., and Vincent, C.A. 2012. Patient involvement in patient safety: The health-care professional's perspective. *J Patient Saf* 8: 182–188.
Day, R.O., Roffe, D., Richardson, K. et al. 2011. Implementing electronic medication management at an Australian teaching hospital. *Med J Aust* 195: 498–502.
De Clifford, J. 1993. Concentrate or kill! *Aust J Hosp Pharm* 23: 72–73.
de Feijter, J.M., de Grave, W.S., Dornan, T., Koopmans, R.P., and Scherpbier, A.J. 2011. Students' perceptions of patient safety during the transition from undergraduate to postgraduate training: An activity theory analysis. *Adv Health Sci Educ Theory Pract* 16: 347–358.
de Lusignan, S., Mold, F., Sheikh, A. et al. 2014. Patients' online access to their electronic health records and linked online services: A systematic interpretative review. *BMJ Open* 4: e006021.
Dean, B., and Barber, N. 1999. A validated, reliable method of scoring the severity of medication errors. *Am J Health Syst Pharm* 56: 57–62.
Dean, B., and Barber, N. 2001. Validity and reliability of observational methods for studying medication administration errors. *Am J Health Syst Pharm* 58: 54–59.

Dean, B., Barber, N., and Schachter, M. 2000. What is a prescribing error? *Qual Health Care* 9: 232–237.

Dean, B., Schachter, M., Vincent, C., and Barber, N. 2002a. Causes of prescribing errors in hospital inpatients—A prospective study. *Lancet* 359: 1373–1378.

Dean, B., Schachter, M., Vincent, C., and Barber, N. 2002b. Prescribing errors in hospital inpatients: Their incidence and clinical significance. *Qual Saf Health Care* 11: 340–344.

Dean, B.S., Allan, E.L., Barber, N.D., and Barker, K.N. 1995. Comparison of medication errors in an American and a British hospital. *Am J Health Syst Pharm* 52: 2543–2549.

Debono, D.S., Greenfield, D., Travaglia, J.F. et al. 2013. Nurses' workarounds in acute health-care settings: A scoping review. *BMC Health Serv Res* 13: 175.

Deming, W.E. 1993. *The New Economics for Industry, Government, Education*. Cambridge, MA: Massachusetts Institute of Technology, Center for Advanced Engineering Study.

Department of Health. 2000. *An Organisation with a Memory*. London: The Stationery Office.

Department of Health. 2008. *High Quality Care For All. NHS Next Stage Review Final Report*. London: The Stationery Office.

Department of Health. 2010. *Equity and Excellence: Liberating the NHS*. London: The Stationery Office.

Department of Health. 2011. *A Framework for Technology Enhanced Learning*. London: Department of Health.

Department of Health. 2013. *Delivering High Quality, Effective, Compassionate Care: Developing the Right People with the Right Skills and the Right Values*. London: Williams Lea.

DesRoches, C.M., Charles, D., Furukawa, M.F. et al. 2013. Adoption of electronic health records grows rapidly, but fewer than half of US hospitals had at least a basic system in 2012. *Health Aff (Millwood)* 32: 1478–1485.

Devine, E.B., Williams, E.C., Martin, D.P. et al. 2010. Prescriber and staff perceptions of an electronic prescribing system in primary care: A qualitative assessment. *BMC Med Inform Decis Mak* 10: 72.

DeYoung, J.L., VanderKooi, M.E., and Barletta, J.F. 2009. Effect of bar-code-assisted medication administration on medication error rates in an adult medical intensive care unit. *Am J Health Syst Pharm* 66: 1110–1115.

Dickinson, A., McCall, E., Twomey, B., and James, N. 2010. Paediatric nurses' understanding of the process and procedure of double-checking medications. *J Clin Nurs* 19: 728–735.

Diller, T., Helmrich, G., Dunning, S. et al. 2014. The Human Factors Analysis Classification System (HFACS) applied to health care. *Am J Med Qual* 29: 181–190.

Dixon-Woods, M., Leslie, M., Bion, J., and Tarrant, C. 2012. What counts? An ethnographic study of infection data reported to a patient safety program. *Milbank Q* 90: 548–591.

Doherty, C., and McDonnell, C. 2012. Tenfold medication errors: 5 years' experience at a university-affiliated pediatric hospital. *Pediatrics* 129: 916–924.

Doll, W.E., Jr., and Trueit, D. 2010. Complexity and the health care professions. *J Eval Clin Pract* 16: 841–848.

Donabedian, A. 1988. The quality of care. How can it be assessed? *JAMA* 260: 1743–1748.

Donaldson, L. 2004. When will health care pass the orange-wire test? *Lancet* 364: 1567–1568.

Donaldson, L.J., and Fletcher, M.G. 2006. The WHO World Alliance for Patient Safety: Towards the years of living less dangerously. *Med J Aust* 184: S69–S72.

Doran, G.T. 1981. There's a S.M.A.R.T. way to write management's goals and objectives. *Manage Rev* 70: 35–38.

Dornan, T., Ashcroft, D., Heathfield, H. et al. 2009. An in depth investigation into causes of prescribing errors by foundation trainees in relation to their medical education. EQUIP study. http://www.gmc-uk.org/FINAL_Report_prevalence_and_causes_of_prescribing_errors.pdf_28935150.pdf (Accessed December 5, 2014).

Dowell, D., Manwell, L.B., Maguire, A. et al. 2005. Urban outpatient views on quality and safety in primary care. *Healthc Q* 8(Suppl): 2–8.

Duncan, E.M., Francis, J.J., Johnston, M. et al. 2012. Learning curves, taking instructions, and patient safety: Using a theoretical domains framework in an interview study to investigate prescribing errors among trainee doctors. *Implement Sci* 7: 86.

Dunning, D., Johnson, K., Ehrlinger, J., and Kruger, J. 2003. Why people fail to recognise their own incompetence. *Curr Dir Psychol Sci* 12: 83–87.

Dwibedi, N., Sansgiry, S.S., Frost, C.P. et al. 2011. Effect of bar-code-assisted medication administration on nurses' activities in an intensive care unit: A time-motion study. *Am J Health Syst Pharm* 68: 1026–1031.

Ebeling, C.E. 1997. *An Introduction to Reliability and Maintainability Engineering*. New York, NY: McGraw-Hill.

Egan, S., Murphy, P.G., Fennell, J.P. et al. 2012. Using Six Sigma to improve once daily gentamicin dosing and therapeutic drug monitoring performance. *BMJ Qual Saf* 21: 1042–1051.

Elwyn, G., Legare, F., van der Weijden, T., Edwards, A., and May, C. 2008. Arduous implementation: Does the Normalisation Process Model explain why it's so difficult to embed decision support technologies for patients in routine clinical practice. *Implement Sci* 3: 57.

Engeström, Y., and Sannino, A. 2010. Studies of expansive learning: Foundations, findings and future challenges. *Educ Res Rev* 5: 1–24.

Entwistle, V.A., McCaughan, D., Watt, I.S. et al. 2010. Speaking up about safety concerns: Multi-setting qualitative study of patients' views and experiences. *Qual Saf Health Care* 19: e33.

Eraut, M. 2004. Transfer of knowledge between education and workplace settings. In *Workplace Learning in Context*, eds H. Rainbird, A. Fuller and A. Munro, 201–221. London: Routledge.

Eraut, M. 2010. Knowledge, working practices, and learning. In *Learning Through Practice*, ed S. Billett, 37–58. New York, NY: Springer.

Escoms, M.C., Cabanas, M.J., Oliveras, M., Hidalgo, E., and Barroso, C. 1996. Errors evolution and analysis in antineoplastic drug preparation during one year. *Pharm World Sci* 18: 178–181.

European Parliament, and Council of European Union. 2009. Council Directive 2001/83/EC on the community code relating to medicinal products for human use. *Official J L113* 174: 1–129.

Eva, K.W., and Regehr, G. 2005. Self-assessment in the health professions: A reformulation and research agenda. *Acad Med* 80: S46–S54.

Evans, J.S. 2008. Dual-processing accounts of reasoning, judgment, and social cognition. *Annu Rev Psychol* 59: 255–278.

Fanikos, J., Fiumara, K., Baroletti, S. et al. 2007. Impact of smart infusion technology on administration of anticoagulants (unfractionated heparin, argatroban, lepirudin, and bivalirudin). *Am J Cardiol* 99: 1002–1005.

Ferner, R.E. 2009. The epidemiology of medication errors: The methodological difficulties. *Br J Clin Pharmacol* 67: 614–620.

Fijn, R., van den Bemt, P.M., Chow, M. et al. 2002. Hospital prescribing errors: Epidemiological assessment of predictors. *Br J Clin Pharmacol* 53: 326–331.

Filik, R., Price, J., Darker, I. et al. 2010. The influence of tall man lettering on drug name confusion. *Drug Saf* 33: 677–687.

Fishbein, M., Triandis, H., Kanfer, F., Becker, M., and Middlestadt, S. 2000. Factors influencing behavior and behavior change. In *Handbook of Health Psychology*, eds A.S. Baum, T.A. Revenson and J.E. Singer, 1–17. Mahwah, NJ: Lawrence Erlbaum.

FitzHenry, F., Doran, J., Lobo, B. et al. 2011. Medication-error alerts for warfarin orders detected by a bar-code-assisted medication administration system. *Am J Health Syst Pharm* 68: 434–441.

Fitzpatrick, R., Cooke, P., Southall, C., Kauldhar, K., and Waters, P. 2005. Evaluation of an automated dispensing system in a hospital pharmacy dispensary. *Pharm J* 274: 763–765.

Fitzsimons, M., Grimes, T., and Galvin, M. 2011. Sources of pre-admission medication information: Observational study of accuracy and availability. *Int J Pharm Pract* 19: 408–416.

Flavell, J.H. 1979. Metacognition and cognitive monitoring: A new area of cognitive–developmental inquiry. *Am Psychol* 34: 906–911.

Flynn, E.A., and Barker, K.N. 2006. Effect of an automated dispensing system on errors in two pharmacies. *J Am Pharm Assoc* 46: 613–615.

Flynn, E.A., Barker, K.N., and Carnahan, B.J. 2003. National observational study of prescription dispensing accuracy and safety in 50 pharmacies. *J Am Pharm Assoc* 43: 191–200.

Flynn, E.A., Barker, K.N., Gibson, J.T. et al. 1996. Relationships between ambient sounds and the accuracy of pharmacists' prescription-filling performance. *Hum Factors* 38: 614–622.

Flynn, E.A., Barker, K.N., Gibson, J.T. et al. 1999. Impact of interruptions and distractions on dispensing errors in an ambulatory care pharmacy. *Am J Health Syst Pharm* 56: 1319–1325.

Flynn, E.A., Dorris, N.T., Holman, G.T., Carnahan, B.J., and Barker, K.N. 2002. Medication dispensing errors in community pharmacies: A nationwide study. *Proc HFES Annu Meeting* 46: 1448–1451.

Flynn, E.A., Pearson, R.E., and Barker, K.N. 1997. Observational study of accuracy in compounding i.v. admixtures at five hospitals. *Am J Health Syst Pharm* 54: 904–912.

Folli, H.L., Poole, R.L., and Benitz, W.E. 1987. Medication error prevention by clinical pharmacists in two children's hospitals. *Pediatrics* 79: 718–722.

Fontan, J.E., Maneglier, V., Nguyen, V.X., Loirat, C., and Brion, F. 2003. Medication errors in hospitals: Computerized unit dose drug dispensing system versus ward stock distribution system. *Pharm World Sci* 25: 112–117.

Food and Drug Administration. 2006. 2006 limited FDA survey of compounded drug products. http://www.fda.gov/drugs/guidancecomplianceregulatoryinformation/pharmacy-compounding/ucm204237.htm (Accessed March 21, 2014).

Ford, D.G., Seybert, A.L., Smithburger, P.L. et al. 2010. Impact of simulation-based learning on medication error rates in critically ill patients. *Intensive Care Med* 36: 1526–1531.

Forrester, S.H., Hepp, Z., Roth, J.A., Wirtz, H.S., and Devine, E.B. 2014. Cost-effectiveness of a computerized provider order entry system in improving medication safety ambulatory care. *Value Health* 17: 340–349.

Forrey, R.A., Pedersen, C.A., and Schneider, P.J. 2007. Interrater agreement with a standard scheme for classifying medication errors. *Am J Health Syst Pharm* 64: 175–181.

Frank, J., and Brien, S. 2009. *The Safety Competencies: Enhancing Patient Safety Across the Health Professions*. Toronto: Canadian Patient Safety Institute.

Franklin, B.D. 2014. Medication errors: Do they occur in isolation? *BMJ Qual Saf* 23: e1.

Franklin, B.D., Birch, S., Savage, I. et al. 2009. Methodological variability in detecting prescribing errors and consequences for the evaluation of interventions. *Pharmacoepidemiol Drug Saf* 18: 992–999.

Franklin, B.D., Birch, S., Schachter, M., and Barber, N. 2010a. Testing a trigger tool as a method of detecting harm from medication errors in a UK hospital: A pilot study. *Int J Pharm Pract* 18: 305–311.

Franklin, B.D., McLeod, M., and Barber, N. 2010b. Comment on 'Prevalence, incidence and nature of prescribing errors in hospital inpatients: A systematic review.' *Drug Saf* 33: 163–165.

Franklin, B.D., and O'Grady, K. 2007. Dispensing errors in community pharmacy; frequency, clinical significance and potential impact of authentication at the point of dispensing. *Int J Pharm Pract* 15: 273–281.

Franklin, B.D., O'Grady, K., Donyai, P., Jacklin, A., and Barber, N. 2007a. The impact of a closed-loop electronic prescribing and administration system on prescribing errors, administration errors and staff time: A before-and-after study. *Qual Saf Health Care* 16: 279–284.

Franklin, B.D., O'Grady, K., Paschalides, C. et al. 2007b. Providing feedback to hospital doctors about prescribing errors; a pilot study. *Pharm World Sci* 29: 213–220.

Franklin, B.D., O'Grady, K., Voncina, L., Popoola, J., and Jacklin, A. 2008. An evaluation of two automated dispensing machines in UK hospital pharmacy. *Int J Pharm Pract* 16: 47–53.

Franklin, B.D., Panesar, S.S., Vincent, C., and Donaldson, L.J. 2014a. Identifying systems failures in the pathway to a catastrophic event: An analysis of national incident report data relating to vinca alkaloids. *BMJ Qual Saf* 23: 765–772.

Franklin, B.D., Reynolds, M., Sadler, S. et al. 2014b. The effect of the electronic transmission of prescriptions on dispensing errors and prescription enhancements made in English community pharmacies: A naturalistic stepped wedge study. *BMJ Qual Saf* 23: 629–638.

Franklin, B.D., Reynolds, M., Shebl, N.A., Burnett, S., and Jacklin, A. 2011. Prescribing errors in hospital inpatients: A three-centre study of their prevalence, types and causes. *Postgrad Med J* 87: 739–745.

Franklin, B.D., Reynolds, M.J., Hibberd, R., Sadler, S., and Barber, N. 2013. Community pharmacists' interventions with electronic prescriptions in England: An exploratory study. *Int J Clin Pharm* 35: 1030–1035.

Franklin, B.D., Vincent, C., Schachter, M., and Barber, N. 2005. The incidence of prescribing errors in hospital inpatients: An overview of the research methods. *Drug Saf* 28: 891–900.

Frankovich, J., Longhurst, C.A., and Sutherland, S.M. 2011. Evidence-based medicine in the EMR era. *N Engl J Med* 365: 1758–1759.

Fraser, S.W., and Greenhalgh, T. 2001. Coping with complexity: Educating for capability. *BMJ* 323: 799–803.

Fuji, K.T., Paschal, K.A., Galt, K.A., and Abbott, A.A. 2010. Pharmacy student attitudes toward an interprofessional patient safety course: An exploratory mixed methods study. *Curr Pharm Teach Learning* 2: 238–247.

Fung, E.Y., and Leung, B. 2009. Do automated dispensing machines improve patient safety? *Can J Hosp Pharm* 62: 516–517.

Furniss, D., Barber, N., Lyons, I., Eliasson, L., and Blandford, A. 2014. Unintentional non-adherence: Can a spoon full of resilience help the medicine go down? *BMJ Qual Saf* 23: 95–98.

Gaba, D.M. 2004. The future vision of simulation in health care. *Qual Saf Health Care* 13(Suppl 1): i2–i10.

Gabriele, S. 2006. The role of typography in differentiating look-alike/sound-alike drug names. *Healthc Q* 9: 88–95.

Galvin, M., Jago-Byrne, M.C., Fitzsimons, M., and Grimes, T. 2012. Clinical pharmacist's contribution to medication reconciliation on admission to hospital in Ireland. *Int J Clin Pharm* 35: 14–21.

Gandhi, T.K., Weingart, S.N., Seger, A.C. et al. 2005. Outpatient prescribing errors and the impact of computerized prescribing. *J Gen Intern Med* 20: 837–841.

Garfield, S., Reynolds, M., Dermont, L., and Franklin, B.D. 2013. Measuring the severity of prescribing errors: A systematic review. *Drug Saf* 36: 1151–1157.

Garg, A.X., Adhikari, N.K., McDonald, H. et al. 2005. Effects of computerized clinical decision support systems on practitioner performance and patient outcomes: A systematic review. *JAMA* 293: 1223–1238.

Gawande, A.A. 2009. *The Checklist Manifesto: How to Get Things Right*. New York, NY: Metropolitan Books.

Gelfand, M.J., Frese, M., and Salmon, E. 2011. Cultural influences on errors: Prevention, detection, and management. In *Errors in Organizations*, eds D.A. Hofman and M. Frese, 273–315. London: Routledge.

General Pharmaceutical Council. 2014. Pharmacy technicians. Accredited courses. http://www.pharmacyregulation.org/education/pharmacy-technician/accredited-courses (Accessed August 28, 2014).

Georgiou, A., Ampt, A., Creswick, N., Westbrook, J.I., and Braithwaite, J. 2009. Computerized provider order entry—What are health professionals concerned about? A qualitative study in an Australian hospital. *Int J Med Inf* 78: 60–70.

Ghaleb, M., Barber, N., Franklin, B.D. et al. 2006. Systematic review of medication errors in pediatric patients. *Ann Pharmacother* 40: 1766–1776.

Ghaleb, M.A., Barber, N., Franklin, B.D., and Wong, I.C.K. 2010. The incidence and nature of prescribing and medication administration errors in paediatric inpatients. *Arch Dis Child* 95: 113–118.

Gilbar, P.J. 2014. Intrathecal chemotherapy: Potential for medication error. *Cancer Nurs* 37: 299–309.

Gladstone, J. 1995. Drug administration errors: A study into the factors underlying the occurrence and reporting of drug errors in a district general hospital. *J Adv Nurs* 22: 628–637.

Gleason, K.M., McDaniel, M.R., Feinglass, J. et al. 2010. Results of the medications at transitions and clinical handoffs (match) study: An analysis of medication reconciliation errors and risk factors at hospital admission. *J Gen Intern Med* 25: 441–447.

Goldzweig, C.L., Orshansky, G., Paige, N.M. et al. 2013. Electronic patient portals: Evidence on health outcomes, satisfaction, efficiency, and attitudes: A systematic review. *Ann Intern Med* 159: 677–687.

Gollwitzer, P.M., and Sheeran, P. 2006. Implementation intentions and goal achievement: A meta-analysis of effects and processes. *Adv Exp Soc Psychol* 38: 69–119.

Gonzales, D.M., Ceruelo, J., Romero, M.V., and Dominquez-Gil, A. 2005. Analysis of prescription, transcription and dispensing quality through the information gathered in a pharmacy service. *Eur J Hosp Pharm Sci* 11: 91–93.

Gonzales, K. 2015. Risk propensity and safe medication administration. *J Patient Saf* (in press).

Gonzales, K.J. 2012. Assessments of safe medication administration in nursing education. *J Nurs Educ Pract* 2: 39–50.

Goodrich, J., and Cornwell, J. 2008. *Seeing the Person in the Patient—The Point of Care Review Paper*. London: The King's Fund.

Goorman, E., and Berg, M. 2000. Modelling nursing activities: Electronic patient records and their discontents. *Nurs Inq* 7: 3–9.

Gordon, M. 2013. Non-technical skills training to enhance patient safety. *Clin Teach* 10: 170–175.

Gothard, A.M., Dade, J.P., Murphy, K., and Mellor, E.J. 2004. Using error theory in the pharmacy dispensary can reduce accidents. *Pharm Prac* 14: 44–48.

Grasso, B.C., Genest, R., Jordan, C.W., and Bates, D.W. 2003. Use of chart and record reviews to detect medication errors in a state psychiatric hospital. *Psychiatr Serv* 54: 677–681.

Green, C.F., Burgul, K., and Armstrong, D.J. 2010. A study of the use of medicine lists in medicines reconciliation: Please remember this, a list is just a list. *Int J Pharm Pract* 18: 116–121.

Greenall, J., U, D., and Lam, R. 2005. An effective tool to enhance a culture of patient safety and assess the risks of medication use systems. *Healthc Q* 8: 53–58.

Greenes, R.A. 2007. *Clinical Decision Support: The Road Ahead.* Burlington, MA: Academic Press.

Greenhalgh, T., Howick, J., Maskrey, N., and Evidence-Based Medicine Renaissance Group. 2014. Evidence based medicine: A movement in crisis? *BMJ* 348: g3725.

Greenwood, M., and Woods, H. 1919. *The Incidence of Industrial Accidents Upon Individuals with Special Reference to Multiple Accidents.* London: HMSO, Her Majesty's Stationery Office.

Greiner, A., and Knebel, E. 2003. *Health Professions Education: A Bridge to Quality.* Washington, DC: The National Academies Press.

Greysen, S.R., Khanna, R.R., Jacolbia, R., Lee, H.M., and Auerbach, A.D. 2014. Tablet computers for hospitalized patients: A pilot study to improve inpatient engagement. *J Hosp Med* 9: 396–399.

Grimshaw, J.M., Thomas, R.E., MacLennan, G. et al. 2004. Effectiveness and efficiency of guideline dissemination and implementation strategies. *Health Technol Assess* 8: iii–iv, 1–72.

Guchelaar, H.J., Colen, H.B., Kalmeijer, M.D., Hudson, P.T., and Teepe-Twiss, I.M. 2005. Medication errors: Hospital pharmacist perspective. *Drugs* 65: 1735–1746.

Gudeman, J., Jozwiakowski, M., Chollet, J., and Randell, M. 2013. Potential risks of pharmacy compounding. *Drugs R D* 13: 1–8.

Guernsey, B.G., Ingrim, N.B., Hokanson, J.A. et al. 1983. Pharmacists' dispensing accuracy in a high-volume outpatient pharmacy service: Focus on risk management. *Drug Intell Clin Pharm* 17: 742–746.

Guldenmund, F.W. 2010. *Understanding and Exploring Safety Culture.* PhD Thesis. TU Delft, The Netherlands.

Gurwitz, J.H., Field, T.S., Judge, J. et al. 2005. The incidence of adverse drug events in two large academic long-term care facilities. *Am J Med* 118: 251–258.

Habraken, M.M., and van der Schaaf, T.W. 2010. If only...: Failed, missed and absent error recovery opportunities in medication errors. *Qual Saf Health Care* 19: 37–41.

Hall, J., Peat, M., Birks, Y. et al. 2010. Effectiveness of interventions designed to promote patient involvement to enhance safety: A systematic review. *Qual Saf Health Care* 19: e10.

Hall, K.W., Ebbeling, P., Brown, B., and Shwortz, I. 1985. A retrospective–prospective study of medication errors: Basis for an ongoing monitoring program. *Can J Hosp Pharm* 38: 141–143, 146.

Hammad, E.A., Wright, D.J., Walton, C., Nunney, I., and Bhattacharya, D. 2014. Adherence to UK national guidance for discharge information: An audit in primary care. *Brit J Clin Pharmacol* 78: 1453–1464.

Hammick, M., Freeth, D., Koppel, I., Reeves, S., and Barr, H. 2007. A best evidence systematic review of interprofessional education: BEME Guide no. 9. *Med Teach* 29: 735–751.

Han, W.H., and Maxwell, S.R. 2006. Are medical students adequately trained to prescribe at the point of graduation? Views of first year foundation doctors. *Scot Med J* 51: 27–32.

Hanna, K.M. 1993. Effect of nurse–client transaction on female adolescents' oral contraceptive adherence. *Image J Nurs Sch* 25: 285–290.

Harden, R.M. 1998. AMEE guide No. 12: Multiprofessional education: Part 1—Effective multiprofessional education: A three-dimensional perspective. *Med Teach* 20: 402–408.

Hardisty, J., Scott, L., Chandler, S., Pearson, P., and Powell, S. 2014. Interprofessional learning for medication safety. *Clin Teach* 11: 290–296.

Hassink, J.J., Essenberg, M.D., Roukema, J.A., and van den Bemt, P.M. 2013. Effect of bar-code-assisted medication administration on medication administration errors. *Am J Health Syst Pharm* 70: 572–573.

Haw, C., and Stubbs, J. 2003. Prescribing errors at a psychiatric hospital. *Pharm Prac* 13: 64–66.

Health and Social Care Information Centre. 2013. Prescriptions dispensed in the community, statistics for England—2002–2012. http://www.hscic.gov.uk/catalogue/PUB11291 (Accessed October 10, 2014).

Health and Social Care Information Centre. 2014. *Prescriptions Dispensed in the Community: England 2003–2013*. Leeds: Health and Social Care Information Centre.

Health Care Professions Council. 2012. *Standards of Education and Training*. London: Health Care Professions Council.

Health and Care Professions Council. 2013. *Service User and Carer Involvement in Education and Training Programmes*. London: Health Care Professions Council.

Hellstrom, L.M., Bondesson, A., Hoglund, P., and Eriksson, T. 2012. Errors in medication history at hospital admission: Prevalence and predicting factors. *BMC Clin Pharmacol* 12: 9.

Helmons, P.J., Wargel, L.N., and Daniels, C.E. 2009. Effect of bar-code-assisted medication administration on medication administration errors and accuracy in multiple patient care areas. *Am J Health Syst Pharm* 66: 1202–1210.

Hendey, G.W., Barth, B.E., and Soliz, T. 2005. Overnight and postcall errors in medication orders. *Acad Emerg Med* 12: 629–634.

Hendrick, H. 1996. The ergonomics of economics is the economics of ergonomics. *Proc HFES Annu Meeting* 40: 1–10.

Henriksen, K., Dayton, E., Keyes, M.A., Carayon, P., and Hughes, R. 2008. Understanding adverse events: A human factors framework. In *Patient Safety and Quality: An Evidence-Based Handbook for Nurses*, ed R.G. Hughes, 67–85. Rockville, MD: Agency for Healthcare Research and Quality (US).

Herring, H., Ripley, T.L., Farmer, K.C., and St Cyr, M. 2012. An intervention to increase safety feature use on smart pumps: A quality improvement initiative. *J Pharm Technol* 28: 119–123.

Hesselink, G., Schoonhoven, L., Barach, P. et al. 2012. Improving patient handovers from hospital to primary care: A systematic review. *Ann Intern Med* 157: 417–428.

Heyworth, L., Clark, J., Marcello, T.B. et al. 2013. Aligning medication reconciliation and secure messaging: Qualitative study of primary care providers' perspectives. *J Med Internet Res* 15: e264.

Heyworth, L., Paquin, A.M., Clark, J. et al. 2014. Engaging patients in medication reconciliation via a patient portal following hospital discharge. *J Am Med Inform Assoc* 21: e157–e162.

Hibbard, J.H., Peters, E., Slovic, P., and Tusler, M. 2005. Can patients be part of the solution? Views on their role in preventing medical errors. *Med Care Res Rev* 62: 601–616.

Hillestad, R., Bigelow, J., Bower, A. et al. 2005. Can electronic medical record systems transform health care? Potential health benefits, savings, and costs. *Health Aff (Millwood)* 24: 1103–1117.

Hines, L.E., Warholak, T.L., Saverno, K.R. et al. 2011. Drug–drug interaction software quality assurance: Lessons learned. *J Am Pharm Assoc* 51: 570–572.

Hodge, M.H. 1990. History of the TDS medical information system. In *A History of Medical Informatics*, eds B.I. Blum and K. Duncan, 328–344. Reading, MA: Addison-Wesley Publishing Company.

Hoffmann, B., Muller, V., Rochon, J. et al. 2014. Effects of a team-based assessment and intervention on patient safety culture in general practice: An open randomised controlled trial. *BMJ Qual Saf* 23: 35–46.

Hofmann, D.A., and Mark, B. 2006. An investigation of the relationship between safety climate and medication errors as well as other nurse and patient outcomes. *Pers Psychol* 59: 847–869.

Holden, R.J. 2010. Physicians' beliefs about using EMR and CPOE: In pursuit of a contextualised understanding of health IT use behaviour. *Int J Med Inf* 79: 71–80.

Holden, R.J., Carayon, P., Gurses, A.P. et al. 2013a. SEIPS 2.0: A human factors framework for studying and improving the work of healthcare professionals and patients. *Ergonomics* 56: 1669–1686.

Holden, R.J., Rivera-Rodriguez, A.J., Faye, H., Scanlon, M.C., and Karsh, B.T. 2013b. Automation and adaptation: Nurses' problem-solving behavior following the implementation of bar coded medication administration technology. *Cogn Technol Work* 15: 283–296.

Horsky, J., Phansalkar, S., Desai, A., Bell, D., and Middleton, B. 2013. Design of decision support interventions for medication prescribing. *Int J Med Inf* 82: 492–503.

Horsky, J., Schiff, G.D., Johnston, D. et al. 2012. Interface design principles for usable decision support: A targeted review of best practices for clinical prescribing interventions. *J Biomed Inform* 45: 1202–1216.

Howard, R.L., Avery, A.J., Slavenburg, S. et al. 2007. Which drugs cause preventable admissions to hospital? A systematic review. *Br J Clin Pharmacol* 63: 136–147.

Hoxsie, D.M., Keller, A.E., and Armstrong, E.P. 2006. Analysis of community pharmacy workflow processes in preventing dispensing errors. *J Pharm Pract* 19: 124–130.

Hrisos, S., Eccles, M., Johnston, M. et al. 2008. Developing the content of two behavioural interventions: Using theory-based interventions to promote GP management of upper respiratory tract infection without prescribing antibiotics #1. *BMC Health Serv Res* 8: 11.

Hsieh, S.H., Hou, I.C., Tan, C.T. et al. 2009. Design and implementation of mobile electronic medication administration record. *Stud Comput Intell* 199: 493–507.

Hsieh, T.C., Kuperman, G.J., Jaggi, T. et al. 2004. Characteristics and consequences of drug allergy alert overrides in a computerized physician order entry system. *J Am Med Inform Assoc* 11: 482–491.

Huckels-Baumgart, S., and Manser, T. 2014. Identifying medication error chains from critical incident reports: A new analytic approach. *J Clin Pharmacol* 54: 1188–1197.

Hudson, P. 2003. Applying the lessons of high risk industries to health care. *Qual Saf Health Care* 12 (Suppl 1): i7–12.

Hulscher, M.E., Grol, R.P., and van der Meer, J.W. 2010. Antibiotic prescribing in hospitals: A social and behavioural scientific approach. *Lancet Infect Dis* 10: 167–175.

Husch, M., Sullivan, C., Rooney, D. et al. 2005. Insights from the sharp end of intravenous medication errors: Implications for infusion pump technology. *Qual Saf Health Care* 14: 80–86.

IMS Institute for Healthcare Informatics. 2013. *Declining Medicines Use and Cost: For Better or Worse? A Review of the Use of Medicines in the United States in 2012*. Parsippany, NJ: IMS Institute for Healthcare Informatics.

Institute for Healthcare Improvement. 2004. Failure Modes and Effects Analysis (FMEA) tool. http://www.ihi.org/resources/Pages/Tools/FailureModesandEffectsAnalysisTool.aspx (Accessed March 18, 2014).

Institute for Healthcare Improvement. 2012. How-to guide: Prevent harm from high alert medications. http://www.ihi.org/resources/Pages/Tools/HowtoGuidePreventHarmfromHighAlertMedications.aspx (Accessed February 27, 2014).

Institute for Safe Medication Practices. 2001. *Medication Safety Self-Assessment for Community/Ambulatory Pharmacy*. Horsham, PA: Institute for Safe Medication Practices.

Institute for Safe Medication Practices. 2006. *Double Key Bounce and Double Keying Errors*. Horsham, PA: Institute for Safe Medication Practices.

Institute for Safe Medication Practices. 2009. Proceedings from the ISMP summit on the use of smart infusion pumps: Guidelines for safe implementation and use. http://www.ismp.org/tools/guidelines/smartpumps/comments/ (Accessed August 28, 2014).

Institute for Safe Medication Practices. 2013. Independent double checks: Undervalued and misused: Selective use of this strategy can play an important role in medication safety. https://www.ismp.org/newsletters/acutecare/showarticle.aspx?id=51 (Accessed September 17, 2014).

Institute for Safe Medication Practices. 2014a. Confused drug names. http://www.ismp.org/Tools/confuseddrugnames.pdf (Accessed September 17, 2014).

Institute for Safe Medication Practices. 2014b. ISMP high-alert medications. https://www.ismp.org/tools/highalertmedicationLists.asp (Accessed September 17, 2014).

Institute for Safe Medication Practices. 2014c. The national medication errors reporting Program (ISMP MERP). https://www.ismp.org/orderforms/reporterrortoismp.asp (Accessed September 17, 2014).

Institute of Medicine. 2000. *To Err Is Human: Building a Safer Health System*. Washington, DC: National Academy Press.

Interprofessional Education Collaborative Expert Panel. 2011. *Core Competencies for Interprofessional Collaborative Practice: Report of an Expert Panel*. Washington, DC: Interprofessional Education Collaborative.

Isaac, T., Weissman, J.S., Davis, R.B. et al. 2009. Overrides of medication alerts in ambulatory care. *Arch Intern Med* 169: 305–311.

Ivers, N., Jamtvedt, G., Flottorp, S. et al. 2012. Audit and feedback: Effects on professional practice and healthcare outcomes. *Cochrane Database Syst Rev* 6: CD000259.

Iyer, S.B., Schubert, C.J., Schoettker, P.J., and Reeves, S.D. 2011. Use of quality-improvement methods to improve timeliness of analgesic delivery. *Pediatrics* 127: e219–e225.

Jackson, M., and Lowey, A. 2010. *Handbook of Extemporaneous Preparation*. London: Pharmaceutical Press.

Jaensch, S.L., Baysari, M.T., Day, R.O., and Westbrook, J.I. 2013. Junior doctors' prescribing work after-hours and the impact of computerized decision support. *Int J Med Inf* 82: 980–986.

James, K.L., Barlow, D., Bithell, A. et al. 2013. The impact of automation on workload and dispensing errors in a hospital pharmacy. *Int J Pharm Pract* 21: 92–104.

James, K.L., Barlow, D., Burfield, R. et al. 2007. Impact of automation on the occurrence of unprevented dispensing incidents at Llandough hospital. *Int J Pharm Pract* 15: B59.

James, K.L., Barlow, D., Burfield, R. et al. 2008. Analysis of unprevented dispensing incidents in Welsh NHS hospitals 2003–2004. *Int J Pharm Pract* 16: 175–188.

James, K.L., Barlow, D., Burfield, R. et al. 2011. Unprevented or prevented dispensing incidents: Which outcome to use in dispensing error research? *Int J Pharm Pract* 19: 36–50.

James, K.L., Barlow, D., McArtney, R. et al. 2009. Incidence, type and causes of dispensing errors: A review of the literature. *Int J Pharm Pract* 17: 9–30.

Jarman, H., Jacobs, E., and Zielinski, V. 2002. Medication study supports registered nurses' competence for single checking. *Int J Nurs Pract* 8: 330–335.

Jha, A.K. 2010. Meaningful use of electronic health records: The road ahead. *JAMA* 304: 1709–1710.

Johnson, C.W. 2006. What are the emergent properties and how do they affect the engineering of complex systems? *Reliab Eng Syst Safe* 91: 1475–1481.

Johnson, J.K., Miller, S.H., and Horowitz, S.D. 2008. Systems-based practice: Improving the safety and quality of patient care by recognizing and improving the systems in which we work. In *Advances in Patient Safety: New Directions and Alternative Approaches*, eds K. Henriksen, J.B. Battles, M.A. Keyes and M. Grady, 1–10. Rockville, MD: Agency for Healthcare Research and Quality (US).

Joint Commission. 2008. Speak up: Help avoid mistakes with your medicines. http://www.jointcommission.org/topics/speakup_brochures.aspx (Accessed October 12, 2014).

Joint Commission. 2011. Taking medicine safely: What can you do to make sure it is safe for your child to take a new medicine? http://www.jointcommission.org/assets/1/6/speakup_peds.pdf (Accessed October 12, 2014).

Joint Commission. 2014. National patient safety goals. http://www.jointcommission.org/standards_information/npsgs.aspx (Accessed April 25, 2014).

Kadmon, G., Bron-Harlev, E., Nahum, E. et al. 2009. Computerized order entry with limited decision support to prevent prescription errors in a PICU. *Pediatrics* 124: 935–940.

Kaiser Permanente of Colorado, and Evergreen Colorado. 2011. SBAR technique for communication: A situational briefing model. http://www.ihi.org/knowledge/Pages/Tools/SBARTechniqueforCommunicationASituationalBriefingModel.aspx (Accessed August 28, 2014).

Kamarudin, G., Penm, J., Chaar, B., and Moles, R. 2013. Educational interventions to improve prescribing competency: A systematic review. *BMJ Open* 3: e003291.

Kamath, J.R., Osborn, J.B., Roger, V.L., and Rohleder, T.R. 2011. Highlights from the third annual Mayo Clinic conference on systems engineering and operations research in health care. *Mayo Clin Proc* 86: 781–786.

Kane-Gill, S.L., Kowiatek, J.G., and Weber, R.J. 2010. A comparison of voluntarily reported medication errors in intensive care and general care units. *Qual Saf Health Care* 19: 55–59.

Kaplan, G., Bo-Linn, G., Carayon, P. et al. 2013. Bringing a systems approach to health. http://www.iom.edu/systemsapproaches (Accessed April 25, 2014).

Karapinar, F., van den Bemt, P.M., Zoer, J., Nijpels, G., and Borgsteede, S.D. 2010. Informational needs of general practitioners regarding discharge medication: Content, timing and pharmacotherapeutic advice. *Pharm World Sci* 32: 172–178.

Karapinar-Çarkıt, F., Borgsteede, S.D., Zoer, J. et al. 2009. Effect of medication reconciliation with and without patient counseling on the number of pharmaceutical interventions among patients discharged from the hospital. *Ann Pharmacother* 43: 1001–1010.

Karapinar-Çarkıt, F., van Breukelen, B.R.L., Borgsteede, S.D. et al. 2014. Completeness of patient records in community pharmacies post-discharge after in-patient medication reconciliation: A before–after study. *Int J Clin Pharm* 36: 807–814.

Karsh, B.T. 2004. Beyond usability: Designing effective technology implementation systems to promote patient safety. *Qual Saf Health Care* 13: 388–394.

Karsh, B.T., Holden, R.J., Alper, S.J., and Or, C.K. 2006. A human factors engineering paradigm for patient safety: Designing to support the performance of the healthcare professional. *Qual Saf Health Care* 15(Suppl 1): i59–i65.

Kaushal, R., Bates, D.W., Landrigan, C. et al. 2001. Medication errors and adverse drug events in pediatric inpatients. *JAMA* 285: 2114–2120.

Kaushal, R., Goldmann, D.A., Keohane, C.A. et al. 2010. Medication errors in paediatric outpatients. *Qual Saf Health Care* 19: e30.

Kawamoto, K., Houlihan, C.A., Balas, E.A., and Lobach, D.F. 2005. Improving clinical practice using clinical decision support systems: A systematic review of trials to identify features critical to success. *BMJ* 330: 765.

Kayne, S. 1996. Negligence and the pharmacist. Part 3. Dispensing and prescribing errors. *Pharm J* 257: 32–35.

Keers, R.N., Williams, S.D., Cooke, J., and Ashcroft, D.M. 2013a. Causes of medication administration errors in hospitals: A systematic review of quantitative and qualitative evidence. *Drug Saf* 36: 1045–1067.

Keers, R.N., Williams, S.D., Cooke, J., and Ashcroft, D.M. 2013b. Prevalence and nature of medication administration errors in health care settings: A systematic review of direct observational evidence. *Ann Pharmacother* 47: 237–256.

Keers, R.N., Williams, S.D., Vattakatuchery, J.J. et al. 2014. Prevalence, nature and predictors of prescribing errors in mental health hospitals: A prospective multicentre study. *BMJ Open* 4: e006084.

Kelly, D.F., Faught, W.J., and Holmes, L.A. 1999. Ovarian cancer treatment: The benefit of patient telephone follow-up post-chemotherapy. *Can Oncol Nurs J* 9: 175–178.

Kelly, D.V., Bishop, L., Young, S. et al. 2013. Pharmacist and physician views on collaborative practice: Findings from the community pharmaceutical care project. *Can Pharm J (Ott)* 146: 218–226.

Kennedy, L.M. 1990. *The Effectiveness of a Self-Care Medication Education Protocol on the Home Medication Behaviors of Recently Hospitalized Elderly*. PhD Thesis. The University of Texas at Austin, TX, USA.

Ker, J., Mole, L., and Bradley, P. 2003. Early introduction to interprofessional learning: A simulated ward environment. *Med Educ* 37: 248–255.

Kerosuo, H., Kajamaa, A., and Engeström, Y. 2010. Promoting innovation and learning through Change laboratory: An example from Finnish healthcare. *Cent Eur J Publ Policy* 4: 110–131.

Kiekkas, P., Karga, M., Lemonidou, C., Aretha, D., and Karanikolas, M. 2011. Medication errors in critically ill adults: A review of direct observation evidence. *Am J Crit Care* 20: 36–44.

King, D., Jabbar, A., Charani, E. et al. 2014. Redesigning the 'choice architecture' of hospital prescription charts. *BMJ Open* 4: e005473.

Kirk, S., Parker, D., Claridge, T., Esmail, A., and Marshall, M. 2007. Patient safety culture in primary care: Developing a theoretical framework for practical use. *Qual Saf Health Care* 16: 313–320.

Kistner, U.A., Keith, M.R., Sergeant, K.A., and Hokanson, J.A. 1994. Accuracy of dispensing in a high-volume, hospital-based outpatient pharmacy. *Am J Hosp Pharm* 51: 2793–2797.

Klein, E.G., Santora, J.A., Pascale, P.M., and Kitrenos, J.G. 1994. Medication cart-filling time, accuracy, and cost with an automated dispensing system. *Am J Hosp Pharm* 51: 1193–1196.

Knudsen, P., Herborg, H., Mortensen, A.R., Knudsen, M., and Hellebek, A. 2007. Preventing medication errors in community pharmacy: Frequency and seriousness of medication errors. *Qual Saf Health Care* 16: 291–296.

Kopp, B.J., Erstad, B.L., Allen, M.E., Theodorou, A.A., and Priestley, G. 2006. Medication errors and adverse drug events in an intensive care unit: Direct observation approach for detection. *Crit Care Med* 34: 415–425.

Koppel, R., Leonard, C.E., Localio, A.R. et al. 2008a. Identifying and quantifying medication errors: Evaluation of rapidly discontinued medication orders submitted to a computerized physician order entry system. *J Am Med Inform Assoc* 15: 461–465.

Koppel, R., Metlay, J.P., Cohen, A. et al. 2005. Role of computerized physician order entry systems in facilitating medication errors. *JAMA* 293: 1197–1203.

Koppel, R., Wetterneck, T., Telles, J.L., and Karsh, B.T. 2008b. Workarounds to barcode medication administration systems: Their occurrences, causes, and threats to patient safety. *J Am Med Inform Assoc* 15: 408–423.

Kozer, E., Scolnik, D., Macpherson, A. et al. 2002. Variables associated with medication errors in pediatric emergency medicine. *Pediatrics* 110: 737–742.

Kratz, K., and Thygesen, C. 1992. A comparison of the accuracy of unit dose cart fill with the Baxter ATC-212 computerized system and manual filling. *Hosp Pharm* 27: 19–20, 22.

Kripalani, S., LeFevre, F., Phillips, C.O. et al. 2007. Deficits in communication and information transfer between hospital-based and primary care physicians: Implications for patient safety and continuity of care. *JAMA* 297: 831–841.

Kripalani, S., Roumie, C.L., Dalal, A.K. et al. 2012. Effect of a pharmacist intervention on clinically important medication errors after hospital discharge: A randomized trial. *Ann Intern Med* 157: 1–10.

Kruse, H., Johnson, A., O'Connell, D., and Clarke, T. 1992. Administering non-restricted medications in hospital: The implications and cost of using two nurses. *Austr Clin Rev* 12: 77–83.

Kugler, K.D., Beekman III, R.H., Rosenthal, G.L. et al. 2009. Development of a pediatric cardiology quality improvement collaborative: From inception to implementation. From the Joint Council on Congenital Heart Disease Quality Improvement Task Force. *Congenit Heart Dis* 4: 318–328.

Kwan, J.L., Lo, L., Sampson, M., and Shojania, K.G. 2013. Medication reconciliation during transitions of care as a patient safety strategy: A systematic review. *Ann Intern Med* 158: 397–403.

Lalonde, L., Lampron, A.M., Vanier, M.C. et al. 2008. Effectiveness of a medication discharge plan for transitions of care from hospital to outpatient settings. *Am J Health Syst Pharm* 65: 1451–1457.

Lalor, D.J., Chen, T.F., Walpola, R. et al. 2015. An exploration of Australian hospital pharmacists' attitudes to patient safety. *Int J Pharm Pract* 23: 67–76.

Lamont, T., and Scarpello, J. 2009. National Patient Safety Agency: Combining stories with statistics to minimise harm. *BMJ* 339: b4489.

Langley, G.J., Nolan, K.M., Nolan, T.W., Norman, C.L., and Provost, L.P. 1996. *The Improvement Guide. A Practical Approach to Enhancing Organizational Performance*. San Francisco, CA: Jossey-Bass.

Lave, J., and Wenger, E. 1991. *Situated Learning: Legitimate Peripheral Participation*. Cambridge: Cambridge University Press.

Lavon, O., Ben-Zeev, A., and Bentur, Y. 2014. Medication errors outside healthcare facilities: A national poison centre perspective. *Basic Clin Pharmacol Toxicol* 114: 288–292.

Lawton, G., and Shields, A. 2005. Bar-code verification of medication administration in a small hospital. *Am J Health Syst Pharm* 62: 2413–2415.

Lawton, R., and Armitage, G. 2012. *The Role of the Patient in Clinical Safety*. London: The Health Foundation.

Lawton, R., and Parker, D. 1998. Individual differences in accident liability: A review and integrative approach. *Hum Factors* 40: 655–671.

Lawton, R., McEachan, R.R., Giles, S.J. et al. 2012. Development of an evidence-based framework of factors contributing to patient safety incidents in hospital settings: A systematic review. *BMJ Qual Saf* 21: 369–380.

Leape, L.L. 2000. Institute of medicine medical error figures are not exaggerated. *JAMA* 284: 95–97.

Leape, L.L. 2002. Reporting of adverse events. *N Engl J Med* 347: 1633–1638.

Leape, L.L., Bates, D.W., Cullen, D.J. et al. 1995. Systems analysis of adverse drug events. ADE Prevention Study Group. *JAMA* 274: 35–43.

Lederman, R.M., and Parkes, C. 2005. Systems failure in hospitals—Using Reason's model to predict problems in a prescribing information system. *J Med Syst* 29: 33–43.

Lee, J.Y., Leblanc, K., Fernandes, O.A. et al. 2010. Medication reconciliation during internal hospital transfer and impact of computerized prescriber order entry. *Ann Pharmacother* 44: 1887–1895.

Leendertse, A.J., Egberts, A.C., Stoker, L.J., and van den Bemt, P.M. 2008. Frequency of and risk factors for preventable medication-related hospital admissions in The Netherlands. *Arch Intern Med* 168: 1890–1896.

Lefeber, G.J., Drenth-van Maanen, A.C., Wilting, I. et al. 2014. Effect of a transitional pharmaceutical care intervention at hospital discharge on registration of changes in medication regimens in primary care. *J Am Geriatr Soc* 62: 565–567.

Legault, K., Ostro, J., Khalid, Z., Wasi, P., and You, J.J. 2012. Quality of discharge summaries prepared by first year internal medicine residents. *BMC Med Educ* 12: 77.

Lent, V., Eckstein, E.C., Cameron, A.S. et al. 2009. Evaluation of patient participation in a patient empowerment initiative to improve hand hygiene practices in a Veterans Affairs medical center. *Am J Infect Control* 37: 117–120.

Leontiev, A.N. 1981. *Problems of the Development of Mind*. Moscow, Russia: Progress.

Leotsakos, A., Zheng, H., Croteau, R. et al. 2014. Standardization in patient safety: The WHO High 5s project. *Int J Qual Health Care* 26: 109–116.

Lépée, C., Klaber, R.E., Benn, J. et al. 2012. The use of a consultant-led ward round checklist to improve paediatric prescribing: An interrupted time series study. *Eur J Pediatr* 171: 1239–1245.

Lesar, T.S., Briceland, L., Delcoure, K. et al. 1990. Medication-prescribing errors in a teaching hospital. *JAMA* 263: 2329–2334.

Lesar, T.S., Briceland, L., and Stein, D.S. 1997a. Factors related to errors in medication prescribing. *JAMA* 277: 312–317.

Lesar, T.S., Lomaestro, B.M., and Pohl, H. 1997b. Medication-prescribing errors in a teaching hospital. A 9-year experience. *Arch Intern Med* 157: 1569–1576.

Lesselroth, B., Adams, S., Felder, R. et al. 2009. Using consumer-based kiosk technology to improve and standardize medication reconciliation in a specialty care setting. *Jt Comm J Qual Patient Saf* 35: 264–270.

Lesselroth, B., Eisenhauer, W., Adams, S. et al. 2011. Simulation modeling of a check-in and medication reconciliation ambulatory clinic kiosk. *J Healthcare Eng* 2011: 197–222.

Leung, A.A., Denham, C.R., Gandhi, T.K. et al. 2015. A safe practice standard for barcode technology. *J Patient Saf* (in press).

Lewis, P.J., Ashcroft, D.M., Dornan, T. et al. 2014. Exploring the causes of junior doctors' prescribing mistakes: A qualitative study. *Br J Clin Pharmacol* 78: 310–319.

Lewis, P.J., Dornan, T., Taylor, D. et al. 2009. Prevalence, incidence and nature of prescribing errors in hospital inpatients. A systematic review. *Drug Saf* 32: 379–389.

Lewis, R., Stachan, A., and Smith, M.M. 2012. High fidelity simulation the most effective method for the development of non-technical skills in nursing. *Open Nurs J* 6: 82–89.

Limat, S., Drouhin, J.P., Demesmay, K. et al. 2001. Incidence and risk factors of preparation errors in a centralized cytotoxic preparation unit. *Pharm World Sci* 23: 102–106.

Lin, L., Vicente, K.J., and Doyle, D.J. 2001. Patient safety, potential adverse drug events, and medical device design: A human factors engineering approach. *J Biomed Inform* 34: 274–284.

Linden, J.V., and Kaplan, H.S. 1994. Transfusion errors: Causes and effects. *Transfus Med Rev* VIII: 169–183.

Linsky, A., and Simon, S.R. 2013. Medication discrepancies in integrated electronic health records. *BMJ Qual Saf* 22: 103–109.

Lo, C., Burke, R., and Westbrook, J.I. 2010. Comparison of pharmacists' work patterns on hospital wards with and without an electronic medication management system (eMMS). *J Pharm Pract Res* 40: 108–112.

Low, D.K., and Belcher, J.V. 2002. Reporting medication errors through computerized medication administration. *Comput Inform Nurs* 20: 178–183.

Luria, J.W., Muething, S.E., Schoettker, P.J., and Kotagal, U.R. 2006. Reliability science and patient safety. *Pediatr Clin North Am* 53: 1121–1133.

Lyons, M. 2009. Towards a framework to select techniques for error prediction: Supporting novice users in the healthcare sector. *Appl Ergon* 40: 379–395.

Macaire, P., Nadhari, M., Greiss, H. et al. 2014. Internet remote control of pump settings for postoperative continuous peripheral nerve blocks: A feasibility study in 59 patients. *Ann Fr Anesth Reanim* 33: e1–e7.

Mager, D.R., and Campbell, S.H. 2013. Home care simulation for student nurses: Medication management in the home. *Nurse Educ Today* 33: 1416–1421.

Mahaguna, V., McDermott, J.M., Zhang, F., and Ochoa, F. 2004. Investigation of product quality between extemporaneously compounded progesterone vaginal suppositories and an approved progesterone vaginal gel. *Drug Dev Ind Pharm* 30: 1069–1078.

Mangoni, A.A., and Jackson, S.H. 2004. Age-related changes in pharmacokinetics and pharmacodynamics: Basic principles and practical applications. *Br J Clin Pharmacol* 57: 6–14.

Manias, E., Aitken, R., and Dunning, T. 2005. How graduate nurses use protocols to manage patients' medications. *J Clin Nurs* 14: 935–944.

Mannion, R., Konteh, F.H., and Davies, H.T. 2009. Assessing organisational culture for quality and safety improvement: A national survey of tools and tool use. *Qual Saf Health Care* 18: 153–156.

Manrique-Rodriguez, S., Sanchez-Galindo, A.C., de Lorenzo-Pinto, A. et al. 2015. Implementation of smart pump technology in a paediatric intensive care unit. *Health Informatics J* 10.1177/1460458213518058.

Mansfield, J., and Jarrett, S. 2013. Using smart pumps to understand and evaluate clinician practice patterns to ensure patient safety. *Hosp Pharm* 48: 942–950.

Maran, N.J., and Glavin, R.J. 2003. Low- to high-fidelity simulation—A continuum of medical education? *Med Educ* 37(Suppl 1): 22–28.

Marciano, N.J., Merlin, T.L., Bessen, T., and Street, J.M. 2014. To what extent are current guidelines for cutaneous melanoma follow up based on scientific evidence? *Int J Clin Pract* 68: 761–770.

Marella, W.M., Edward, F., Thomas, A., and Clarke, J.R. 2007. Healthcare consumers' inclination to engage in selected patient safety practices. *J Patient Saf* 3: 184–189.

Margolis, P., Provost, L.P., Schoettker, P.J., and Britto, M.T. 2009. Quality improvement, clinical research, and quality improvement research—Opportunities for integration. *Pediatr Clin North Am* 56: 831–841.

Martin, H.M., Navne, L.E., and Lipczak, H. 2013. Involvement of patients with cancer in patient safety: A qualitative study of current practices, potentials and barriers. *BMJ Qual Saf* 22: 836–842.

Mason, J.J., Roberts-Turner, R., Amendola, V., Sill, A.M., and Hinds, P.S. 2014. Patient safety, error reduction, and pediatric nurses' perceptions of smart pump technology. *J Pediatr Nurs* 29: 143–151.

Mason, M. 2008. *Complex Theory and the Philosophy of Education (Educational Philosophy and Theory Special Issues).* Chichester: Wiley-Blackwell.

Masotti, P., McColl, M.A., and Green, M. 2010. Adverse events experienced by homecare patients: A scoping review of the literature. *Int J Qual Health Care* 22: 115–125.

Massaro, T.A. 1993. Introducing physician order entry at a major academic medical center: I. Impact on organizational culture and behavior. *Acad Med* 68: 20–25.

Mattheos, N., Nattestad, A., Falk-Nilsson, E., and Attstrom, R. 2004. The interactive examination: Assessing students' self-assessment ability. *Med Educ* 38: 378–389.

Mayo, A.M., and Duncan, D. 2004. Nurse perceptions of medication errors: What we need to know for patient safety. *J Nurs Care Qual* 19: 209–217.

McAlearney, A.S., Vrontos, J., Schneider, P.J., Curran, C., and Czerwinski, B.S. 2007. Strategic work-arounds to accommodate new technology: The case of smart pumps in hosptial care. *J Patient Saf* 3: 75–81.

McDowell, S.E., Mt-Isa, S., Ashby, D., and Ferner, R.E. 2010. Where errors occur in the preparation and administration of intravenous medicines: A systematic review and Bayesian analysis. *Qual Saf Health Care* 19: 341–345.

McKee, L., West, M., Flin, R. et al. 2010. Understanding the dynamics of organizational culture change: Creating safe places for patients and staff. Report to the National Institute for Health Research Service Delivery and Organisation Programme, 08/1501/92. http://www.nets.nihr.ac.uk/__data/assets/pdf_file/0005/64499/FR-08-1501-092.pdf (Accessed August 28, 2014).

McKeon, L.M., Oswaks, J.D., and Cunningham, P.D. 2006. Safeguarding patients: Complexity science, high reliability organizations, and implications for team training in healthcare. *Clin Nurse Spec* 20: 298–304.

McKibbon, K.A., Lokker, C., Handler, S.M. et al. 2011. *Enabling Medication Management through Health Information Technology*. Rockville, MD: Agency for Healthcare Research and Quality.

McKibbon, K.A., Lokker, C., Handler, S.M. et al. 2012. The effectiveness of integrated health information technologies across the phases of medication management: A systematic review of randomized controlled trials. *J Am Med Inform Assoc* 19: 22–30.

McLellan, L., Tully, M.P., and Dornan, T. 2012. How could undergraduate education prepare new graduates to be safer prescribers? *Br J Clin Pharmacol* 74: 605–613.

McLeod, M., Ahmed, Z., Barber, N., and Franklin, B.D. 2014. A national survey of inpatient medication systems in English NHS hospitals. *BMC Health Serv Res* 14: 93.

McLeod, M.C., Barber, N., and Franklin, B.D. 2013. Methodological variations and their effects on reported medication administration error rates. *BMJ Qual Saf* 22: 278–289.

McNally, K.M., and Sunderland, V.B. 1998. No-blame medication administration error reporting by nursing staff at a teaching hospital in Australia. *Int J Pharm Pract* 6: 67–71.

Meadows, S., Baker, K., and Butler, J. 2005. The incident decision tree: Guidelines for action following patient safety incidents. In *Advances in Patient Safety: From Research to Implementation*, eds K. Henriksen, J.B. Battles, E.S. Marks, and D.I. Lewin, 387–399. Rockville, MD: Agency for Healthcare Research and Quality.

Medical Schools Council. 2008. Outcomes of the Medical Schools Council Safe Prescribing Working Group. http://www.medschools.ac.uk/Publications/Pages/Safe-Prescribing-Working-Group-Outcomes.aspx (Accessed October 8, 2014).

Medicines and Healthcare Products Regulatory Agency. 2010. MHRA Drug Safety Advice. Intravenous paracetamol (Perfalgan): Risk of accidental overdose, especially in infants and neonates. *Drug Safety Update* 3: 2–3.

Medicines and Healthcare Products Regulatory Agency. 2014. *Rules and Guidance for Pharmaceutical Manufacturers and Distributors*. London: Pharmaceutical Press.

MedMARx. 2002. MedMARx report may aid in error prevention. *Healthcare Benchmarks Qual Improv* 1: 8–9.

Mennin, S. 2010. Self-organisation, integration and curriculum in the complex world of medical education. *Med Educ* 44: 20–30.

Menon, L., Taylor, Z., and Tuthill, D.P. 2006. Can paediatric junior hospital doctors prescribe competently? *Paediatr Perinat Drug Ther* 7: 118–120.

Meredith, S., Feldman, P.H., Frey, D. et al. 2001. Possible medication errors in home healthcare patients. *J Am Geriatr Soc* 49: 719–724.

Meyer, C., Stern, M., Woolley, W., Jeanmonod, R., and Jeanmonod, D. 2012. How reliable are patient-completed medication reconciliation forms compared with pharmacy lists? *Am J Emerg Med* 30: 1048–1054.

Meyer-Massetti, C., Kaiser, E., Hedinger-Grogg, B., Luterbacher, S., and Hersberger, K. 2012. Medication safety in the home care setting: Error-prone process steps. *Pflege* 25: 261–269.

Michie, S., Johnson, M., Abraham, C. et al. 2005. Making psychological theory useful for implementing evidence based practice: A consensus approach. *Qual Saf Health Care* 14: 26–33.

Milligan, F.J. 2007. Establishing a culture for patient safety—The role of education. *Nurse Educ Today* 27: 95–102.

Mills, N., Bachmann, M.O., Harvey, I., Hine, I., and McGowan, M. 1999. Effect of a primary-care-based epilepsy specialist nurse service on quality of care from the patients' perspective: Quasi-experimental evaluation. *Seizure* 8: 1–7.

Mitchell, D., Usher, J., Gray, S. et al. 2004a. Evaluation and audit of a pilot of electronic prescribing and drug administration. *J Inform Technol Healthcare* 2: 19–29.

Mitchell, J., Hayhurst, C., and Robinson, S.M. 2004b. Can a senior house officer's time be used more effectively? *Emerg Med J* 21: 545–547.

Mohsin-Shaikh, S., Garfield, S., and Franklin, B.D. 2014. Patient involvement in medication safety in hospital: An exploratory study. *Int J Clin Pharm* 36: 657–666.

Moll, L.C. 1992. *Vygotsky and Education: Instructional Implications and Applications of Sociohistorical Psychology*. Cambridge: Cambridge University Press.

Moore, P., Armitage, G., Wright, J. et al. 2011. Medicines reconciliation using a shared electronic health care record. *J Patient Saf* 7: 148–154.

Moreland, P.J., Gallagher, S., Bena, J.F., Morrison, S., and Albert, N.M. 2012. Nursing satisfaction with implementation of electronic medication administration record. *Comput Inform Nurs* 30: 97–103.

Morimoto, T., Gandhi, T.K., Seger, A.C., Hseih, T.C., and Bates, D.W. 2004. Adverse drug events and medication errors: Detection and classification methods. *Qual Saf Health Care* 13: 306–314.

Morriss, F.H., Abramowitz, P.W., Nelson, S.P. et al. 2009. Effectiveness of a barcode medication administration system in reducing preventable adverse drug events in a neonatal intensive care unit: A prospective cohort study. *J Pediatr* 154: 363–8, 368.

Mueller, S.K., Sponsler, K.C., Kripalani, S., and Schnipper, J.L. 2012. Hospital-based medication reconciliation practices: A systematic review. *Arch Intern Med* 172: 1057–1069.

Muething, S.E., Conway, P.H., Kloppenborg, E. et al. 2010. Identifying causes of adverse events detected by an automated trigger tool through in-depth analysis. *Qual Saf Health Care* 19: 435–439.

Muething, S.E., Goudie, A., Schoettker, P.J. et al. 2012. Quality improvement initiative to reduce serious safety events and improve patient safety culture. *Pediatrics* 130: e423–e431.

National Coordinating Council for Medication Error Reporting and Prevention. 2011. NCC MERP taxonomy of medication errors. http://www.nccmerp.org/sites/default/files/taxonomy2001-07-31.pdf (Accessed May 26, 2015).

National Patient Safety Agency. 2004. Seven steps to patient safety. http://www.nrls.npsa.nhs.uk/resources/collections/seven-steps-to-patient-safety/?entryid45=59787 (Accessed August 24, 2014).

National Patient Safety Agency. 2007. Patient safety alert 20—Promoting safer use of injectable medicines. http://www.nrls.npsa.nhs.uk/resources/?entryid45=59812 (Accessed November 12, 2014).

National Patient Safety Agency. 2008. Reporting anaesthetic patient safety incidents. http://www.npsa.nhs.uk/nrls/improvingpatientsafety/anaesthesia-and-surgery/anaesthesia-current-projects/anaesthetic-incident-reporting/ (Accessed August 24, 2014).

National Patient Safety Agency. 2010. NHS places reporting and learning from patient safety incidents at top of its agenda—England. http://www.npsa.nhs.uk/corporate/news/nhs-places-reporting-and-learning-from-patient-safety-incidents-at-top-of-its-agenda-england/ (Accessed August 24, 2014).

National Patient Safety Agency. 2014. The incident decision tree. Information and advice on use. http://www.suspension-nhs.org/Resources/Safety%20-%20IDT%20(info%20and%20advice%20on%20use).pdf (Accessed September 9, 2014).

National Reporting and Learning System. 2014. Reducing treatment dose errors with low molecular weight heparins. http://www.nrls.npsa.nhs.uk/alerts/?entryid45=75208 (Accessed August 24, 2014).

Nau, D.P., and Erickson, S.R. 2005. Medication safety: Patients' experiences, beliefs, and behaviors. *J Am Pharm Assoc* 45: 452–457.

Neafsey, P.J., Strickler, Z., Shellman, J., and Chartier, V. 2002. An interactive technology approach to educate older adults about drug interactions arising from over-the-counter self-medication practices. *Public Health Nurs* 19: 255–262.

Ness, J.E., Sullivan, S.D., and Stergachis, A. 1994. Accuracy of technicians and pharmacists in identifying dispensing errors. *Am J Hosp Pharm* 51: 354–357.

Neuenschwander, M. 1996. Limiting or increasing opportunities for errors with dispensing automation. *Hosp Pharm* 31: 1102–1106.

New York State Education Department. 2004. Education Law, Article 137, Pharmacy. Office for the Professions. http://www.op.nysed.gov/prof/pharm/article137.htm#sect6829 (Accessed October 10, 2014).

Newgreen, D.B., Presley, J.A., and Marty, S.H. 2005. A survey of dispensing errors reported to the Pharmacy Board of Victoria July 1998 to December 2004. *Aust Pharmacist* 28: 644–649.

Nguyen, H.T., Pham, H.T., Vo, D.K. et al. 2014. The effect of a clinical pharmacist-led training programme on intravenous medication errors: A controlled before and after study. *BMJ Qual Saf* 23: 319–324.

NHS Connecting for Health. 2009. *Electronic Prescribing in Hospitals: Challenges and Lessons Learned*. London: National Health Service.

Nichols, P., Copeland, T.S., Craib, I.A., Hopkins, P., and Bruce, D.G. 2008. Learning from error: Identifying contributory causes of medication errors in an Australian hospital. *Med J Aust* 188: 276–279.

Nieva, V.F., and Sorra, J. 2003. Safety culture assessment: A tool for improving patient safety in healthcare organizations. *Qual Saf Health Care* 12: ii17–ii23.

Nirodi, P., and Mitchell, A. 2002. The quality of psychotropic drug prescribing in patients in psychiatric units for the elderly. *Aging Ment Health* 6: 191–196.

Nisbet, G., Hendry, G.D., Rolls, G., and Field, M.J. 2008. Interprofessional learning for prequalification health care students: An outcomes-based evaluation. *J Interprof Care* 22: 57–68.

Noble, D.J., and Donaldson, L.J. 2010. The quest to eliminate intrathecal vincristine errors: A 40-year journey. *Qual Saf Health Care* 19: 323–326.

Noble, D.J., Panesar, S.S., and Pronovost, P.J. 2011. A public health approach to patient safety reporting systems is urgently needed. *J Patient Saf* 7: 109–112.

Nolan, T., Resar, R., Haraden, C., and Griffin, F.A. 2004. *Improving the Reliability of Healthcare*. Cambridge, MA: Institute for Healthcare Improvement.

Noott, A., and Phipps, G.C. 2003. Monitoring and preliminary analysis of internal dispensing errors within a hospital trust. *Pharm World Sci* 25: A42–A43.

Norden-Hagg, A., Sexton, J.B., Kalvemark-Sporrong, S., Ring, L., and Kettis-Lindblad, A. 2010. Assessing safety culture in pharmacies: The psychometric validation of the Safety Attitudes Questionnaire (SAQ) in a national sample of community pharmacies in Sweden. *BMC Clin Pharmacol* 10: 8.

Nordic Pharmacy Association. 2008. The added value of Nordic pharmacies documentation examples. http://www.pharmakon.dk/data/files/Aps/Afdaekning/PAPA%20-%20Added%20value%20doc%202008%20(final)%20(2)%20(4).pdf (Accessed June 20, 2014).

Novak, L.L., Holden, R.J., Anders, S.H., Hong, J.Y., and Karsh, B.T. 2013. Using a sociotechnical framework to understand adaptations in health IT implementation. *Int J Med Inform* 82: e331–e344.

Novak, L.L., and Lorenzi, N.M. 2008. Barcode medication administration: Supporting transitions in articulation work. *AMIA Annu Symp Proc* 515–519, Washington, DC, USA.

Nuckols, T., Smith-Spangler, C., Morton, S. et al. 2014. The effectiveness of computerized order entry at reducing preventable adverse drug events and medication errors in hospital settings: A systematic review and meta-analysis. *Syst Rev* 3: 56.

Nursing and Midwifery Council. 2007. *Standards for Medicines Management*. London: Nursing and Midwifery Council.

Odukoya, O.K., Stone, J.A., and Chui, M.A. 2014. How do community pharmacies recover from e-prescription errors? *Res Social Adm Pharm* 10: 837–852.

Ohashi, K., Dykes, P., McIntosh, K. et al. 2013. Evaluation of intravenous medication errors with smart infusion pumps in an academic medical center. *AMIA Annu Symp Proc* 1089–1098, Washington, DC, USA.

Ojeleye, O., Avery, A., Gupta, V., and Boyd, M. 2013. The evidence for the effectiveness of safety alerts in electronic patient medication record systems at the point of pharmacy order entry: A systematic review. *BMC Med Inform Decis Mak* 13: 69.

O'Leary, K., Burke, R., and Kirsa, S. 2006. SHPA standards of practice for the distribution of medicines in Australian Hospital. *J Pharm Pract Res* 36: 143–149.

Olvey, E.L., Clauschee, S., and Malone, D.C. 2010. Comparison of critical drug–drug interaction listings: The Department of Veterans Affairs medical system and standard reference compendia. *Clin Pharmacol Ther* 87: 48–51.

Or, C.K.L., and Chan, A.H.S. 2014. Effects of text enhancements on the differentiation performance of orthographically similar drug names. *Work* 48: 521–528.

Orser, B.A., Chen, R.J., and Yee, D.A. 2001. Medication errors in anesthetic practice: A survey of 687 practitioners. *Can J Anaesth* 48: 139–146.

Osheroff, J.A., Pifer, E.A., Teich, J.M., Sittig, D.F., and Jenders, R.A. 2005. *Improving Outcomes with Clinical Decision Support: An Implementer's Guide*. Chicago, IL: Healthcare Information and Management Systems Society.

Oxford University Press. 2014. Oxford English Dictionary. http://www.oed.com/ (Accessed October 6, 2014).

Panesar, S.S., Cleary, K., Bhandari, M., and Sheikh, A. 2009a. To cement or not in hip fracture surgery? *Lancet* 374: 1047–1049.

Panesar, S.S., Cleary, K., and Sheikh, A. 2009b. Reflections on the National Patient Safety Agency's database of medical errors. *J R Soc Med* 102: 256–258.

Pang, R.K.Y., Kong, D.C.M., deClifford, J., Lam, S.S., and Leung, B.K. 2011. Smart infusion pumps reduce intravenous medication administration errors at an Australian teaching Hospital. *J Pharm Pract Res* 41: 192–195.

Paoletti, R.D., Suess, T.M., Lesko, M.G. et al. 2007. Using bar-code technology and medication observation methodology for safer medication administration. *Am J Health Syst Pharm* 64: 536–543.

Parke, J. 2006. Risk analysis of errors in prescribing, dispensing and administering medications within a district hospital. *J Pharm Pract Res* 36: 21–24.

Parker, D. 2009. Managing risk in healthcare: Understanding your safety culture using the Manchester Patient Safety Framework (MaPSaF). *J Nurs Manag* 17: 218–222.

Parker, D., Lawrie, M., and Hudson, P. 2006. A framework for understanding the development of organisational safety culture. *Saf Sci* 44: 551–562.

Parker, D., Manstead, A.S.R., Stradling, S.G., Reason, J.T., and Baxter, J.S. 1992. Intention to commit driving violations—An application of the theory of planned behavior. *J Appl Psychol*: 94–101.

Parker, D., Reason, J.T., Manstead, A.S.R., and Stradling, S.G. 1995a. Driving errors, driving violations and accident involvement. *Ergonomics*: 1036–1048.

Parker, D., West, R., Stradling, S., and Manstead, A.S. 1995b. Behavioural characteristics and involvement in different types of traffic accident. *Accid Anal Prev* 27: 571–581.

Paterno, M.D., Maviglia, S.M., Gorman, P.N. et al. 2009. Tiering drug–drug interaction alerts by severity increases compliance rates. *J Am Med Inform Assoc* 16: 40–46.

Patey, R., Flin, R., Fletcher, G., Maran, N.J., and Glavin, R.J. 2005. Developing a taxonomy of anesthetists' nontechnical skills (ANTS). In *Advances in Patient Safety: From Research to Implementation (Volume 4: Programs, Tools, and Products)*, eds K. Henriksen, J.B.

Battles, E.S. Marks and D.I. Lewin, 325–336. Rockville, MD: Agency for Healthcare Research and Quality (US).

Paton, C., and Gill-Banham, S. 2003. Prescribing errors in psychiatry. *Psychiatr Bull* 27: 208–2010.

Patterson, E.S., Cook, R.I., Woods, D.D., and Render, M.L. 2004. Examining the complexity behind a medication error: Generic patterns in communication. *IEEE Trans Syst Man Cybern A Syst Humans* 34: 749–756.

Patterson, E.S., Rogers, M.L., Chapman, R.J., and Render, M.L. 2006. Compliance with intended use of bar code medication administration in acute and long-term care: An observational study. *Hum Factors* 48: 15–22.

Pearson, P., Steven, A., and on behalf of the Patient Safety Education Study Group. 2009. Patient safety in health care professional educational curricula: Examining the learning experience. http://nrl.northumbria.ac.uk/id/eprint/594 (Accessed October 8, 2014).

Pedersen, C.A., Schneider, P.J., and Scheckelhoff, D.J. 2012. ASHP national survey of pharmacy practice in hospital settings: Dispensing and administration—2011. *Am J Health Syst Pharm* 69: 768–785.

Pellegrino, E.D. 1976. Prescribing and drug ingestion symbols and substances. *Drug Intell Clin Pharm* 10: 624–630.

Pennsylvania Patient Safety Authority. 2010. Annual reports. http://patientsafetyauthority.org/PatientSafetyAuthority/Pages/AnnualReports.aspx (Accessed August 24, 2014).

Peterson, G.M., Wu, M.S., and Bergin, J.K. 1999. Pharmacist's attitudes towards dispensing errors: Their causes and prevention. *J Clin Pharm Therap* 24: 57–71.

Peterson, J.F., and Bates, D.W. 2001. Preventable medication errors: Identifying and eliminating serious drug interactions. *J Am Pharm Assoc* 41: 159–160.

Phansalkar, S., Desai, A., Choksi, A. et al. 2013. Criteria for assessing high-priority drug–drug interactions for clinical decision support in electronic health records. *BMC Med Inform Decis Mak* 13: 65.

Phansalkar, S., Desai, A.A., Bell, D.S. et al. 2012. High-priority drug–drug interactions for use in electronic health records. *J Am Med Inform Assoc* 19: 735–743.

Phansalkar, S., Edworthy, J., Hellier, E. et al. 2010. A review of human factors principles for the design and implementation of medication safety alerts in clinical information systems. *J Am Med Inform Assoc* 17: 493–501.

Phansalkar, S., Zachariah, M., Seidling, H.M. et al. 2014. Evaluation of medication alerts in electronic health records for compliance with human factors principles. *J Am Med Inform Assoc* 21(e2): 332–340.

Pharmaceutical Society of Australia. 2010. *Professional Practice Standards.* Deakin West: Pharmaceutical Society of Australia.

Phelps, P.K. 2011. *Smart Infusion Pumps: Implementation, Management, and Drug Libraries.* Bethesda, MD: American Society of Health-System Pharmacists.

Phipps, D., Noyce, P., Walshe, K., Parker, D., and Ashcroft, D. 2011. Assessing risk associated with contemporary pharmacy practice in Northern Ireland. http://www.psni.org.uk/documents/803/UOMrevalreport.pdf (Accessed August 28, 2014).

Phipps, D.L., and Ashcroft, D.M. 2011. Psychosocial influences on safety climate: Evidence from community pharmacies. *BMJ Qual Saf* 20: 1062–1068.

Phipps, D.L., and Ashcroft, D.M. 2012. An investigation of occupational subgroups with respect to patient safety culture. *Saf Sci* 50: 1290–1298.

Phipps, D.L., and Ashcroft, D.M. 2014. Looking behind patient safety culture: Organisational dynamics, job characteristics and the work domain. In *Patient Safety Culture: Theory, Methods and Application*, ed P. Waterson, 99–118. Farnham: Ashgate.

Phipps, D.L., de Bie, J., Herborg, H. et al. 2012. Evaluation of the pharmacy safety climate questionnaire in European community pharmacies. *Int J Qual Health Care* 24: 16–22.

Phipps, D.L., Noyce, P.R., Parker, D., and Ashcroft, D.M. 2009. Medication safety in community pharmacy: A qualitative study of the sociotechnical context. *BMC Health Serv Res* 9: 158.

Pippins, J.R., Gandhi, T.K., Hamann, C. et al. 2008. Classifying and predicting errors of inpatient medication reconciliation. *J Gen Intern Med* 23: 1414–1422.

Pirmohamed, M., James, S., Meakin, S. et al. 2004. Adverse drug reactions as cause of admission to hospital: Prospective analysis of 18,820 patients. *BMJ* 329: 15–19.

Pittet, D., Hugonnet, S.P., Harbarth, S. et al. 2000. Effectiveness of a hospital-wide programme to improve compliance with hand hygiene. *Lancet* 356: 1307–1312.

Poissant, L., Pereira, J., Tamblyn, R., and Kawasumi, Y. 2005. The impact of electronic health records on time efficiency of physicians and nurses: A systematic review. *J Am Med Inform Assoc* 12: 505–516.

Poon, E.G., Cina, J.L., Churchill, W. et al. 2006. Medication dispensing errors and potential adverse drug events before and after implementing bar code technology in the pharmacy. *Ann Intern Med* 145: 426–434.

Poon, E.G., Keohane, C.A., Bane, A. et al. 2008. Impact of barcode medication administration technology on how nurses spend their time providing patient care. *J Nur Adm* 38: 541–549.

Poon, E.G., Keohane, C.A., Yoon, C.S. et al. 2010. Effect of bar-code technology on the safety of medication administration. *N Engl J Med* 362: 1698–1707.

Power, M., Fogarty, M., Madsen, J. et al. 2014. Learning from the design and development of the NHS Safety Thermometer. *Int J Qual Health Care* 26: 287–297.

Prakash, V., Koczmara, C., Savage, P. et al. 2014. Mitigating errors caused by interruptions during medication verification and administration: Interventions in a simulated ambulatory chemotherapy setting. *BMJ Qual Saf* 23: 884–892.

Presseau, J., Johnston, M., Heponiemi, T. et al. 2014. Reflective and automatic processes in health care professional behaviour: A dual process model tested across multiple behaviours. *Ann Behav Med* 48: 347–358.

Pronovost, P., and Sexton, B. 2005. Assessing safety culture: Guidelines and recommendations. *Qual Saf Health Care* 14: 231–233.

Pronovost, P., Weast, B., Rosenstein, B. et al. 2005. Implementing and validating a comprehensive unit-based safety program. *J Patient Saf* 1: 33–40.

Pronovost, P.J., Goeschel, C.A., Olsen, K.L. et al. 2009. Reducing health care hazards: Lessons from the commercial aviation safety team. *Health Aff (Millwood)* 28: w479–w489.

Qulennec, B., Beretz, L., Paya, D. et al. 2013. Potential clinical impact of medication discrepancies at hospital admissions. *Eur J Intern Med* 24: 530–535.

Raban, M.Z., and Westbrook, J.I. 2014. Are interventions to reduce interruptions and errors during medication administration effective?: A systematic review. *BMJ Qual Saf* 23: 414–421.

Rack, L.L., Dudjak, L.A., and Wolf, G.A. 2012. Study of nurse workarounds in a hospital using bar code medication administration system. *J Nurs Care Qual* 27: 232–239.

Radecki, R.P., McCoy, A.B., Sirajuddin, A.M., Murphy, R.E., and Sittig, D.F. 2012. Effectiveness of bar coded medication alerts for elevated potassium. *AMIA Annu Symp Proc* 1360–1365, Chicago, Illinois, USA.

Ramasamy, S., Baysari, M.T., Lehnbom, E.C., and Westbrook, J.I. 2013. Evidence briefings on interventions to improve medication safety: Double checking medication administration. http://www.safetyandquality.gov.au/wp-content/uploads/2013/12/Evidence-briefings-on-interventions-to-improve-medication-safety-Double-checking-medication-administration-PDF-888KB.pdf (Accessed November 11, 2014).

Ramsay, M.A., Savege, T.M., Simpson, B.R., and Goodwin, R. 1974. Controlled sedation with alphaxalone–alphadolone. *Br Med J* 2: 656–659.

Ranchon, F., Salles, G., Spath, H.M. et al. 2011. Chemotherapeutic errors in hospitalised cancer patients: Attributable damage and extra costs. *BMC Cancer* 11: 478.

Rasmussen, J. 1982. Human errors: A taxonomy for describing human malfunction in industrial installations. *J Occup Acc* 4: 311–333.

Rawlins, M.D. 2013. Comment: Of snarks, boojums and national drug charts. *J R Soc Med* 106: 9.

Reason, J. 1990. *Human Error.* Cambridge: University of Cambridge.

Reason, J. 2000. Human error: Models and management. *BMJ* 320: 768–770.

Reason, J., Parker, D., and Lawton, R. 1998. Organisational controls and safety: The varieties of rule-related behavior. *J Occup Organ Psych* 71: 289–304.

Reason, J.T. 1995. Understanding adverse events: Human factors. In *Clinical Risk Management*, ed C.A. Vincent, 31–54. London: BMJ.

Reason, J.T. 1997. *Managing the Risks of Organizational Accidents.* Burlington, VT: Ashgate.

Reason, J.T. 2001. A systems approach to organizational error. *Ergonomics* 38: 1708–1721.

Reckmann, M., Westbrook, J.I., Koh, Y., Lo, C., and Day, R.O. 2009. Does computerized provider order entry reduce prescribing errors for hospital inpatients? A systematic review. *J Am Med Inform Assoc* 16: 613–623.

Redwood, S., Rajakumar, A., Hodson, J., and Coleman, J. 2011. Does the implementation of an electronic prescribing system create unintended medication errors? A study of the sociotechnical context through the analysis of reported medication incidents. *BMC Med Inform Decis Mak* 11: 29.

Reeve, J.F., and Allinson, M. 2005. High-risk medication alert: Intravenous potassium chloride. *Aust Prescr* 28: 14–16.

Relihan, E., O'Brien, V., O'Hara, S., and Silke, B. 2010. The impact of a set of interventions to reduce interruptions and distractions to nurses during medication administration. *Qual Saf Health Care* 19: e52.

Resar, R. 2005. Why we need to learn standardisation. *Aust Fam Physician* 34: 67–68.

Resar, R.K., Rozich, J.D., and Classen, D.C. 2003. Methodology and rationale for the measurement of harm with trigger tools. *Qual Saf Health Care* 12(Suppl 2): ii39–ii45.

Reynolds, M., Jheeta, S., and Franklin, B.D. 2014. A plan-do-study-act approach to increasing the prevalence of prescribers' names on individual inpatient medication orders. *Int J Pharm Pract* 22: 95.

Ridley, S.A., Booth, S.A., Thompson, C.M., and Intensive Care Society's Working Group on Adverse Incidents. 2004. Prescription errors in UK critical care units. *Anaesthesia* 59: 1193–1200.

Rijpma, J.A. 1997. Complexity, tight-coupling and reliability: Connecting normal accidents theory and high reliability theory. *J Contingencies Crisis Manage* 5: 15–23.

Ringold, D.J., Santell, J.P., and Schneider, P.J. 2000. ASHP national survey of pharmacy practice in acute care settings: Dispensing and administration—1999. *Am J Health Syst Pharm* 57: 1759–1775.

Roberts, K.H. 1990a. Managing high reliability organizations. *Calif Manage Rev* 32: 101–113.

Roberts, K.H. 1990b. Some characteristics of one type of high reliability organization. *Organ Sci* 1: 160–176.

Robertson, N., Baker, R., and Hearnshaw, H. 1996. Changing the clinical behavior of doctors: A psychological framework. *Qual Health Care* 5: 51–54.

Rogers, R.W. 1975. A protection motivation theory of fear appeals and attitude change. *J Psychol* 91: 93–114.

Romero, C.M., Salazar, N., Rojas, L. et al. 2013. Effects of the implementation of a preventive interventions program on the reduction of medication errors in critically ill adult patients. *J Crit Care* 28: 451–460.

Rosenthal, M.B., Fernandopulle, R., Song, H.R., and Landon, B. 2004. Paying for quality: Providers' incentives for quality improvement. *Health Aff (Millwood)* 23: 127–141.

Ross, S., Hamilton, L., Ryan, C., and Bond, C. 2012. Who makes prescribing decisions in hospital inpatients? An observational study. *Postgrad Med J* 88: 507–510.

Ross, S., and Loke, Y.K. 2009. Do educational interventions improve prescribing by medical students and junior doctors? A systematic review. *Brit J Clin Pharmacol* 67: 662–670.

Ross, S., Patey, R., and Flin, R. 2014. Is it time for a nontechnical skills approach to prescribing? *Br J Clin Pharmacol* 78: 681–683.

Rothschild, J.M., Keohane, C.A., Cook, E.F. et al. 2005. A controlled trial of smart infusion pumps to improve medication safety in critically ill patients. *Crit Care Med* 33: 533–540.

Rothschild, J.M., Mann, K., Keohane, C.A. et al. 2007. Medication safety in a psychiatric hospital. *Gen Hosp Psychiatry* 29: 156–162.

Rowe, C., Koren, T., and Koren, G. 1998. Errors by paediatric residents in calculating drug doses. *Arch Dis Child* 79: 56–58.

Royal Pharmaceutical Society. 2014. Foundation Pharmacy Framework: A framework for professional development in foundation practice across pharmacy. http://www.rpharms.com/development/foundation-practice.asp (Accessed October 8, 2014).

Royal Pharmaceutical Society of Great Britain. 1997. *From Compliance to Concordance: Achieving Shared Goals in Medicine Taking.* London: Royal Pharmaceutical Society of Great Britain.

Rozich, J.D., Haraden, C.R., and Resar, R.K. 2003. Adverse drug event trigger tool: A practical methodology for measuring medication related harm. *Qual Saf Health Care* 12: 194–200.

Rozich, J.D., Howard, R.J., Justeson, J.M. et al. 2004. Standardization as a mechanism to improve safety in health care. *Jt Comm J Qual Saf* 30: 5–14.

Runciman, W.B., Sellen, A., Webb, R.K. et al. 1993. The Australian Incident Monitoring Study. Errors, incidents and accidents in anaesthetic practice. *Anaesth Intensive Care* 21: 506–519.

Russ, A.L., Fairbanks, R.J., Karsh, B.T. et al. 2013. The science of human factors: Separating fact from fiction. *BMJ Qual Saf* 22: 802–808.

Russ, A.L., Weiner, M., Saleem, J.J., and Wears, R.L. 2012a. When technically preventable alerts occur, the design—not the prescriber—has failed. *J Am Med Inform Assoc* 19: 1119.

Russ, A.L., Zillich, A.J., McManus, M.S., Doebbeling, B.N., and Saleem, J.J. 2012b. Prescribers' interactions with medication alerts at the point of prescribing: A multi-method, *in situ* investigation of the human-computer interaction. *Int J Med Inf* 81: 232–243.

Russ, A.L., Zillich, A.J., Melton, B.L. et al. 2014. Applying human factors principles to alert design increases efficiency and reduces prescribing errors in a scenario-based simulation. *J Am Med Inform Assoc* 21: e287–e296.

Ryan, C., Ross, S., Davey, P. et al. 2014. Prevalence and causes of prescribing errors: The prescribing outcomes for trainee doctors engaged in clinical training (PROTECT) study. *PLoS One* 9: e79802.

Ryckman, F.C., Schoettker, P.J., Hays, K.R. et al. 2009a. Reducing surgical site infections at a pediatric academic medical center. *Jt Comm J Qual Patient Saf* 35: 192–198.

Ryckman, F.C., Yelton, P.A., Anneken, A.M. et al. 2009b. Redesigning intensive care unit flow using variability management to improve access and safety. *Jt Comm J Qual Patient Saf* 35: 535–543.

Sacks, G.S., Rough, S., and Kudsk, K.A. 2009. Frequency and severity of harm of medication errors related to the parenteral nutrition process in a large university teaching hospital. *Pharmacotherapy* 29: 966–974.

Sagan, S.D. 2004. The problem of redundancy problem: Why more nuclear forces may produce less nuclear security. *Risk Analysis* 24: 935–946.

Sakowski, J., Leonard, T., Colburn, S. et al. 2005. Using a bar-coded medication administration system to prevent medication errors in a community hospital network. *Am J Health Syst Pharm* 62: 2619–2625.

Salemi, C.S., and Singleton, N. 2007. Decreasing medication discrepancies between outpatient and inpatient care through the use of computerized pharmacy data. *Perm J* 11: 31–34.

Sanghera, I.S., Franklin, B.D., and Dhillon, S. 2007. The attitudes and beliefs of healthcare professionals on the causes and reporting of medication errors in a UK intensive care unit. *Anaesthesia* 62: 53–61.

Santell, J.P., Hicks, R.W., McMeekin, J., and Cousins, D.D. 2003. Medication errors: Experience of the United States Pharmacopeia (USP) MEDMARX reporting system. *J Clin Pharmacol* 43: 760–767.

Sari, A.B., Sheldon, T.A., Cracknell, A., and Turnbull, A. 2007. Sensitivity of routine system for reporting patient safety incidents in an NHS hospital: Retrospective patient case note review. *BMJ* 334: 79.

Sawyer, D. 1996. Do it by design. An introduction to human factors in medical devices. http://www.fda.gov/MedicalDevices/DeviceRegulationandGuidance/GuidanceDocuments/ucm094957.htm (Accessed April 25, 2014).

Schectman, G., Hiatt, J., and Hartz, A. 1994. Telephone contacts do not improve adherence to niacin or bile acid sequestrant therapy. *Ann Pharmacother* 28: 29–35.

Schedlbauer, A., Prasad, V., Mulvaney, C. et al. 2009. What evidence supports the use of computerized alerts and prompts to improve clinicians' prescribing behavior? *J Am Med Inform Assoc* 16: 531–538.

Schnipper, J.L., Gandhi, T.K., Wald, J.S. et al. 2012. Effects of an online personal health record on medication accuracy and safety: A cluster-randomized trial. *J Am Med Inform Assoc* 19: 728–734.

Schnipper, J.L., Hamann, C., Ndumele, C.D. et al. 2009. Effect of an electronic medication reconciliation application and process redesign on potential adverse drug events: A cluster-randomized trial. *Arch Intern Med* 169: 771–780.

Schnipper, J.L., Kirwin, J.L., Cotugno, M.C. et al. 2006. Role of pharmacist counseling in preventing adverse drug events after hospitalization. *Arch Intern Med* 166: 565–571.

Schroeder, M.E., Wolman, R.L., Wetterneck, T.B., and Carayon, P. 2006. Tubing misload allows free flow event with smart intravenous infusion pump. *Anesthesiology* 105: 434–435.

Schulman, P.R. 1993. The negotiated order of organizational reliability. *Adm Soc* 25: 353–372.

Schwappach, D.L., and Wernli, M. 2010a. Chemotherapy patients' perceptions of drug administration safety. *J Clin Oncol* 28: 2896–2901.

Schwappach, D.L., and Wernli, M. 2010b. Predictors of chemotherapy patients' intentions to engage in medical error prevention. *Oncologist* 15: 903–912.

Schwappach, D.L., and Wernli, M. 2010c. Am I (un)safe here? Chemotherapy patients' perspectives towards engaging in their safety. *Qual Saf Health Care* 19: e9.

Schwappach, D.L., and Wernli, M. 2011. Barriers and facilitators to chemotherapy patients' engagement in medical error prevention. *Ann Oncol* 22: 424–430.

Schwartz, B., Wasserman, E.A., and Robbins, S.J. 2001. *Psychology of Learning and Behavior*. New York, NY: W.W. Norton & Co.

Schwartz, R.K., Soumerai, S.B., and Avorn, J. 1989. Physician motivations for nonscientific drug prescribing. *Soc Sci Med* 28: 577–582.

Schwarz, H.O., and Brodowy, B.A. 1995. Implementation and evaluation of an automated dispensing system. *Am J Health Syst Pharm* 52: 823–828.

Scott, G.P.T., Shah, P.A., Wyatt, J.C., Makubate, B., and Cross, F.W. 2011. Making electronic prescribing alerts more effective: Scenario-based experimental study in junior doctors. *J Am Med Inform Assoc* 18: 789–798.

Scott-Cawiezell, J., Madsen, R.W., Pepper, G. et al. 2009. Medication safety teams' guided implementation of electronic medication administration records in five nursing homes. *Jt Comm J Qual Patient Saf* 35: 29–35.

Sears, K., Goldsworthy, S., and Goodman, W.M. 2010. The relationship between simulation in nursing education and medication safety. *J Nurs Educ* 49: 52–55.

Seibert, H.H., Maddox, R.R., Flynn, E.A., and Williams, C.K. 2014. Effect of barcode technology with electronic medication administration record on medication accuracy rates. *Am J Health Syst Pharm* 71: 209–218.

Seidling, H.M., Storch, C.H., Bertsche, T. et al. 2009. Successful strategy to improve the specificity of electronic statin–drug interaction alerts. *Eur J Clin Pharmacol* 65: 1149–1157.

Seifert, S.A., and Jacobitz, K. 2002. Pharmacy prescription dispensing errors reported to a regional poison control center. *J Toxicol Clin Toxicol* 40: 919–923.

Serrano-Fabia, A., Albert-Mari, A., Almenar-Cubells, D., and Jimenez-Torres, N.V. 2010. Multidisciplinary system for detecting medication errors in antineoplastic chemotherapy. *J Oncol Pharm Pract* 16: 105–112.

Sexton, J.B., Helmreich, R.L., Neilands, T.B. et al. 2006. The Safety Attitudes Questionnaire: Psychometric properties, benchmarking data, and emerging research. *BMC Health Serv Res* 6: 44.

Shah, S.N.H., Aslam, M., and Avery, A.J. 2001. A survey of prescription errors in general practice. *Pharmaceutical J* 267: 860–862.

Sharek, P.J., McClead, R.E., Jr., Taketomo, C.K. et al. 2008. An intervention to decrease narcotic-related adverse drug events in children's hospitals. *Pediatrics* 122: e861–e866.

Sheikh, A., and Hurwitz, B. 1999. A national database of medical error. *J R Soc Med* 92: 554–555.

Shillito, J., Arfanis, K., and Smith, A. 2010. Checking in healthcare safety: Theoretical basis and practical application. *Int J Health Care Qual Assur* 23: 699–707.

Sikka, R., Sweis, R., Kaucky, C., and Kulstad, E. 2012. Automated dispensing cabinet alert improves compliance with obtaining blood cultures before antibiotic administration for patients admitted with pneumonia. *Jt Comm J Qual Patient Saf* 38: 224–228.

Silva, A., Miasso, A., Oliveira, R. et al. 2008. The process of drug dispensing and distribution at four Brazilian hospitals: A multicentre descriptive study. *Latin Amer J Pharm* 27: 446–453.

Sirois, P., Fournier, H.Ã., Lebouthilier, A. et al. 2013. Nurses' perceptions and attitudes towards new ADU technology and use. *Technol Health Care* 21: 41–47.

Sittig, D.F., and Ash, J.S. 2011. *Clinical Information Systems: Overcoming Adverse Consequences*. Sudbury, MA: Jones and Bartlett.

Sittig, D.F., Krall, M.A., Dykstra, R.H., Russell, A., and Chin, H.L. 2006. A survey of factors affecting clinician acceptance of clinical decision support. *BMC Med Inform Decis Mak* 6: 6.

Slight, S.P., Howard, R., Ghaleb, M. et al. 2013. The causes of prescribing errors in English general practices: A qualitative study. *Br J Gen Pract* 63: e713–e720.

Smetzer, J., Baker, C., Byrne, F.D., and Cohen, M.R. 2010. Shaping systems for better behavioral choices: Lessons learned from a fatal medication error. *Jt Comm J Qual Patient Saf* 36: 152–163.

Smetzer, J.L., Vaida, A.J., Cohen, M.R. et al. 2003. Findings from the ISMP Medication Safety Self-Assessment for hospitals. *Jt Comm J Qual Saf* 29: 586–597.

Smith, E.R., and DeCoster, J. 2000. Dual-process models in social and cognitive psychology: Conceptual integration and links to underlying memory systems. *Pers Soc Psychol Rev* 4: 108–131.

Smithburger, P.L., Buckley, M.S., Bejian, S., Burenheide, K., and Kane-Gill, S.L. 2011. A critical evaluation of clinical decision support for the detection of drug–drug interactions. *Expert Opin Drug Saf* 10: 871–882.

Snodgrass, R.D. 2005. Smart pump technology. *Biomed Instrum Technol* 39: 444–446.

Spencer, M.G., and Smith, A.P. 1993. A multi-centre study of dispensing in British hospitals. *Int J Pharm Pract* 2: 142–146.

Spinewine, A., Claeys, C., Foulon, V., and Chevalier, P. 2013. Approaches for improving continuity of care in medication management: A systematic review. *Int J Qual Health Care* 25: 403–417.

Spivey, P. 2012. Ensuring good dispensing practices. In *MDS3—Managing Access to Medicines and Health Technologies,* Management Sciences for Health, Inc, 30.1–30.17. Arlington, VA: Management Sciences for Health.

Spooner, S.H., and Emerson, P.K. 1994. Using hospital pharmacy technicians to check unit dose carts. *Hosp Pharm* 29: 433–437.

Staroselsky, M., Volk, L.A., Tsurikova, R. et al. 2008. An effort to improve electronic health record medication list accuracy between visits: Patients' and physicians' response. *Int J Med Inform* 77: 153–160.

Steven, K., Wenger, E., Boshuizen, H., Scherpbier, A., and Dornan, T. 2014. How clerkship students learn from real patients in practice settings. *Acad Med* 89: 469–476.

Stevenson, F.A., Cox, K., Britten, N., and Dundar, Y. 2004. A systematic review of the research on communication between patients and health care professionals about medicines: The consequences for concordance. *Health Expect* 7: 235–245.

Stewart, M., Purdy, J., Kennedy, N., and Burns, A. 2010. An interprofessional approach to improving paediatric medication safety. *BMC Med Educ* 10: 19.

Stratton, K.M., Blegen, M.A., Pepper, G., and Vaughn, T. 2004. Reporting of medication errors by pediatric nurses. *J Pediatr Nurs* 19: 385–392.

Stubbs, J., Haw, C., and Cahill, C. 2004. Auditing prescribing errors in a psychiatric hospital. Are pharmacists' interventions effective? *Hosp Pharmacist* 11: 203–206.

Stubbs, J., Haw, C., and Taylor, D. 2006. Prescription errors in psychiatry—A multi-centre study. *J Psychopharmacol* 20: 553–561.

Sturmberg, J.P., and Martin, C. 2013. *Handbook of Systems and Complexity in Health.* New York: Springer.

Suresh, G., Horbar, J.D., Plsek, P. et al. 2004. Voluntary anonymous reporting of medical errors for neonatal intensive care. *Pediatrics* 113: 1609–1618.

Sutton, J., and Weiss, M. 2008. Involving patients as advisors in pharmacy practice research: What are the benefits? *Int J Pharm Pract* 16: 231–238.

Swanson, D. 2004. Automated dispensing—An overview of the types of system available. *Hosp Pharm* 11: 66–68.

Sweidan, M., Reeve, J., Brien, J.E., Jayasuria, P., and Vernon, G. 2009. Quality of drug interaction alerts in prescribing and dispensing software. *Med J Aust* 190: 251–254.

Szczepura, A., Wild, D., and Nelson, S. 2011. Medication administration errors for older people in long-term residential care. *BMC Geriatrics* 11: 82.

Tague, N.R. 2005. *The Quality Toolbox*. Milwaukee, WI: ASQ Quality Press.

Takata, G.S., Mason, W., Taketomo, C., Logsdon, T., and Sharek, P.J. 2008. Development, testing, and findings of a pediatric-focused trigger tool to identify medication-related harm in US children's hospitals. *Pediatrics* 121: e927–e935.

Tallentire, V.R., Hale, R.L., Dewhurst, N.G., and Maxwell, S.R. 2013. The contribution of prescription chart design and familiarity to prescribing error: A prospective, randomised, cross-over study. *BMJ Qual Saf* 22: 864–869.

Tam, V.C., Knowles, S.R., Cornish, P.L. et al. 2005. Frequency, type and clinical importance of medication history errors at admission to hospital: A systematic review. *CMAJ* 173: 510–515.

Tamblyn, R., Huang, A., Taylor, L.K. et al. 2008. A randomized trial of the effectiveness of on-demand versus computer-triggered drug decision support in primary care. *J Am Med Inform Assoc* 15: 430–438.

Taxis, K., and Barber, N. 2003a. Causes of intravenous medication errors: An ethnographic study. *Qual Saf Health Care* 12: 343–347.

Taxis, K., and Barber, N. 2003b. Ethnographic study of incidence and severity of intravenous drug errors. *BMJ* 326: 684–687.

Taxis, K., and Barber, N. 2004. Causes of intravenous medication errors: Observation of nurses in a German hospital. *J Publ Health* 12: 132–138.

Taxis, K., Dean, B., and Barber, N. 1999. Hospital drug distribution systems in the UK and Germany—A study of medication errors. *Pharm World Sci* 21: 25–31.

Taxis, K., Dean, B., and Barber, N. 2002. The validation of an existing method of scoring the severity of medication administration errors for use in Germany. *Pharm World Sci* 24: 236–239.

Taxis, K., Wirtz, V., and Barber, N. 2004. Variations in aseptic techniques during preparation and administration of intravenous drugs—An observation-based study in the UK and in Germany. *J Hosp Infect* 56: 79–81.

Taylor, D., Yuen, S., Hunt, L., and Emond, A. 2012. An interprofessional pediatric prescribing workshop. *Am J Pharm Educ* 76: 111.

Taylor, N., Lawton, R., Slater, B., and Foy, R. 2013. The demonstration of a theory-based approach to the design of localized patient safety interventions. *Implement Sci* 8: 123.

Taylor-Adams, S., and Vincent, C. 2004. Systems analysis of clinical incidents: The London protocol. *Clin Risk* 10: 211–220.

Teagarden, J.R., Nagle, B., Aubert, R.E. et al. 2005. Dispensing error rate in a highly automated mail-service pharmacy practice. *Pharmacotherapy* 25: 1629–1635.

Teich, J.M., Osheroff, J.A., Pifer, E.A. et al. 2005. Clinical decision support in electronic prescribing: Recommendations and an action plan: Report of the Joint Clinical Decision Support Workgroup. *J Am Med Inform Assoc* 12: 365–376.

Temple, J., and Ludwig, B. 2010. Implementation and evaluation of carousel dispensing technology in a university medical center pharmacy. *Am J Health Syst Pharm* 67: 821–829.

Tham, E., Calmes, H.M., Poppy, A. et al. 2011. Sustaining and spreading the reduction of adverse events in a multicenter collaborative. *Pediatrics* 128: e438–e445.

The Health Foundation. 2010. *Complex Adaptive Systems*. London: The Health Foundation.

The Health Foundation. 2013. Measuring harm. A summary of learning from a Health Foundation roundtable. http://www.health.org.uk/public/cms/75/76/313/4470/Measuring%20harm.pdf?realName=687aQB.pdf (Accessed July 1, 2014).

The Pharmacy Guild of Australia. 2013. Dispensing your prescription medicine: More than sticking a label on a bottle. http://www.guild.org.au/docs/default-source/public-documents/issues-and-resources/Fact-Sheets/the-dispensing-process.pdf?sfvrsn=0 (Accessed August 24, 2014).

The Pharmacy Guild of Australia. 2014. Training courses. http://www.guild.org.au/sa_branch/training/courses (Accessed June 20, 2014).

Thistlethwaite, J., and Nisbet, G. 2007. Interprofessional education: What's the point and where we're at … *Clin Teach* 4: 67–72.

Thomsen, L.A., Winterstein, A.G., Sondergaard, B., Haugbolle, L.S., and Melander, A. 2007. Systematic review of the incidence and characteristics of preventable adverse drug events in ambulatory care. *Ann Pharmacother* 41: 1411–1426.

Thornton, P.D., Goh, S.F., and Tasker, J.L. 1990. Pharmacists don't make mistakes: A review of pharmacy-originated errors occurring in a major Australian teaching hospital. *Aust J Hosp Pharm* 20: 133.

Tofani, B.F., Rineair, S.A., Gosdin, C.H. et al. 2012. Quality improvement project to reduce infiltration and extravasation events in a pediatric hospital. *J Pediatr Nurs* 27: 682–689.

Toft, B. 2001. *External Inquiry into the Adverse Incident that Occurred at Queen's Medical Centre, Nottingham, 4th January 2001*. London: Department of Health.

Tompa, E., Dolinschi, R., and Natale, J. 2013. Economic evaluation of a participatory ergonomics intervention in a textile plant. *Appl Ergon* 44: 480–487.

Trbovich, P.L., Pinkney, S., Colvin, C., and Easty, A.C. 2014. Increased complexity of medical technology and the need for human factors informed design and training. In *XIII Mediterranean Conference on Medical and Biological Engineering and Computing 2013*, ed L.M. Roa Romero, 1163–1165. Cham, Switzerland: Springer International Publishing.

Tsai, S.L., Sun, Y.C., and Taur, F.M. 2010. Comparing the working time between bar-code medication administration system and traditional medication administration system: An observational study. *Int J Med Inf* 79: 681–689.

Tully, M.P., Ashcroft, D.M., Dornan, T. et al. 2009. The causes of and factors associated with prescribing errors in hospital inpatients: A systematic review. *Drug Saf* 32: 819–836.

Tully, M.P., Dornan, T.L., Lewis, P.J. et al. 2014. Pharmacist-led feedback to junior doctors to reduce suboptimal prescribing of antimicrobials. *Res Social Adm Pharm* 10: e40.

Tully, M.P., Kettis, A., Hoglund, A.T. et al. 2013. Transfer of data or re-creation of knowledge—Experiences of a shared electronic patient medical records system. *Res Social Adm Pharm* 9: 965–974.

Turvey, C.L., Zulman, D.M., Nazi, K.M. et al. 2012. Transfer of information from personal health records: A survey of veterans using My HealtheVet. *Telemed J E Health* 18: 109–114.

Ulanimo, V., O'Leary-Kelley, C., and Connolly, P. 2007. Nurses' perceptions of causes of medication errors and barriers to reporting. *J Nurs Care Qual* 22: 28–33.

Unver, V., Tastan, S., and Akbayrak, N. 2012. Medication errors: Perspectives of newly graduated and experienced nurses. *Int J Nurs Pract* 18: 317–324.

Urban, R., Armitage, G., Morgan, J. et al. 2014. Custom and practice: A multi-center study of medicines reconciliation following admission in four acute hospitals in the UK. *Res Social Adm Pharm* 10: 355–368.

U.S. Food and Drug Administration. 2013a. BeSafeRx: Know your online pharmacy. http://www.fda.gov/Drugs/ResourcesForYou/Consumers/BuyingUsingMedicineSafely/BuyingMedicinesOvertheInternet/BeSafeRxKnowYourOnlinePharmacy/ucm20027044.htm (Accessed November 16, 2014).

U.S. Food and Drug Administration. 2013b. Name differentiation project. http://www.fda.gov/Drugs/DrugSafety/MedicationErrors/ucm164587.htm (Accessed November 15, 2014).

Uzuner, O., Solti, I., and Cadag, E. 2010. Extracting medication information from clinical text. *J Am Med Inform Assoc* 17: 514–518.

van den Bemt, P.M., Idzinga, J.C., Robertz, H., Kormelink, D.G., and Pels, N. 2009a. Medication administration errors in nursing homes using an automated medication dispensing system. *J Am Med Inform Assoc* 16: 486–492.

van den Bemt, P.M., van den Broek, S., van Nunen, A.K., Harbers, J.B., and Lenderink, A.W. 2009b. Medication reconciliation performed by pharmacy technicians at the time of preoperative screening. *Ann Pharmacother* 43: 868–874.

van den Bemt, P.M., van der Schrieck-de Loos EM, van der Linden, C., Theeuwes, A.M., and Pol, A.G. 2013. Effect of medication reconciliation on unintentional medication discrepancies in acute hospital admissions of elderly adults: A multicenter study. *J Am Geriatr Soc* 61: 1262–1268.

van den Bemt, P.M.L.A., Postma, M.J., van Roon, E.N. et al. 2002. Cost–benefit analysis of the detection of prescribing errors by hospital pharmacy staff. *Drug Saf* 25: 135–143.

van der Geest, S., Whyte, S.R., and Hardon, A. 1996. The anthropology of pharmaceuticals: A biographical approach. *Annu Rev Anthropol* 25: 153–178.

van der Linden, C.M., Kerskes, M.C., Bijl, A.M. et al. 2006. Represcription after adverse drug reaction in the elderly: A descriptive study. *Arch Intern Med* 166: 1666–1667.

van der Sijs, H., Aarts, J., van Gelder, T., Berg, M., and Vulto, A. 2008. Turning off frequently overridden drug alerts: Limited opportunities for doing it safely. *J Am Med Inform Assoc* 15: 439–448.

van der Sijs, H., Aarts, J., Vulto, A., and Berg, M. 2006. Overriding of drug safety alerts in computerized physician order entry. *J Am Med Inform Assoc* 13: 138–147.

van der Sijs, H., Bouamar, R., van Gelder, T. et al. 2010. Functionality test for drug safety alerting in computerized physician order entry systems. *Int J Med Inform* 79: 243–251.

van der Sijs, H., Mulder, A., van Gelder, T. et al. 2009. Drug safety alert generation and overriding in a large Dutch university medical centre. *Pharmacoepidemiol Drug Saf* 18: 941–947.

van Merriënboer, J.J.G., and Kirschner, P.A. 2007. *Ten Steps to Complex Learning: A Systematic Approach to Four-Component Instructional Design*. Abingdon, Oxon, UK: Routledge.

van Mil, J.W.F., Westerlund, L.O.T., Hersberger, K.E., and Schaefer, M.A. 2004. Drug-related problem classification systems. *Ann Pharmacother* 38: 859–867.

van Onzenoort, H.A., van de Plas, A., Kessels, A.G. et al. 2008. Factors influencing barcode verification by nurses during medication administration in a Dutch hospital. *Am J Health Syst Pharm* 65: 644–648.

van Roon, E.N., Flikweert, S., le Comte, M. et al. 2005. Clinical relevance of drug–drug interactions: A structured assessment procedure. *Drug Saf* 28: 1131–1139.

van Vuuren, W., Shea, C.E., and van der Schaaf, T.W. 1997. *The Development of An Incident Analysis Tool for the Medical Field*. Eindhoven, The Netherlands: Faculty of Technology Management, University of Eindhoven.

Varadarajan, R., Barker, K.N., Flynn, E.A., and Thomas, R.E. 2008. Comparison of two error-detection methods in a mail service pharmacy serving health facilities. *J Am Pharm Assoc* 48: 371–378.

Varkey, P., Cunningham, J., and Bisping, D.S. 2007. Improving medication reconciliation in the outpatient setting. *Jt Comm J Qual Patient Saf* 33: 286–292.

Vermaire, D., Caruso, M.C., Lesko, A. et al. 2011. Quality improvement project to reduce perioperative opioid oversedation events in a paediatric hospital. *BMJ Qual Saf* 20: 895–902.

Vermeulen, K.M., van Doormaal, J.E., Zaal, R.J. et al. 2014. Cost-effectiveness of an electronic medication ordering system (CPOE/CDSS) in hospitalized patients. *Int J Med Inf* 83: 572–580.

Verrue, C.L., Mehuys, E., Somers, A. et al. 2010. Medication administration in nursing homes: Pharmacists' contribution to error prevention. *J Am Med Dir Assoc* 11: 275–283.

Vincent, C. 2010. *Patient Safety*, 2nd Edition. Chichester: Wiley-Blackwell.

Vincent, C., Aylin, P., Franklin, B.D. et al. 2008. Is health care getting safer? *BMJ* 337: a2426.

Vincent, C., Burnett, S., and Carthey, J. 2013. *The Measurement and Monitoring of Safety. Drawing Together Academic Evidence and Practical Experience to Produce a Framework for Safety Measurement and Monitoring*. London: The Health Foundation.

Vincent, C., Taylor-Adams, S., Chapman, E.J. et al. 2000. How to investigate and analyse clinical incidents: Clinical risk unit and association of litigation and risk management protocol. *BMJ* 320: 777–781.

Vincent, C., Taylor-Adams, S., and Stanhope, N. 1998. Framework for analysing risk and safety in clinical medicine. *BMJ* 316: 1154–1157.

Vira, T., Colquhoun, M., and Etchells, E. 2006. Reconcilable differences: Correcting medication errors at hospital admission and discharge. *Qual Saf Health Care* 15: 122–126.

Virtanen, M., Kurvinen, T., Terho, K. et al. 2009. Work hours, work stress, and collaboration among ward staff in relation to risk of hospital-associated infection among patients. *Med Care* 47: 310–318.

Vogelsmeier, A., Pepper, G.A., Oderda, L., and Weir, C. 2013. Medication reconciliation: A qualitative analysis of clinicians' perceptions. *Res Social Adm Pharm* 9: 419–430.

Vogelsmeier, A.A., Halbesleben, J.R., and Scott-Cawiezell, J.R. 2008. Technology implementation and workarounds in the nursing home. *J Am Med Inform Assoc* 15: 114–119.

Wahr, J.A., Shore, A.D., Harris, L.H. et al. 2014. Comparison of intensive care unit medication errors reported to the United States' MedMarx and the United Kingdom's National Reporting and Learning System: A cross-sectional study. *Am J Med Qual* 29: 61–69.

Walker, A.E., Grimshaw, J., Johnston, M. et al. 2003. PRIME—PRocess modelling in ImpleMEntation research: Selecting a theoretical basis for interventions to change clinical practice. *BMC Health Serv Res* 3: 22.

Walsh, K.E., Roblin, D.W., Weingart, S.N. et al. 2013. Medication errors in the home: A multisite study of children with cancer. *Pediatrics* 131: e1405–e1414.

Warrick, C., Naik, H., Avis, S. et al. 2011. A clinical information system reduces medication errors in paediatric intensive care. *Intensive Care Med* 37: 691–694.

Waterman, A.D., Gallagher, T.H., Garbutt, J. et al. 2006. Brief report: Hospitalized patients' attitudes about and participation in error prevention. *J Gen Intern Med* 21: 367–370.

Waterson, P. 2009. A critical review of the systems approach within patient safety research. *Ergonomics* 52: 1185–1195.

Waterson, P. 2014. *Patient Safety Culture: Theories, Methods and Applications*. Farnham: Ashgate.

Watson, R.J. 1977. A large-scale professionally oriented medical information system—Five years later. *J Med Syst* 1: 3–21.

Watt, I.S., Birks, Y., Entwistle, V. et al. 2009. A review of strategies to promote patient involvement, a study to explore patient's views and attitudes and a pilot study to evaluate the acceptability of selected patient involvement strategies. http://www.birmingham.ac.uk/Documents/college-mds/haps/projects/cfhep/psrp/finalreports/PS034-Finalreport2009.pdf (Accessed November 11, 2014).

Wears, R.L., Cook, R.I., and Perry, S.J. 2006. Automation, interaction, complexity and failure: A case study. *Reliab Eng Syst Safe* 91: 1494–1501.

Weaver, S.J., Lubomksi, L.H., Wilson, R.F. et al. 2013. Promoting a culture of safety as a patient safety strategy: A systematic review. *Ann Intern Med* 158: 369–374.

Webbe, D., Dhillon, S., and Roberts, C.M. 2007. Improving junior doctor prescribing—The positive impact of a pharmacist intervention. *Pharm J* 278: 136–139.

Weick, K.E. 1987. Organizational culture as a source of high reliability. *Calif Manage Rev* 29: 112–127.

Weick, K.E., Sutcliffe, K.M., and Obstfeld, D. 1999. Organizing for high reliability: Processes of collective mindfulness. *Res Organ Behav* 21: 81–123.

Weingart, S.N., Hamrick, H.E., Tutkus, S. et al. 2008. Medication safety messages for patients via the web portal: The MedCheck intervention. *Int J Med Inf* 77: 161–168.

Weingart, S.N., Simchowitz, B., Padolsky, H. et al. 2009. An empirical model to estimate the potential impact of medication safety alerts on patient safety, health care utilization, and cost in ambulatory care. *Arch Intern Med* 169: 1465–1473.

Weingart, S.N., Toth, M., Eneman, J. et al. 2004. Lessons from a patient partnership intervention to prevent adverse drug events. *Int J Qual Health Care* 16: 499–507.

Weller, J.M. 2004. Simulation in undergraduate medical education: Bridging the gap between theory and practice. *Med Educ* 38: 32–38.

Wenger, E. 1998. *Communities of Practice: Learning, Meaning and Identity*. Cambridge: Cambridge University Press.

Wenger, E. 2010. Social learning systems and communities of practice: The career of a concept. In *Social Learning Systems and Communities of Practice*, ed C. Blackmore, 179–198. New York, NY: Springer.

Wertsch, J.V. 1985. *Vygotsky and the Social Formation of Mind*. Boston, MA: Harvard University Press.

Westbrook, J.I. 2014. Interruptions and multi-tasking: Moving the research agenda in new directions. *BMJ Qual Saf* 23: 877–879.

Westbrook, J.I., and Ampt, A. 2009. Design, application and testing of the work observation method by activity timing (WOMBAT) to measure clinicians' patterns of work and communication. *Int J Med Inf* 78S: S25–S33.

Westbrook, J.I., Baysari, M.T., Li, L. et al. 2013a. The safety of electronic prescribing: Manifestations, mechanisms, and rates of system-related errors associated with two commercial systems in hospitals. *J Am Med Inform Assoc* 20: 1159–1167.

Westbrook, J.I., and Braithwaite, J. 2010. Will ICT disrupt the health system and deliver on its promise? *Med J Aust* 193: 399–400.

Westbrook, J.I., Coiera, E., Dunsmuir, W.T.M. et al. 2010a. The impact of interruptions on clinical task completion. *Qual Saf Health Care* 19: 284–289.

Westbrook, J.I., and Li, L. 2013. Changes in medication administration errors following the implementation of electronic medication management systems in hospitals. In *30th International Society for Quality in Healthcare (ISQua) Conference Proceedings*, ed ISQua, 2444. Edinburgh: Scotland.

Westbrook, J.I., Li, L., Georgiou, A., Paoloni, R., and Cullen, J. 2013b. Impact of an electronic medication management system on hospital doctors' and nurses' work: A controlled pre–post, time and motion study. *J Am Med Inform Assoc* 20: 1150–1158.

Westbrook, J.I., Reckmann, M., Li, L. et al. 2012. Effects of two commercial electronic prescribing systems on prescribing error rates in hospital inpatients: A before and after study. *PLoS Med* 9: e1001164.

Westbrook, J.I., Rob, M.I., Woods, A., and Parry, D. 2011. Errors in the administration of intravenous medications in hospital and the role of correct procedures and nurse experience. *BMJ Qual Saf* 20: 1027–1034.

Westbrook, J.I., Woods, A., Rob, M.I., Dunsmuir, W.T., and Day, R.O. 2010b. Association of interruptions with an increased risk and severity of medication administration errors. *Arch Intern Med* 170: 683–690.

Westrum, R. 2004. A typology of organisational cultures. *Qual Saf Health Care* 13(Suppl 2): ii22–ii27.

Wetterneck, T.B. 2012. Error recovery in health care. In *Handbook of Human Factors and Ergonomics in Health Care and Patient Safety*, ed. P. Carayon, 763–774. Boca Raton, FL: Taylor & Francis.

Wetterneck, T.B., and Karsh, B.-T. 2011. Human factors applications—Understanding and using close calls to improve health care. In *Using Near Misses and Close Calls to Improve Patient Safety*, ed A. Wu, 39–54. Oakbrook Terrace, IL: Joint Commission Resources.

Wetterneck, T.B., Skibinski, K.A., Roberts, T.L. et al. 2006. Using failure mode and effects analysis to plan implementation of smart i.v. pump technology. *Am J Health Syst Pharm* 63: 1528–1538.

Wetterneck, T.B., Walker, J.M., Blosky, M.A. et al. 2011. Factors contributing to an increase in duplicate medication order errors after CPOE implementation. *J Am Med Inform Assoc* 18: 774–782.

Wheeler, D.S., Giaccone, M.J., Hutchinson, N. et al. 2011. A hospital-wide improvement collaborative to reduce catheter-associated bloodstream infections. *Pediatrics* 128: e995–e1007.

White, C.M., Schoettker, P.J., Conway, P.H. et al. 2011. Utilising improvement science methods to optimise medication reconciliation. *BMJ Qual Saf* 20: 372–380.

White, C.M., Statile, A.M., Conway, P.H. et al. 2012. Utilizing improvement science methods to improve physician compliance with proper hand hygiene. *Pediatrics* 129: e1042–e1050.

White, C.M., Statile, A.M., White, D.L. et al. 2014. Using quality improvement to optimise paediatric discharge efficiency. *BMJ Qual Saf* 23: 428–436.

Whyte, S.R., Geest, S.V.D., and Hardon, A. 2002. *Social Lives of Medicines*. Cambridge; New York, NY: Cambridge University Press.

Wiegmann, D., and Shappell, S. 2003. *A Human Error Approach to Aviation Accident Analysis*. Burlington, VT: Ashgate.

Wilder-Smith, C.H., and Schuler, L. 1992. Postoperative analgesia: Pain by choice? The influence of patient attitudes and patient education. *Pain* 50: 257–262.

Williams, S.D., Phipps, D.L., and Ashcroft, D.M. 2013. Understanding the attitudes of hospital pharmacists to reporting medication incidents: A qualitative study. *Res Social Adm Pharm* 9: 80–89.

Wilson, J.R. 2000. Fundamentals of ergonomics in theory and practice. *Appl Ergon* 31: 557–567.

Wilson, J.R. 2014. Fundamentals of systems ergonomics/human factors. *Appl Ergon* 45: 5–13.

Wilson, R.M., Runciman, W.B., Gibberd, R.W. et al. 1995. The quality in Australian Health Care Study. *Med J Aust* 163: 458–471.

Wilson-Donnelly, K.A., Priest, H.A., Salas, E., and Burke, C.S. 2005. The impact of organizational practices on safety in manufacturing: A review and reappraisal. *Hum Factor Ergon Man* 15: 135–176.

Winson, G. 1991. A survey of nurses' attitudes towards single administration of medicines. *Nurse Pract* 4: 20–23.

Winters, B., Gurses, A., Lehmann, H. et al. 2009. Clinical review: Checklists—Translating evidence into practice. *Crit Care* 13: 210.

Winterstein, A.G., Johns, T.E., Rosenberg, E.I. et al. 2004. Nature and causes of clinically significant medication errors in a tertiary care hospital. *Am J Health Syst Pharm* 61: 1908–1916.

Wirtz, V., Taxis, K., and Barber, N.D. 2003. An observational study of intravenous medication errors in the United Kingdom and in Germany. *Pharm World Sci* 25: 104–111.

Wolf, Z.R., and Hughes, R.G. 2008. Chapter 35: Error Reporting and Disclosure. In *Patient Safety and Quality: An Evidence-Based Handbook for Nurses*, ed R.G. Hughes, 333–379. Rockville, MD: Agency for Healthcare Research and Quality (US).

Wong, I.C., Ghaleb, M.A., Franklin, B.D., and Barber, N. 2004. Incidence and nature of dosing errors in paediatric medications: A systematic review. *Drug Saf* 27: 661–670.

Wong, J.D., Bajcar, J.M., Wong, G.G. et al. 2008. Medication reconciliation at hospital discharge: Evaluating discrepancies. *Ann Pharmacother* 42: 1373–1379.

Wood, J.L., and Burnette, J.S. 2012. Enhancing patient safety with intelligent intravenous infusion devices: Experience in a specialty cardiac hospital. *Heart Lung* 41: 173–176.

Woodehouse, S., Burney, B., and Coste, K. 2004. To err is human: Improving patient safety through failure mode and effect analysis. *Clin Leadership Manag Rev* 18: 32–36.

Woods, D.D., and Hollnagel, E. 2006. Prologue: Resilience engineering concepts. In *Resilience Engineering—Concepts and Precepts*, eds E. Hollnagel, D.D. Woods and N. Leveson, 1–7. Hampshire: Ashgate.

World Alliance for Patient Safety. 2005. *WHO Draft Guidelines for Adverse Event Reporting and Learning Systems. From Information to Action*. Geneva, Switzerland: World Health Organization.

World Health Organization. 2002. World Health Assembly Resolution WHA55.18. http://www.who.int/patientsafety/about/wha_resolution/en/index.html (Accessed August 24, 2014).

World Health Organization. 2009a. A Guide to the implementation of the WHO multimodal hand hygiene improvement strategy. http://whqlibdoc.who.int/hq/2009/WHO_IER_PSP_2009.02_eng.pdf (Accessed July 21, 2014).

World Health Organization. 2009b. *Conceptual Framework for the International Classification for Patient Safety: Final Technical Report version 1.1. WHO*. Geneva, Switzerland: World Health Organization. http://www.who.int/patientsafety/taxonomy/icps_full_report.pdf (Accessed June 23, 2015).

World Health Organization. 2011. *Patient Safety Curriculum Guide: Multi-Professional Edition*. Geneva, Switzerland: World Health Organization. http://www.who.int/patientsafety/education/curriculum/tools-download/en/ (Accessed June 23, 2015).

Management Sciences for Health. 2012. *MDS3—Managing Access to Medicines and Health Technologies*. Arlington, VA: Management Sciences for Health. https://www.msh.org/resources/mds-3-managing-access-to-medicines-and-health-technologies (Accessed June 23, 2015).

World Health Organization, and EuroPharm Forum. 1993. Questions to ask about your medicines.http://www.apotheker.or.at/internet/oeak/newspresse.nsf/e02b9cd11265691ec1256a7d005209ee/90f6ba615dd5603ac1256ab8004148be/$FILE/QaM%20Guidelines%20WHO%20Europharm.pdf (Accessed September 9, 2014).

Wu, A.W., Lipshutz, A.K., and Pronovost, P.J. 2008. Effectiveness and efficiency of root cause analysis in medicine. *JAMA* 299: 685–687.

Wu, P.F. 2000. Measuring potential error rates in the pharmacy dispensary. *Pharm World Sci* 22: B31.

Xiao, Y., and Fairbanks, R.J. 2011. Speaking systems engineering: Bilingualism in health care delivery organizations. *Mayo Clin Proc* 86: 719–720.

Xie, A. 2013. A participatory ergonomics approach to healthcare system redesign: The case of family-centered rounds in a pediatric hospital. http://search.library.wisc.edu/catalog/ocn857671113 (Accessed April 25, 2014).

Yano, E.M. 2008. The role of organizational research in implementing evidence-based practice: QUERI Series. *Implement Sci* 3: 29.

Young, H.M., Sikma, S.K., Reinhard, S.C., McCormick, W.C., and Cartwright, J.C. 2013. Strategies to promote safe medication administration in assisted living settings. *Res Gerontol Nurs* 6: 161–170.

Young, J., Slebodnik, M., and Sands, L. 2010. Bar code technology and medication administration error. *J Patient Saf* 6: 115–120.

Zachariah, M., Phansalkar, S., Seidling, H.M. et al. 2011. Development and preliminary evidence for the validity of an instrument assessing implementation of human-factors principles in medication-related decision-support systems—MeDeSA. *J Am Med Inform Assoc* 18: i62–i72.

Zheng, K., Haftel, H.M., Hirschl, R.B., O'Reilly, M., and Hanauer, D.A. 2010. Quantifying the impact of health IT implementations on clinical workflow: A new methodological perspective. *J Am Med Inform Assoc* 17: 454–461.

Zohar, D. 2010. Thirty years of safety climate research: Reflections and future directions. *Accid Anal Prev* 42: 1517–1522.

Zwart-van Rijkom, J.E., Uijtendaal, E.V., ten Berg, M.J., van Solinge, W.W., and Egberts, A.C. 2009. Frequency and nature of drug–drug interactions in a Dutch university hospital. *Br J Clin Pharmacol* 68: 187–193.

Index

A

B

C

F

G

H

I

K

L

M

N

O

P

Q

R

T

U

V

W